# Solid State Chemistry

# Solid State Chemistry

*by*

**R.C. Ropp**
**Warren**
**New Jersey**
**USA**

2003

ELSEVIER

AMSTERDAM – BOSTON – HEIDELBERG – LONDON – NEW YORK – OXFORD
PARIS – SAN DIEGO – SAN FRANCISCO – SINGAPORE – SYDNEY – TOKYO

ELSEVIER SCIENCE B.V.
Sara Burgerhartstraat 25
P.O. Box 211, 1000 AE Amsterdam, The Netherlands



First edition 2003

Library of Congress Cataloging in Publication Data
A catalog record from the Library of Congress has been applied for.

British Library Cataloguing in Publication Data
A catalogue record from the British Library has been applied for.

ISBN: 0 444 51436 8

♾ The paper used in this publication meets the requirements of ANSI/NISO Z39.48-1992 (Permanence of Paper).
Printed in The Netherlands.

# Preface

Most of the books concerning "Solid State Chemistry" that I have examined have been written for the specialist or for the advanced student. I have composed this book with the novice in mind. That is, those who have little background but wish to begin to learn about the solid state and what it entails can start at the beginning and build upon their own knowledge and experience. This includes those working in the biotechnology, pharmaceutical and proteomics fields. At this point in time, they have sequenced the human genome and are trying to define the structure of proteins. It is my opinion that the material in this book will be helpful to them as well.

With this in mind, I have presented the information in this book in a form that can be easily understood. I think that it is quite important that any student of the body of knowledge that we call "science" needs to be cognizant of the history and effort that has been made by those who preceded us. It was Newton who said: "If I have seen further, it is because I have stood on the shoulders of giants". Thus, I have tried to give a short history of each particular segment of solid state theory and technology.

I have enjoyed composing the material in this book and trust that the reader can use it as a self-study to build upon his (her) store of usable knowledge.

I thank my wife, Francisca Margarita, profusely for her support during the time that it has taken to finish this composition.

R.C. Ropp

March 2003

# Table of Contents

# Chapter 1

## The Phase Chemistry of Solids

To understand the solid state, we must first understand how matter exists. That is, matter was originally defined as "anything that occupies space and has weight". This definition was made in the eighteenth century when little was known about "matter" . Thus, it was defined in terms of its general shape and physical characteristics. We will delve into the "building-blocks" of matter and how solids originate. It is assumed that you have some initial knowledge concerning physics and chemistry as well as some ability in mathematics. The more difficult mathematical treatments are reserved to the appendices of certain chapters.

Originally, in the 18th century, when the "Scientific Revolution" was just getting started, science was referred to as "Natural Philosophy". Researchers were interested in natural science, including what we call today: Physics, Chemistry, Biology and Biochemistry. All of these scientific disciplines deal with molecules having various physical aspects and properties (even proteins). The earliest work dealt with air, water and solids since these were the easiest to manipulate in those times. Note that we regard all of their scientific apparatus to be crude by our standards. Yet, even the most sophisticated instruments (by their criteria) had to be built by hand. Even so, early workers knew that matter existed in three forms, i.e.- solids, liquids and gases. Water was studied extensively because it was easy to freeze and boil. As we will show, many of our basic measurements and scientific standards resulted from such studies.

Reversible transformations between the three forms (phases) of matter, are called "Changes of State". The science related to these phase changes lies in the realm of "Physical Chemistry" while their actual "chemistry", or "reactive-properties", lies either in the realm of "Inorganic" or "Organic" chemistry. Inorganic materials include both elements and compounds. Organic materials have a carbon backbone. Nearly all inorganics are solids at ambient temperature. At elevated temperatures, they transform or melt to form a liquid and then a gas. Most of these phase changes are reversible. A few are gases at room temperature, but liquids are rare. In contrast, organic compounds can be either solid, liquid, or gaseous. Although organic compounds do undergo some Changes-of-State easily and reversibly, their thermal stability is a problem. If one tries to induce a change of state in an organic compound, one usually finds that liquid organic compounds will freeze and/or sublime. Notwithstanding, most organic solids decompose if the gaseous state is sought. This is due to their relatively low thermal stability, i.e.- "they burn". Organic hydrocarbon compounds usually react with oxygen and convert to carbon dioxide and water at rather low temperatures of 200 °C to 600 °C, while many inorganic compounds are stable at ranges of 1500 °C to 2000 °C. and even higher. We will be more interested in inorganics because of their greater thermal stability. Additionally, inorganic compounds have physical properties that cannot be duplicated in the organic domain of chemistry, and vice-versa.

## **1.1.**- PHASE CHANGES OF SOLIDS, LIQUIDS AND GASES

The best way to understand phase changes is to examine those observed with water, i.e.- those involving the molecule, $H_2O$. We are, of course, familiar with the three states of water, namely ice, water and steam, since these forms are encountered daily. The actual difference between them is a matter of energy. To illustrate this fact, let us calculate how many calories are required to change ice into water into steam. This problem is typical of those concerning Changes-of-State. In solving such a problem, we will also define certain concepts and constants relating to its solution.

For example, we know that water (a liquid) will change to ice (a solid) if its internal temperature falls below a certain temperature. Likewise, if its internal temperature rises above a certain point, water changes to steam (a gas). Because water is so abundant on the Earth, it was used in the past to define Changes of State and even to define **Temperature Scales**. However, the concept of "heat" is also involved, and we need to also define the perception of heat as it is used in this context. Note that defining heat implies that we have a reproducible way to measure temperature. A great deal of work was required in the past to reach that stage. First, you have to establish that certain liquids expand when heated. Then you must establish how much they expand. You can do this by making a glass capillary (a glass tube with a uniform small-bore hole in it) and then measuring how far the liquid moves from a given temperature point to another. The easiest way to do this is to establish the freezing point of water and its boiling point on your "thermo" meter. Lastly, you then determine if your chosen liquid expands in a linear manner between these two thermal points (For a history, see p. 401).

There are two heat factors involved in any phase change. They are: "heat capacity", i.e.- $C_p$ or $C_v$ , and "heat of transformation", usually denoted as **H**. The former factor is concerned with **internal temperature change** within any given material (solid, liquid or gas) whereas the latter is involved in **changes of state** of that material. The actual names we use to describe **H** depend upon the direction in which the phase change occurs, vis:

1.1.1.- Changes of State in Water

| CHANGE OF STATE | H, HEAT OF: | TEMPERATURE °C. | TEMPERATURE °F |
|---|---|---|---|
| ice to water | Fusion | 0 | 32 |
| water to ice | Solidification | 0 | 32 |
| water to steam | Vaporization | 100 | 212 |
| steam to water | Condensation | 100 | 212 |

The terms, heat, heat capacity and heat of transformation were originally defined in terms of water. The earliest concept was that of "heat", but it was soon found that every material had its own "heat capacity". That is, the unit of "heat" (or internal energy) itself was originally defined as the amount of energy required to raise the temperature of 1.00 gram of water by 1.00 degree Celsius, the unit of heat energy being set equal to one (1.0) calorie. Heat capacity itself was originally defined as the amount of heat required to raise the temperature of one (1) cubic centimeter (cc.) of water *(whose density was later defined as 0.9999 @ 3.98 °C.) by one (1.00) degree.* However, after this concept had been defined, it was found that the same amount of heat **did not** raise the internal temperature of other materials to the same degree. This led to the concept of heat capacity. Heat of transformation was discovered somewhat later. How these concepts arose can be illustrated as follows.

Suppose you decide to solve the problem of defining the concepts of heat capacity and heat of transformation. The former is the amount of energy required to change the temperature of a material while the latter relates to how much energy is needed to change from one phase to another. Defining these concepts is an intellectual research exercise in itself. In doing this work, it is clear that you must be able to define some sort of measure of energy, or heat, itself (even though you have not yet defined a usable scale of heat). Since water is plentiful (it covers about 75% of the Earth's surface), you decide to use water as a standard. You begin by distilling water to obtain a pure phase. You will find that this is difficult since even the apparatus you use will contaminate the distilled water if you are not careful. After observing its behavior as its temperature is changed, you decide to set up a temperature scale in terms of water. In this work, you define the melting point of ice as 0.000 °C, and the boiling point of water as 100.000 °C. (C is "Celcius", named after the researcher who did the original work in 1734). You do this because these are easily observable points in an arbitrary scale of temperature as related to heat (and more importantly are reproducible and reversible points). As we said, the most important part of this work is to obtain a sample of pure water. Much effort was expended in the past in accomplishing this goal.

Taking your pure phase of water, you then measure its density as a function of temperature. Note that you have already defined a "specific volume" for use in your work, ie- cubic centimeter = cc. You determine that the point of maximum density of water occurs at 3.980 °C and that the density of water decreases both as you approach 0.00 °C, and its boiling point of 100.00 °C. You then assign the arbitrary value of 1.000 gram/1.000 ml. of water (or 0.9998 gram/cc, as later more precise measurements have revealed) as the density of water at its temperature of maximum density. By then measuring the amount of heat required to raise the temperature of 1.000 cubic centimeter of water by 100th of your arbitrary temperature range, i.e. by 1.000 °C, you have also defined your standard unit of heat (and called it 1.000 calorie).

You have now defined "heat" itself as the amount of energy required to raise 1 cc. of water by one degree as one calorie. However, the temperature at which you measured your standard calorie was not specified. For this reason, you choose the temperature range of 14.5 °C to 15.5 °C because you have determined that this range is one where the density of water remains relatively uniform. Thus, the standard calorie was originally defined as the amount of heat required to raise the temperature of **one gram of water** from 14.5 °C to 15.5 °C at a constant pressure of one (1) atmosphere. Note that in your research to define heat, heat capacity and heat of transformation, you have discovered several anomalies of water that had not been known before your work was accomplished.

Next, you now measure other materials to determine how much energy is required to raise their temperature by one degree and find that the amount of energy varies from material to material. In this way, you establish the notion of heat capacity. However, you also note that each material has its own internal energy (heat capacity) at a given temperature in relation to that of water.

The next step that you take is to show that the temperature of ice does not change as it melts (or that the temperature of water does not change as it freezes). This takes several very careful measurements to establish this fact. It is this concept that established that there are two kinds of "heat" involved in Changes-of-State, namely that of heat of tranformation and heat capacity. Thus, the internal temperature of a material changes in a linear manner as energy is added until the point of transformation (phase change) occurs. Then, the internal temperature does not change until the phase change is complete.

Reiterating, heat of transformation, or **H,** is involved in change of form of the material while heat capacity relates to its internal change in temperature as it approaches another point of change. Both of these constants are based upon the standard of energy, or heat, of one (1.00) calorie. Heat capacity is also known as thermal capacity.

We thus label heat capacity as $C_p$, meaning the thermal capacity at **constant pressure.** Sometimes, it is also called specific heat, meaning the ratio of thermal capacity of any given material to that of water, defined as 1.000.

This is the amount of heat in calories that it takes to raise any given material 1.00 °C. Originally, $C_v$ , the thermal capacity at **constant volume** was also used, but its use is rare nowadays. The reason for this is that most materials expand over a given temperature range, and this complicates any accurate measurement (since you have to compensate for this change in volume).

We have now demonstrated that there are two types of "heat" involved in the physical changes of any given material, one involved in change of internal temperature of the material and the other during its tranformation. We specify its internal energy as the "heat" of a material, using $H_{S,L,G}$. where S, L, or G refer to either solid, liquid or gas. Note that this *internal energy* differs from the heat of transformation, **H.** (This concept is perhaps one of the more difficult ones for one to grasp. However, the internal energy is that energy the substance has at a *given* temperature). Thus, the relation between $H_{S,L,G}$ and $C_p$ is:

1.1.2.- $\{H = C_p \cdot T\}_{S,L,G,}$ or $\Delta H_{S,L,G} = C_{p(S,L,G)} \, \Delta T_{S,L,G}$

where the **actual** state of matter is either S, L, or G. (Note that $C_p$ will differ depending upon the state of matter involved).

In contrast, **the internal temperature of a material does not change as the material undergoes a change of state.** (Thus, its internal energy does not change at that point). Therefore, for a change of state between solid and liquid, we would have:

1.1.3- $\{\mathbf{H_S} - \mathbf{H_L}\} = C_p\,(T_{(S)} - T_{(L)})$ or $\mathbf{\Delta H} = C_p \, \Delta T$

Note that $\mathbf{\Delta H}$ here is a heat of transformation involved in a **change of state** whereas $\Delta H_{S,L,G}$ refers to **change in internal heat** for a given state of matter. *In the case of water, we have defined its heat capacity, whereas its heats of transformation were measured.* These constants represent the relative amounts of energy needed to cause changes in form of the molecule, $H_2O$, from one phase to another state. Their values are given in the following values as shown in 1.1.4. on the next page:

1.1.4.-

| | $C_p$ | **ΔH** |
|---|---|---|
| Ice: | 0.5 cal/gm/°C<br>⇓ | 80 cal./gm. (S ⇒ L) |
| Water: | 1.00 cal/gm/°C<br>⇓ | 540 cal./gm. (L ⇒ G) |
| Steam: | 0.5 cal./gm/°C. | |

We can now calculate the amount of energy required to raise **one gram of ice** at - 10 °C. to form **one gram of steam** at 110 °C. This is shown in the following:

1.1.5.- CALORIES REQUIRED TO CHANGE 1 Gram OF ICE TO STEAM

| FORM | ΔT | $C_p$ | Cp ΔT | $\Delta H_X$ | ADDED |
|---|---|---|---|---|---|
| Ice | -10 to 0 | 0.5 | 5.0 cal. | --- | 5.0 cal. |
| Ice | 0 | --- | --- | 80 cal.(fusion) | 85.0 cal. |
| Water | 0 to 100 | 1.0 | 100 cal. | --- | 100.0 cal. |
| Water | 0 | --- | --- | 540 cal. | 540.0 cal. |
| to Steam | | | | (vaporization) | ======== |
| | | | | Total = | 730.0 cal. |

In all cases, $\mathbf{\Delta H_X}$ is specified in terms of the type of change of state occurring, while $C_p$ ΔT is the **change in internal heat** which occurs as the temperature rises (or falls). At a given change of state, all of the energy goes to the change of state and the temperature does not change until the transformation is complete. In the next section, we shall see where the energy goes.

The same transformations take place regardless of the nature of the atoms composing the compound, unless, as we said above, the compound decomposes to some other form. Let us now examine the thermal changes we have described above in terms of a graph showing the change in actual temperature. This is illustrated in the following diagram, given on the next page as 1.1.6.

Note that we have plotted the calories added on a shortened scale so as to

them (J is the number of atoms in the molecule, having quantized vibrational states).

1.2.1-

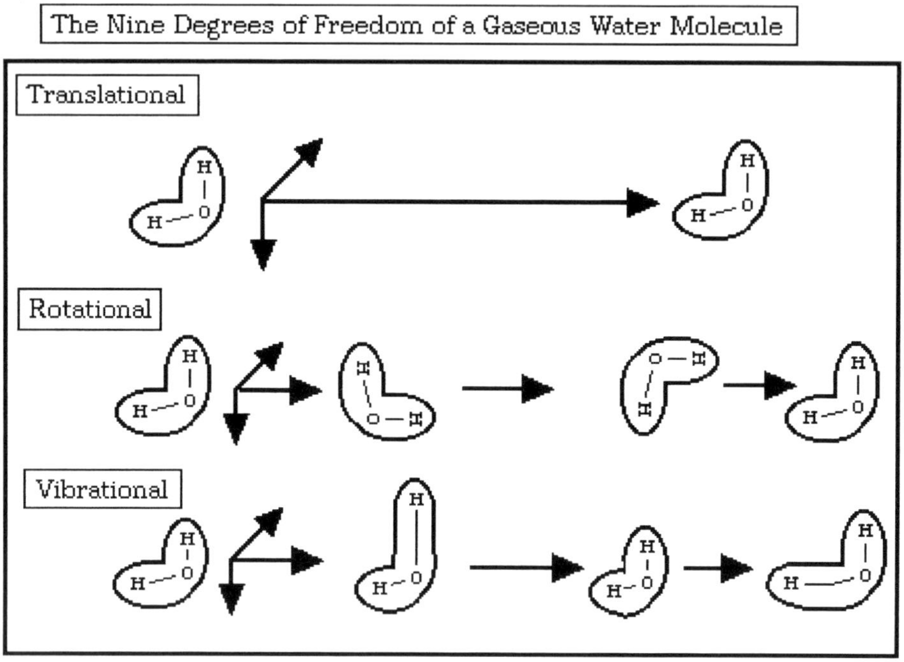


1.2.2.-

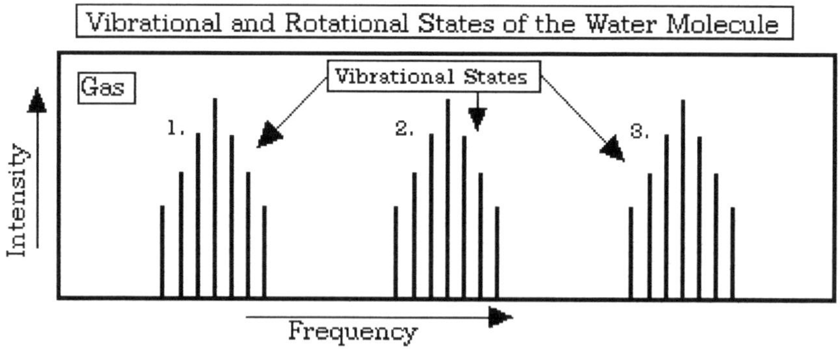


If the molecule happened to be $NH_3$, then the expected number of vibrational states would be nine. If we measure the absorption spectra of any molecule, $MX_2$, in the infra-red region of the spectrum, we will obtain results similar to those shown in the diagram given above, the exact region of the spectrum depending upon the type of molecule present (i.e.- molecular weight). It thus should be clear that each gaseous molecule is free to move in any of the three dimensions until it collides with either the walls of the container, or with some other molecule.

The average distance that each molecule moves before collision is called the "mean free path". The mean free path will be a function of both the temperature and the pressure of the gas. This concept arose from the Kinetic Theory of Gases which in turn arose from Avagadro's Hypothesis. In 1811, Avogadro postulated that equal volumes of gases contain equal numbers of molecules (at a given temperature and pressure). Following this, Joule explained in 1843 that the pressure of a gas is caused by the intense motion of the molecules which bombard the walls of the container. The exact pressure was proposed to be proportional to the speed and momentum of these molecules. Both Avogadro's and Joule's theories were disputed over a number of years on several grounds until 1906 when Jean Perrin directly observed, and Einstein explained, the Brownian movement in gases. These arguments, coupled with direct scientific observation, finally served to establish these theories as Laws.

However, it was Maxwell in 1848 who showed that molecules have a distribution of velocities and that they do not travel in a direct line. One experimental method used to show this was that ammonia molecules are not detected in the time expected, as derived from their calculated velocity, but arrive much later. This arises from the fact that the ammonia molecules *interdiffuse* among the air molecules by intermolecular collisions. The molecular velocity calculated for $NH_3$ molecules from the work done by Joule in 1843 was $5.0 \times 10^2$ meters/sec. at room temperature. This implied that the odor of ammonia ought to be detected in 4 millisec at a distance of 2.0 meters from the source. Since Maxwell observed that it took several minutes, it was fully obvious that the molecules did not travel in a direct path.

Analysis by Clausius in 1849 showed that the ammonia molecules travel only some 0.001 cm. between collisions with air molecules at atmospheric pressure, in time intervals of about $10^{-10}$ sec. between collisions. This meant that they describe a long and intricate path in the process of acquiring a displacement of several meters. The mathematics involved are intricate and we will not present them here. However, they **are** available for those who wish to study them (2). Nevertheless, the fact that there is an average distance that molecules travel between collisions has given us the concept of the Mean Free Path (which is the average distance between collisions).

A by-product of the work by Perrin and Einstein was the first reliable evaluation of Avogadro's number, the number of molecules in a mole. The best current value is believed to be: $6.022045_{31} \times 10^{23}$ molecules per kilogram mole. It is well to note, at this point, that all of these observations are the result of many hours of work by prior investigators from the past. Thus, we have their experience and intuition to draw upon for any work that we may carry forward to the benefit of mankind.

b. <u>The Liquid State</u>

In the liquid state, the molecules are still free to move in three dimensions but still have to be confined in a container in the same manner as the gaseous state if we expect to be able to measure them. However, there are important differences. Since the molecules in the liquid state have had energy removed from them in order to get them to condense, the translational degrees of freedom are found to be restricted. This is due to the fact that the molecules are much closer together and can interact with one another. It is this interaction that gives the liquid state its unique properties. Thus, the molecules of a liquid are not free to flow in any of the three directions, but are bound by intermolecular forces. These forces depend upon the electronic structure of the molecule. In the case of water, which has two electrons on the oxygen atom which do not participate in the bonding structure, the molecule has an electronic moment, i.e.- is a "dipole".

This results in a property which we call fluid viscosity since the moment of each molecule interacts with all of its nearest neighbors. Yet, the same vibrational and rotational states are still present but in a different form. That is, they are mutated forms of the same energy levels that we found in the gaseous state. This is illustrated in the following diagram:

1.2.3.-

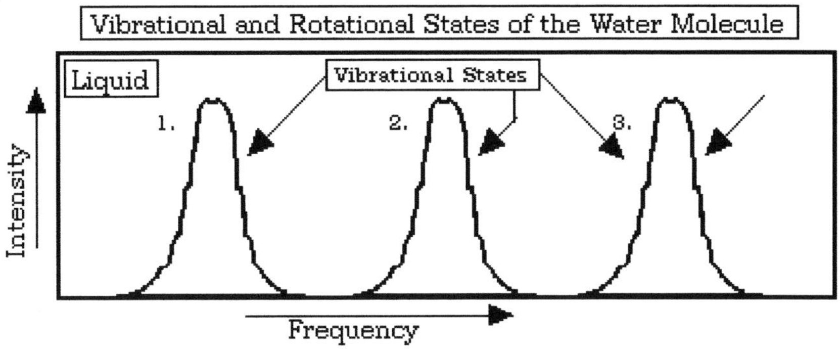


Note that in the above diagram there are still vibrational states but that the rotational states are "smeared" one into the other. There is little translational motion for the water molecules within the interior of the liquid unless they escape from the liquid phase. If they do so, we call this "evaporation" (This may be contrasted to the escape of molecules from a solid which we call "sublimation").

Most liquids do have a defined vapor pressure which means that molecules can and do escape from the surface of the liquid to form a gas. This is another reason that the properties of a liquid vary from those of the gaseous state. Hence, we still have the vibrational and rotational degrees of freedom left in the liquid, but not those of the translational mode. A representation of water molecules in the liquid state is presented in the following diagram, shown as 1.2.4. on the next page.

(Ignore the open spaces since this is merely a simulation). As shown in this diagram, the water molecules fill the container and also have a free surface from which they can escape.

Thus, we conclude that the molecules of a liquid are free to slide past one another but the overall assemblage of molecules does not have a definitive form, except that of the container used to hold it. For this reason, a liquid has been defined as "a substance or state of matter which has the capacity to flow under extremely small shear stesses to conform to the shape of any confining vessel, but is relatively incompressible and lacks the capacity to expand without limit" (1).

1.2.4.-

Representation of the Liquid State Using Water Molecules as Examples

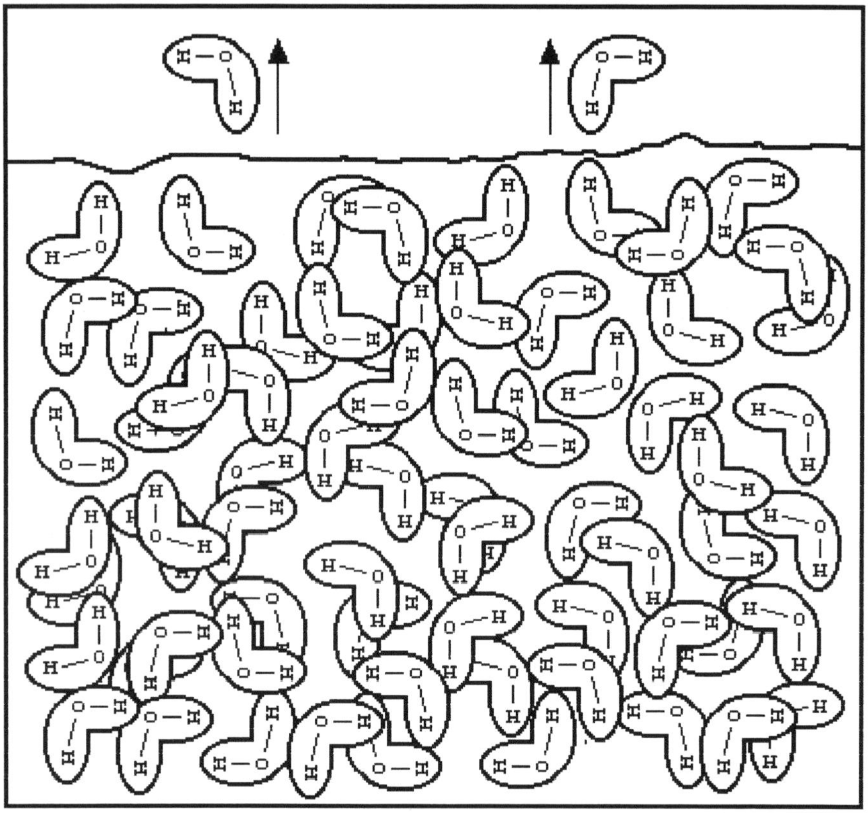

Therefore, as we change the state of matter, the translational degrees of freedom in liquids become severely restricted in relation to those of the gaseous state. And, the vibrational and rotational degrees of freedom appear to be somewhat restricted, even though many of the liquid vibrational and rotational states have been found to be quite similar to those of the gaseous state.

c. The Solid State

If we now remove more energy from the liquid, it finally reaches a temperature where it "freezes", that is - it converts to a solid. What happens, in a molecular sense, is that the molecules become **ordered.** Another way to say this is that they form a lattice-like framework. A representation of the solid state is shown in the following diagram:

1.2.5.-

Representation of the Solid State Using Water Molecules as an Example

In this case, the molecules are found to have arranged themselves in orderly rows. Although we can see only the top layer, there are several layers below with the exact same arrangement (This is not the exact arrangement found in "ice" but is a stylized representation of the solid state of water).

It should thus be clear that as we change the state of matter, the translational degrees of freedom present in gases become restricted in liquids and disappear in solids. For gaseous molecules, both vibrational and rotational degrees of freedom are present while those of the liquid state are modified to the point where only vibrational states can be said to truly free states. The same cannot be said for molecules in the solid state. In the solid, only the vibrational states remain, but they do not resemble those of the gaseous and liquid states. This is illustrated as follows:

1.2.6.-

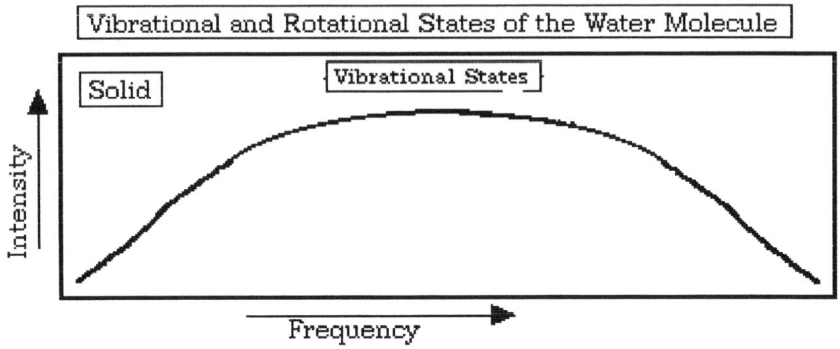


Note that these vibrational states in the solid are not recognizable in terms of those of the gaseous or liquid states. And, the rotational states appear to be completely absent. It has been determined that solids have quite different vibrational states which are called "phonon modes". These vibrational states are **quantized** vibrational modes within the solid structure wherein the atoms all vibrate **together** in a specific pattern. That is, the vibrations have clearly defined energy modes in the solid.

The number of phonon modes are limited and have been described as "phonon branches" where two types are present, "optical " and "acoustical". (These names arose due to the original methods used to study them in solids).

For the solid state, there will be a specific number of phonon branches found in the vibrational spectrum of any given solid, which depends upon the number of atoms composing the solid. The number of branches found in the phonon modes can be found from the following equations, given in 1.2.7. on the next page. Contrast this situation with those of both the liquid state and the gaseous state. (What do we mean by "quantized phonon modes in the solid?- the vibrations have specific amounts of energy and these modes appear only as resonant vibrations- i.e.- the molecules or atoms vibrate together only at certain frequencies, depending upon the mass or the molecules (atoms or ions) and the chemical bonds holding the structure together).

1.2.7.- Number of Branches of Phonon Dispersion:

Equations: Acoustical = y atoms/molecule Optical = 3 y -3

Phonon States: For water with 3 atoms per molecule:
Acoustical = 3
Optical = 6 (i.e.- [3x3] -3)

The major difference, then, between the 3 phases we have discussed is that the solid consists of an assemblage of ***close-packed*** molecules which we have shown to have arisen when we removed enough **energy** from the molecules so as to cause them to condense and to form the solid state. Let us now examine the properties of atoms or molecules when they are crowded together to form a "close-packed" solid.

**1.3.**- THE CLOSE PACKED SOLID

We have already said that the solid differs from the other states of matter in that a long range ordering of atoms or molecules has appeared. To achieve long range order in any solid, one must stack atoms or molecules in a symmetrical way **to completely fill** all of the space available. This is not a trivial matter since solids require that all of the atoms be arranged in a symmetical pattern **in three dimensions**. Thus, if we could actually see these atoms in a solid, we would find that they are composed of specific "building blocks", which we shall call "propagation models" or "Units". (Actually, it is now possible to directly observe the packing of atoms in solids, but that is another story, that of the atomic-force microscope).

Such models must be entirely symmetrical in three-dimensional space so that we can arrange them properly to form a 3-dimensional solid. Because of this limitation, we find that only certain types of propagation models will work. And, in doing so, we can gain further insight into the properties of a solid. To understand this, consider the following discussion.

Of the three-dimensional models available to us, only certain shapes can be used to form a symmetrical solid. These are **even-numbered** sets of atoms, arranged in either of the following forms:

1.3.1.- Atomic Forms Suitable for Assembling Long Distance Arrangements

Tetragonal = 4 atoms per Unit

Hexagonal = 6 atoms per Unit

Cubic = 8 atoms per Unit

These specific shapes are shown in the following diagram:

1.3.2.-

Propagation Models Useful for Forming the Solid State

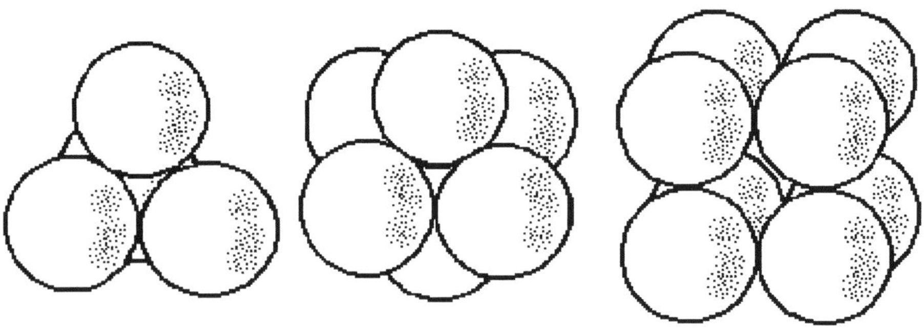

Tetragonal
4 Atoms per Unit

Hexagonal
6 Atoms per Unit

Cubic
8 Atoms per Unit

Note that there are 4, 6 or 8 atoms per Unit, but odd-numbered units of 5, 7, or 9 atoms per Unit are not used. If such Units are tried, one finds that they cannot be fit together in a three-dimensional pattern which has long range order and symmetry (If you do not believe this, try it yourself. You will find that a five-atom Unit, which is the hexagonal Unit given above minus one atom, cannot be stacked together without losing part of the three-dimensional space. In other words, there will be "holes" in the structure). You might wonder why we did not specify either 1, 2 or 3 atoms per Unit. The reason lies in the fact that 1 atom, or 2 atoms are not three-dimensional but are two-dimensional (a fact that you can ascertain by glueing some ping-pong balls together to make individual models- these components can be used to form the three Units given in 1.3.2.). Note that we have now established that only specific shapes can be used to assemble a solid. The next question that needs to be answered is how the solid would appear if we could see the atoms directly.

There are, in general, two kinds of solids, **homogeneous** and **heterogeneous**. The former is composed from atoms that are all the same and the latter from atoms not the same. If the atoms are all of one kind, i.e.- one of the elements, the problem is straight forward. Sets of eight atoms, each set arranged as a cube, will generate a cubic structure. Two sets of three atoms, each set of three arranged in a triangle, will propagate a hexagonal pattern with three dimensional symmetry.

Elemental solids having a tetragonal structure are very few and it is easy to ascertain that most of the elements form structures that are either cubic or hexagonal, but rarely tetragonal. The reason for this is that tetragonal units are more conducive for the case where not all of the atoms are the same, i.e.- the heterogeneous case. One example of this is the case where we have 1 phosphorous atom, combined with 4 oxygen atoms to form the phosphate Unit, i.e.- $PO_4$. Another case might be where we combine 1 carbon atom with 3 oxygen atoms to form a carbonate Unit. In the first case, we have a tetragonal Unit, with the phosphorous atom at the center of the 4 atoms composing a tetrahedron. In the second case, the carbon atom aligned with 3 oxygen atoms which lie in the form of a tetrahedron. In the first case, the tetrahedron arises because of spatial preferences, whereas in the second case, we know that the carbon atom prefers to form tetrahedral bonds. Thus, it should be clear that the specific structure found in a solid arises either because of spatial considerations, or because of bonding preferences of certain atoms comprising the structure.

In forming a solid from atoms or molecules, part of the problem lies in the fact that we must handle a very large number of atoms or molecules. For example, $6.023 \times 10^{23}$ (60.23 septillion) atoms comprise one mole and we must stack each of these in a symmetrical manner to form the close-packed solid. (As a matter of comparison, calcium carbonate, whose molecular weight is about 100 grams/mole, can be held in both of your cupped hands. This amount contains the 6.0 septillion ($10^{23}$) molecules).

There is also another important factor. That is, in building a solid structure one finds that **solid structures are based on the largest atom present**, as well as how it stacks together (its valence) in space filling-form. For most inorganic compounds, this is the **oxygen atom**, e.g.- oxides, silicates, phosphates, sulfates, borates, tungstates, vanadates, etc. The few exceptions involve chalcogenides, halides, hydrides, etc., but even in those compounds, the structure is based upon aggregation of the largest atom, e.g..- the sulfur atom in ZnS. Zinc sulfide exhibits two structures, sphalerite- a cubic arrangement of the sulfide atoms, and wurtzite- a hexagonal arrangement of the sulfide atoms. Divalent zinc atoms have essentially the same coordination in both structures. The following diagram shows how such a solid would look on an atomistic scale:

1.3.3.-

Rows of Atoms Comprising a Solid Lattice

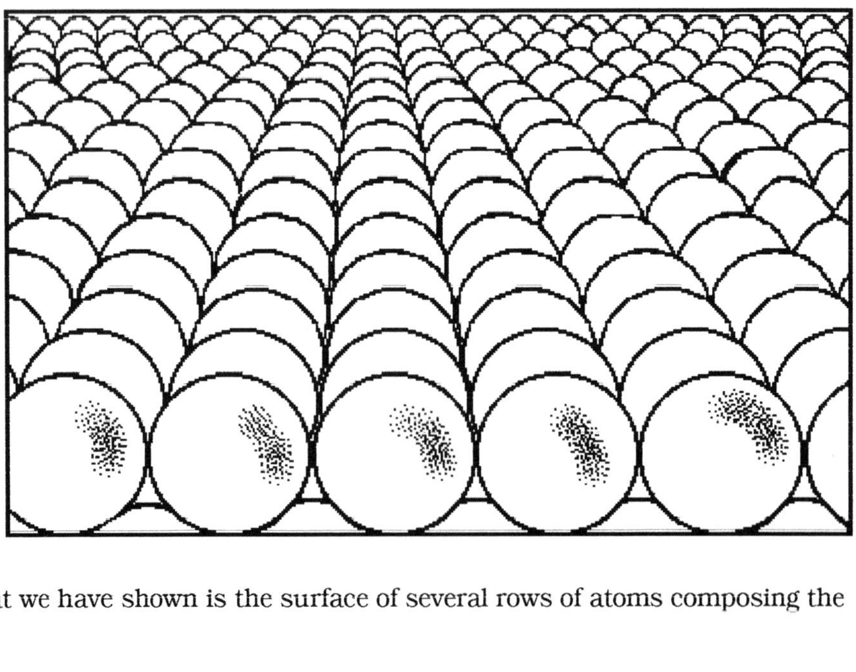

What we have shown is the surface of several rows of atoms composing the solid structure. We see the outermost layer of the structure which is likely to be the oxygen atoms composing the compound.

Let us now consider propagation units in their space-filling aspects. As we defined them above, they are solid state building blocks that we can stack in a symmetrical form to infinity. Thus, 4-atoms will form a 3-dimensional **tetrahedron** (half a cube is only 2-dimensional) which is a valid propagation unit. This means that we can take tetrahedrons and fit them together 3-dimensionally to form a symmetrical structure which extends to infinity. However, the same is not true for 5-atoms, which forms a four-sided pyramid. This shape cannot be completely fitted together in a symmetrical and space-filling manner, because when we stack these pyramids containing 5 atoms, we find that their translational properties preclude formation of a symmetrical structure because there is lost space between the pyramids which results in holes in the long-range structure.

However, if one more atom is added to the pyramid, we then have an **octahedron** which is space-filling with translational properties. This results in a hexagonal structure. Going further, combinations of seven atoms are asymmetrical, but eight atoms form a cube which can be propagated to infinity to form a cubic structure. Note that by turning the top layer of four atoms by 45° (see 1.3.1.), we have a hexagonal unit which is related to the hexagonal unit composed of 6 atoms, two triangles atop of each other. By taking ping-pong balls and gluing them together to form the propagation units shown above, one can get a better perspective between cubic and hexagonal close-packing. Although we can continue with more atoms per propagation unit, it is easy to show that all of those are related to the four basic propagation units found in the solid state, to wit:

1.3.4.- <u>Propagation Units Usually Found in Solids</u>

| | |
|---|---|
| Tetrahedron | (4) |
| Octahedron | (6) |
| Hexagon | (6 or 8) |
| Cube | (8) |

We thus conclude that structures of solids are based, in general, upon these four (4) **basic** propagation units, which can be stacked in a symmetrical and space-filling form to near infinity. The symmetry will be that of the largest atom in the structure, usually oxygen in inorganic solids. Variation of structure depends upon whether the other atoms forming the structure are larger or smaller than the basic propagation units composing the structure.

In many cases they are smaller and will fit into the **interstice** of the propagation unit, illustrated by the $PO_4$ - tetrahedron mentioned above. In this case, the $P^{5+}$ atom is small enough to fit into the center (interstice) of the tetrahedron formed by the four oxygen atoms. If we combine it with $Er^{3+}$ (which is slightly smaller than $PO_4$), we obtain a tetragonal structure, a sort of elongated cube of high symmetry. But if we combine it with $La^{3+}$, which is larger than $PO_4^{3-}$, a monoclinic structure with low symmetry results.

There still is one aspect of phase chemistry that we have not yet addressed. That is the case where more than one solid phase exists. The basic properties of a solid include two factors, namely composition and structure. We will address structures of solids in the next chapter. The composition of solids is one where the individual constituents will vary if the solid is heterogenous. That is, the two types of inorganic solids vary according to whether they are homogeneous or heterogeneous. This is shown in the following:

1.3.4.- Properties of Solids

| | Type | Variation In |
|---|---|---|
| Homogeneous Solids: | Elemental | Structure only |
| Heterogeneous Solids: | Compounds | Structure and Composition |

What this means is that elements (e.g.- metals) can have more than one structure. For example, Fe exists in 4 structures, i.e.- Fe is tetramorphous. $\alpha$-Fe is the one stable at room temperature. But, it transforms to $\beta$-Fe, then $\gamma$-Fe and finally $\delta$ -Fe as the temperature is raised. These changes are all reorganizations in packing density before the melting temperature is reached. In constrast, heterogeneous solids usually exist in one strucural modification (a few change structure as temperature is increased) but they exist in many compositions, depending upon how they were formed. For example, a large number of calcium silicates are known, including:

1.3.5.- Known Calcium Silicates

$Ca_2SiO_3$ $Ca_3Si_2O_7$ $CaSi_2O_5$ $Ca_2SiO_4$ $Ca_4(Si_4O_{17})(OH)_2$
$Ca_4Si_2O_7(OH)_2$ $Ca_9Si_6O_{21}(OH)$ $CaSiO_3$

This is only a partial list (I counted 58 known compounds of varying composition). Each has its own composition and structure. In order to differentiate among such complicated systems, i.e.- oxygenated compounds of calcium and silicon, we resort to what is called a "phase-diagram". A phase diagram shows those compounds which are formed when varying molar ratios of CaO and $SiO_2$ are reacted together.

## **1.-4.** Phase Relations Between Individual Solids

To differentiate and to be able to determine the differences between the phases that may arise when two compounds are present (or are made to react together), we use what are termed "phase-diagrams" to illustrate the nature of the interactions between two solid phase compositions.

Consider the following. Suppose we have two solids, "A" and "B". It does not matter what the exact composition of each may be. A will have a specifc melting point (if it is stable and does not decompose at the M.P.) and likewise for the compound, B. We further suppose that the M.P. of B is higher than that of A. Furthermore, we suppose that A and B form a **solid solution** at all variations of composition. What this means is that from 100% A- 0% B to 50% A-50% B to 0% A-100% B, the two solids dissolve in one another to form a completely homogenous single phase. If we plot the M.P. of the system, we find that three different curves could result, as shown in the following diagram, given as 1.4.1. on the next page.

In this case, the melting point of the ideal solid solution should increase linearly as the ration of B/A increases. However, it usually does not.

1.4.1.-

Possible Changes in Melting Point as a Function of Composition

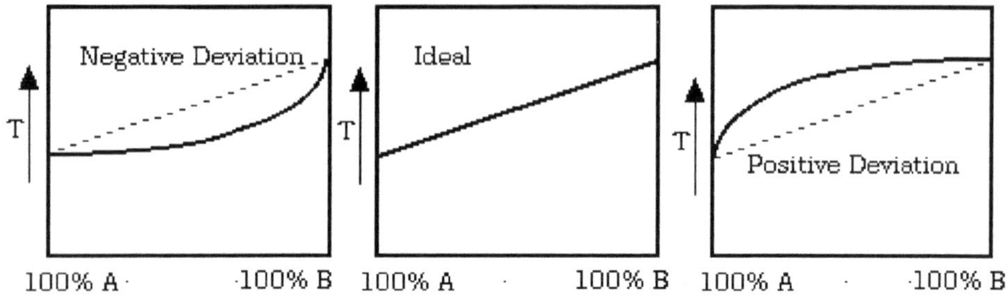


Either a negative deviation or a positive deviation is regularly observed. In any phase diagram, composition is plotted against temperature. In this way, we can see how the interactions between phases change as the temperature changes and the behavior as each solid phase then melts. Either two-phase or three phase systems can be illustrated. This is shown in the following:

1.4.2.-

Two Phase System

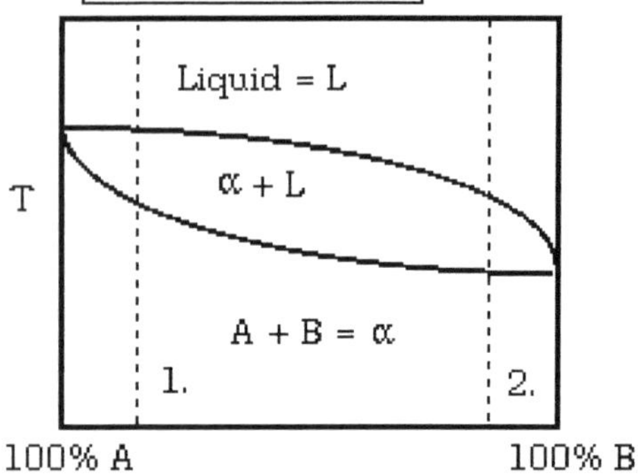


Three Phase System

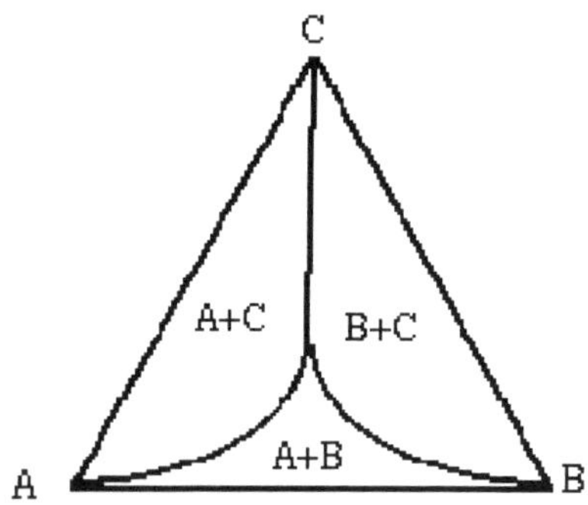


Here, the two-phase diagram is simplified to show a hypothetical phase involving "A" and "B" compounds which form a solid solution from 100% A to 100% B. The solid solution is labelled as "α". The melting temperature of A is higher than that of B. Therefore, the melting temperature of α drops as the composition becomes richer in B. At specific temperatures on the diagram (see 1. & 2.), a two-phase system appears, that of a liquid plus that of α. Finally, as the temperature rises, the melt is homogenous and the solid, α, has melted. In the three-phase system, only the relationship between A, B and C can be illustrated on a two-dimensional drawing. A three-dimensional diagram would be required to show the effect of temperature as well.

The phase diagrams we have shown are based upon the fact that A and B form solid mutually soluble solid state solutions. If they do not, i.e.- they are not mutually soluble in the solid state, then the phase diagram becomes more complicated. As an example, consider the following, which is the case of limited solid solubility between A and B. (N.B.- study the following diagrams carefully)"

1.4.3.-

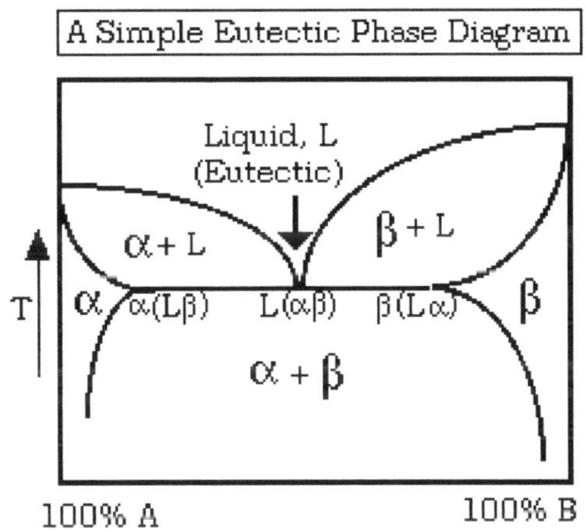


Here, we have the case where A & B form two slightly different componds, α and β. The composition of α is $A_xB_y$ while that of β is $A_uB_v$ , where the subscripts indicate the ratio of A to B (these numbers may be whole numbers or they may be fractional numbers). At low B concentrations, α exists as a solid (the left side of the diagram). As the temperature increases, a melt is obtained and α remains as a solid in the melt (L) {This is indicated by the details given just below the straight line in the diagram}. As the temperature is increased, then α melts to form a uniform liquid. In the middle concentrations, α and β exist as two separate phases. At 75% A-25% B, β melts to form a liquid plus solid α whereas just the opposite occurs at 25% A and 75% B. When A equals B, then both α and β melt together to form the liquid melt, i.e.- the "eutectic point".

In two similar cases but differing cases, i.e.- a system with a melting point minimum and another with a different type of limited solid solubility, the behavior differs as shown in the following diagram:

1.4.5.-

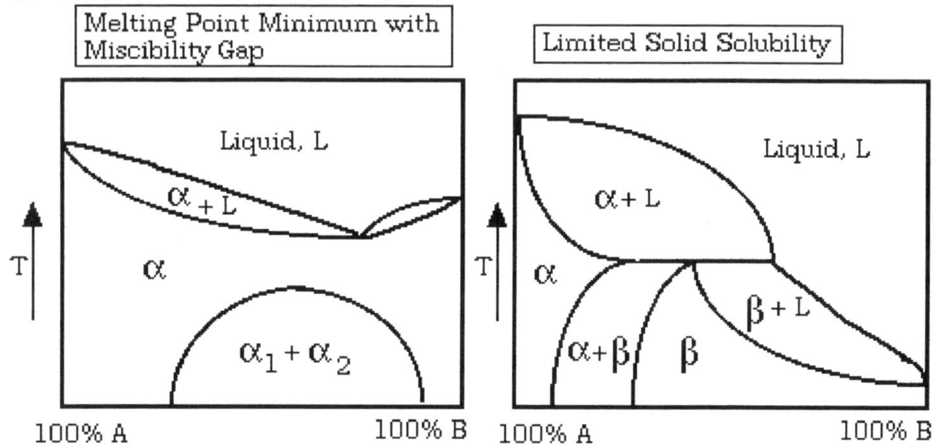


In the case on the left, a composition exists which in which a minimum melting point exists. Again, A and B form α which partially melts to form α + L. However, two forms of α also exist, a compositional area known as a "miscibility gap", i.e.- " $\alpha_1$ and $\alpha_2$". But at higher temperatures, both of these melt into the single phase, α. Finally, we obtain the melt plus α. Note that at about 80% $\alpha_1$-20% $\alpha_2$ , a melting point minimum is seen where it melts directly instead of forming the two-phase system, α + L.

In the case of limited solid solubility, the phase behavior becomes more complicated. Here, both α and β form (where the actual composition of α is $A_xB_y$ while that of β is $A_uB_v$ , as given before. Note that the values of x and y change, but that we still have the α phase. The same holds for u and v of the β phase). At low A concentrations, α exists alone while β exists alone at higher B concentrations. A region exists where the two phase system, + exists. We notew that as the temperature is raised, a two phase system is also seen consisting of the melt liquid phus a solid phase. The phase-behavior shown on the right side of the diagram arises because the two phases, A and B, have limited solubility in each other.

Now, let us consider the case where three (3) separate phases appear in the phase diagram, shown as follows:

1.4.6.-

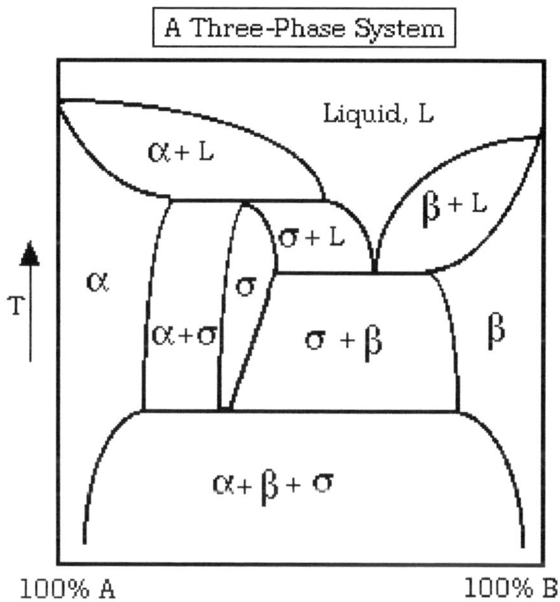


In this case, we have three (3) separate phases that appear in the phase diagram. These phases are α , β , and σ , whose compositions are: $\alpha = A_xB_y$ , $\beta = A_uB_v$ and $\sigma = A_cB_d$ , respectively (the values of x, y, u, v, c, and d all differ from each other so that $A_xB_y$ is a specific compound as are the others). By studying this phase diagram carefully, you can see how the individual phases relate to each other. As you can see, a phase diagram can become quite complicated. However, in most cases involving real compounds, the phase diagrams are usually simple. Those involving compounds like silicates can be complex, but those involving alloys of metals show simple behavior like limited solubility.

## REFERENCES CITED

1. "Dictionary of Scientific and Technical Terms" - D.N. Lapedes- Editor in Chief, McGraw-Hill, New York (1978)

(2) "Enclyclopedia of Physics- 2nd Ed." - Edited by R.G. Lerner & G.L. Trigg, VCH Publishers, NY (1990).

## PROBLEMS FOR CHAPTER 1

1. Given 10.0 grams of ice at - 65 °C, calculate the number of calories required to change it to steam at 240 °C.

2. Do the same for 24.0 grams of ice at -11 °C converted to boiling water.

3. Look up the following heat capacities:

| | |
|---|---|
| $Br_2$ $CO_2$ $PCl_3$ $CH_4$ | - gases |
| $NH_3$ $C_2H_4$ | - liquids |
| $NH_3$ $CO_2$ | - solids |

4. List all of the metallic elements and their crystal structures.

5. List the number of expected vibrational and rotational modes for the following gaseous molecules:

$Br_2$
$CO_2$
$PCl_3$
$CH_4$
$C_2H_4$
$C_2H_6$

6. Draw the following planes for the cubic lattice (see Chapter 2 for help):

{101}
{222}
{101}
{301}.

7. Given the following phase diagram, label the individual phases present, assuming that three phases are present. Use α , β , and σ as symbols for the three phases:

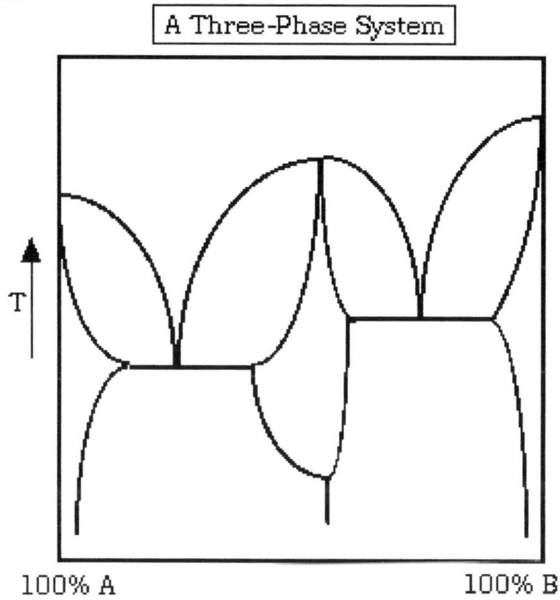

# Chapter 2

## Determining the Structure of Solids

This chapter will present more advanced topics than those of the first chapter in terms of determining the structure of solids. Consequently, you will gain some knowledge of how one goes about determining the structure of a solid, even if you never have to do it.

### 2.1- SCIENTIFIC DETERMINATION OF THE STRUCTURE OF SOLIDS

In this section, we will present the basis developed to explain the structure of solids. That is, the concepts that were perfected in order to accurately describe how atoms or ions fit together to form a solid phase. This work was accomplished by many prior workers who established the rationale used to define the structure of a symmetrical solid. As you will recall, we said that the basic difference between a gas, liquid and that of a solid lay in the orderliness of the solid, compared to the other phases of the same material.

We have already indicated that solids can have several forms or symmetries. To elucidate the structure of solids in more detail, at least three postulates apply:

**First**, the formation of a solid results from a symmetrical "stacking" of atoms to near infinity from atoms or molecules with spacings is much smaller than those found in the liquid or gaseous state.

**Secondly**, if we wish to gain an insight into how these atoms are arranged in the solid, we need to determine what kind of pattern they form while in close proximity.

**Thirdly**, we can then relate our pattern to other structures and thus define the symmetry of solids in general.

One way to approach a solution of the last two postulates is to define the structure of any given solid in terms of its **lattice points.** What this means is that by substituting a point for each atom(ion) composing the structure, we find that these points constitute a latticework, i.e.- three-dimensional solid, having certain symmetries (Examples of the symmetries to which we refer are given in 1.3.2. of Chapter 1).

A lattice is not a structure per se. **A lattice is defined as a set of three-dimensional points**, having a certain symmetry. These points may, or may not, be totally occupied by the atoms composing the structure. Consider a cubic structure such as that given in the following diagram:

2.1.1.- A Cubic Lattice

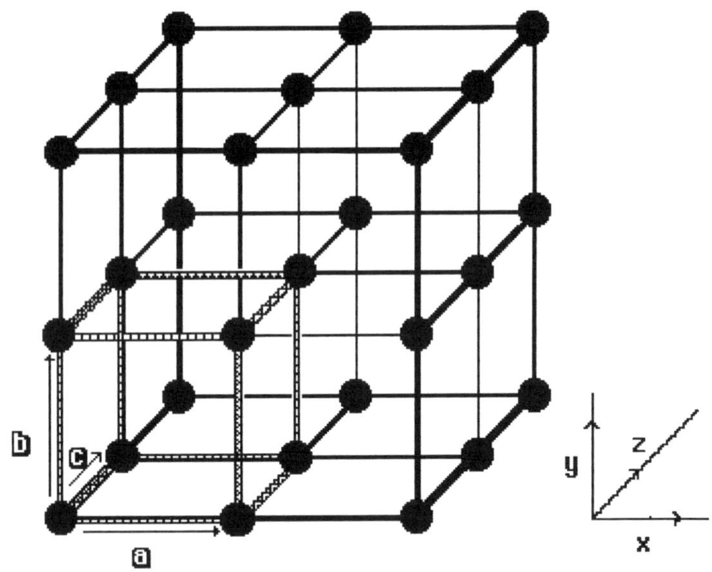


Here, we have a set of points occupied by atoms (ions) arranged in a simple three-dimensional cubic pattern. The lattice **directions** are defined, **by convention**, as x , y & z. Note that there are eight (8) cubes in our example.

In order to further define our cubic pattern, we need to analyze both the smallest Unit and how it fits into the lattice. We call the smallest Unit a "unit-cell" and find that the **unit-cell** is the smallest cube in the diagram.

We also need to define how large the unit-cell is in terms of both the length of its sides and its volume. We do so by defining the unit-cell directions in terms of its "lattice unit-vectors". That is, we define it in terms of the x, y, & z directions of the unit cell with specific vectors having directions corresponding to:

$$x \cong \mathbb{x}$$

$$y \cong \mathbb{y}$$

$$z \cong \mathbb{z}$$

with the length of each unit-vector being equal to 1.0. (Our notation for a **vector** henceforth is a letter which is "outlined", e.g.- $\mathbb{x}$ is defined here as the unit-cell vector in the "x" direction). As you probably remember, a "vector" is specified as a line which has both direction and duration from a given point.

We can now define a set of vectors called "translation" vectors, i.e.- $\mathbb{a}$ , $\mathbb{b}$, and $\mathbb{c}$ in terms of the following:

2.1.2.- $$\mathbb{a} = a_x\,\mathbb{x} + a_y\,\mathbb{y} + a_z\,\mathbb{z}$$

$$\mathbb{b} = b_x\,\mathbb{x} + b_y\,\mathbb{y} + b_z\,\mathbb{z}$$

$$\mathbb{c} = c_x\,\mathbb{x} + c_y\,\mathbb{y} + c_z\,\mathbb{z}$$

Here, the $\mathbb{a}$ , $\mathbb{b}$, and $\mathbb{c}$ vectors are now defined in terms of the a, b, and c lattice constants in each of the x, y, and z directions of 2.1.1., where a, b, and c are lattice constants. They are real numbers corresponding to ångstrom (Å) distance units, i.e.- $10^{-8}$ cm.

Note that we now have: $\mathbb{a}$ as a vector in terms of x, y and z vectors as a function of the lattice distances in the three directions, $a_x$, $a_y$ & $a_z$ . Here, $\mathbb{a} = a_x$, $a_y$ & $a_z$ , i.e.- $\mathbb{a}$ is composed of components in each of the three directions. The same holds for $\mathbb{b}$, and $\mathbb{c}$ vectors.

In general, we use only the lattice constants to define the solid structure (unless we are attempting to determine its symmetry). We can then define a structure factor known as the translation vector. It is a element related to the unit cell and defines the basic unit of the structure. We will call it $\mathbb{T}$ . It is defined according to the following equation:

2.1.3.- $$\mathbb{T} = n_1 \mathbb{a} + n_2 \mathbb{b} + n_3 \mathbb{c}$$

where $n_1$ , $n_2$ , and $n_3$ are intercepts of the unit-vectors, $\mathbb{x}$ , $\mathbb{y}$ , $\mathbb{z}$ , on the x, y, & z - directions in the lattice, respectively. The **unit cell volume** is then defined as:

2.1.4.- $\mathbb{V} = \{ \mathbb{x} \cdot \mathbb{y} \times \mathbb{z} \}$ (This is a "dot - cross" vector product).

Vector notation is being used here because this is the easiest way to define the unit-cell. The reason for using both unit lattice vectors and translation vectors lies in the fact that we can now specify unit-cell parameters in terms of a, b, and c (which are the intercepts of the translation vectors on the lattice). These cell parameters are very useful since they specify the actual length and size of the unit cell, usually in Å., as we shall see.

Although the cubic structure looks like the simplest one of those possible, it actually is the most complicated in terms of symmetry. What this means is that we can "spin" the lattice by holding it at a certain point and rotating it around this axis while still maintaining the same arrangement of atoms in space. (Take a cube and do this for yourself). The cubic lattice gives rise to a great many symmetry elements in contrast to less symmetrical lattices (As we will see, if the lattice is elongated and distorted from cubic, some of these symmetry elements do **not** arise).

In 1895, Röntgen experimentally discovered "x-rays" and produced the first picture of the bones of the human hand. This was followed by work by von Laue in 1912 who showed that solid crystals could act as diffraction gratings to form symmetrical patterns of "dots" whose arrangement depended upon how the atoms were arranged in the solid. It was soon realized that the atoms formed "planes" within the solid. In 1913, Sir William Henry Bragg and his son, William Lawrence Bragg, analyzed the manner in which such x-rays were reflected by planes of atoms in the solid. They showed that these reflections would be most intense at certain angles, and that the values would depend upon the distance between the planes of atoms in the crystal and upon the wavelength of the x-ray. This resulted in the Bragg equation:

2.1.5.- $$n \lambda = 2d \sin \theta$$

where d is the distance in angstroms (Å = $10^{-8}$ cm) between planes and θ is the angle in degrees of the reflection. In Bragg's x-ray diffraction equation, i.e.- the angle , θ , is actually the angle between a given **plane** of atoms in the structure and the **path** of the x-ray beam. The unit, "d", is defined as the distance between planes of the lattice and λ is the wavelength of the radiation. It was Georges Friedel who in 1913 determined that the intensities of reflections of x-rays from these planes in the solid could be used to determine the symmetry of the solid. Thus, by convention, we usually define planes, not points, in the lattice.

We have introduced the term "symmetry" in the solid. Let us now examine exactly what is meant by that term. To illustrate this concept, examine the symmetry elements of our cube, given in the following diagram, given on the next page as 2.1.6.

In this case, we can show that the planes of the cubic lattice are defined by moving various distances in the lattice from the origin. What is meant by this terminology is that as we move from the (0,0,0) origin just 1.00 unit-cell distance in the x" direction to the (1,0,0) point in the cubic lattice, we have defined the {100} plane (Note that we are using the $n_1$ intercept of the unit-cell).

In a like manner, if we move in the "y" direction a distance of just 1 unit-cell, we have defined the {010} plane. Using the the "z" directions gets us a set of one-dimensional planes (See line A in 2.1.6.).

2.1.6.-

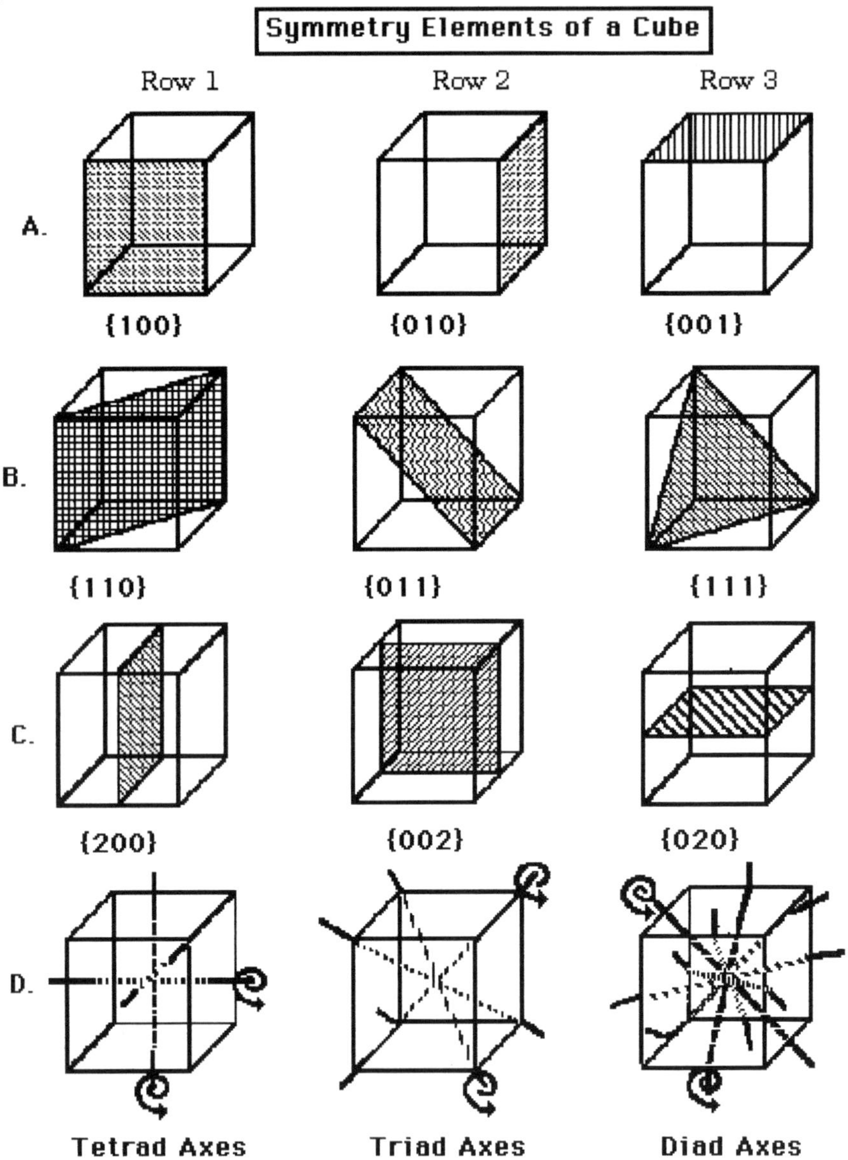


By moving 1 unit-cell distance in the both the x- and y-directions, the {110) has been defined, etc. (Line B above- note that we have not illustrated the {101} plane). Moving 1 unit-cell in all three directions then gives us the {111} plane. In a like manner, we can obtain the {200}, {020}

and {002} planes. A set of planes such as the {003} and {004} series are not as common, but do exist in some solids. Also shown are the rotational axes of the cubic lattice, shown as the diad, triad and tetrad axes. What this means is if we spin the cube on its face, we use the tetrad axis to determine that it takes a total of 4 turns to bring any specific corner back to its original position. Likewise, the triad axis uses the corner of the cube where the lattice is moved a total of three times, and the diad axis employs the diagonal **across the cube** to perform the symmetry operation.

Thus, the planes of the lattice are found to be important and can be defined by moving along one or more of the lattice directions of the unit-cell to define them. Also important are the symmetry operations that can be performed within the unit-cell, as we have illustrated in the preceding diagram. These give rise to a total of 14 different lattices as we will show below. But first, let us confine our discussion to just the simple cubic lattice.

It turns out that the method used to decribe the planes given above for the cubic lattice can also be used to define the planes of all of the known lattices, by use of the so-called "Miller Indices", which are represented by:

2.1.7.- $$\{\mathbf{h}, \mathbf{k}, \mathbf{l}\}$$

The way that Miller Indices arose can be understood by considering the history of crystal structure work, accomplished by many investigators.

In 1921, Ewald developed a method of calculating the sums of diffraction intensities from different planes in the lattice by considering what is called the "Reciprocal Lattice". The reciprocal lattice is obtained by drawing perpendiculars to each plane in the lattice, so that the axes of the reciprocal lattice are perpendicular to those of the crystal lattice. This has the result that the planes of the reciprocal lattice are at right angles (90°) to the real planes in the unit-cell. For our cubic lattice, the reciprocal lattice would look as shown in the following diagram, given as 2.1.8 on the next page.

2.1.8.-

The Reciprocal Lattice within a Simple Cube

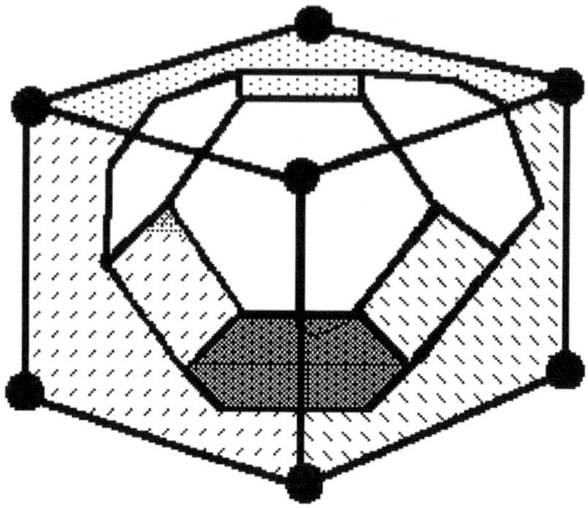

In this diagram, a series of hexagon-shaped planes are shown which are orthogonal, or 90 degrees, to each of the corners of the cubic cell. Each plane connects to another plane (here shown as a rectangle) on each **face** of the unit-cell. Thus, the faces of the lattice unit-cell and those of the reciprocal unit-cell can be seen to lie on the same plane while those at the corners lie at right angles to the corners.

The notion of a reciprocal lattice arose from Ewald who used a sphere to represent how the x-rays interact with any given lattice plane in three dimensional space. He employed what is now called the *Ewald Sphere* to show how reciprocal space could be utilized to represent diffractions of x-rays by lattice planes. Ewald originally **rewrote** the Bragg equation as:

2.1.9.- $$\sin \theta = \frac{n \lambda /}{2d \{hkl\}} = \frac{1/d \{hkl\}}{2/\lambda}$$

Using this equation, Ewald applied it to the case of the diffraction sphere which we show in the following diagram as 2.1.10. on the next page. Study this diagram carefully. In this case, the x-ray beam enters the sphere enters from the left and encounters a lattice plane, L. It is then diffracted by the angle 2θ to the point on the sphere, P, where it

2.1.10.-

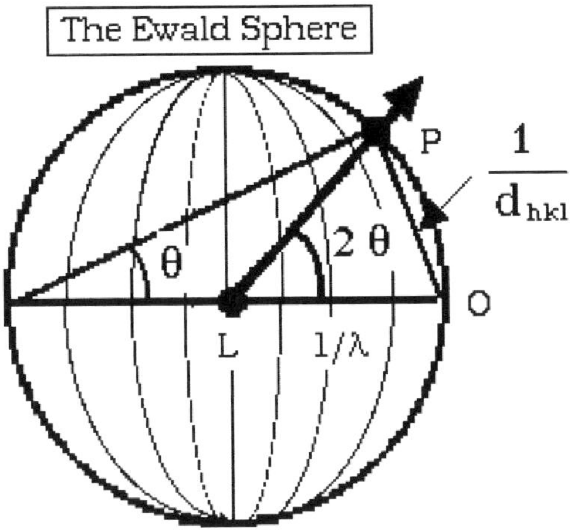


is registered as a diffaction point on the reciprocal lattice. The distance between planes in the reciprocal lattice is given as $1/d_{hkl}$ which is readily obtained from the diagram. It is for these rcasons, we can use the Miller Indices to indicate planes in the real lattice.

The reciprocal lattice is useful in defining some of the electronic properties of solids. That is, when we have a semi-conductor (or even a conductor like a metal), we find that thc electrons are confined in a band, defined by the reciprocal lattice. This has important effects upon thc conductivity of any solid and is known as the "band theory" of solids. It turns out that the reciprocal lattice is also the site of the Brillouin zones, i.e.- the "allowed" electron energy bands in the solid. How this originates is explained as follows.

The free electron resides in a quantized energy well, defined by k (in wave-numbers). This result can be derived from the Schroedinger wave-equation. However, in the presence of a periodic array of electromagnetic potentials arising from the atoms confined in a crystalline lattice, the energies of the electrons from all of the atoms are severely limited in orbit and are restricted to specific allowed energy bands. This potential originates from attraction and repulsion of the electron clouds from the periodic array of atoms in the structure. Solutions to this problem were made by Bloch in 1930 who showed they had the form (for a one-dimensional lattice):

2.1.11.- $\Psi = e^{ikx}\,u(x)$ - one dimensional

$\Psi_k(a) = e^{ika}\,u_k(x)$ - three dimensional

where k is the wave number of the allowed band as modified by the lattice, a may be x, y or z, and $u_k(x)$ is a periodical function with the same periodicity as the potential. One representation is shown in the following:

2.1.12.- The Allowed Energy Bands (Brillouin Zones) in a Crystal

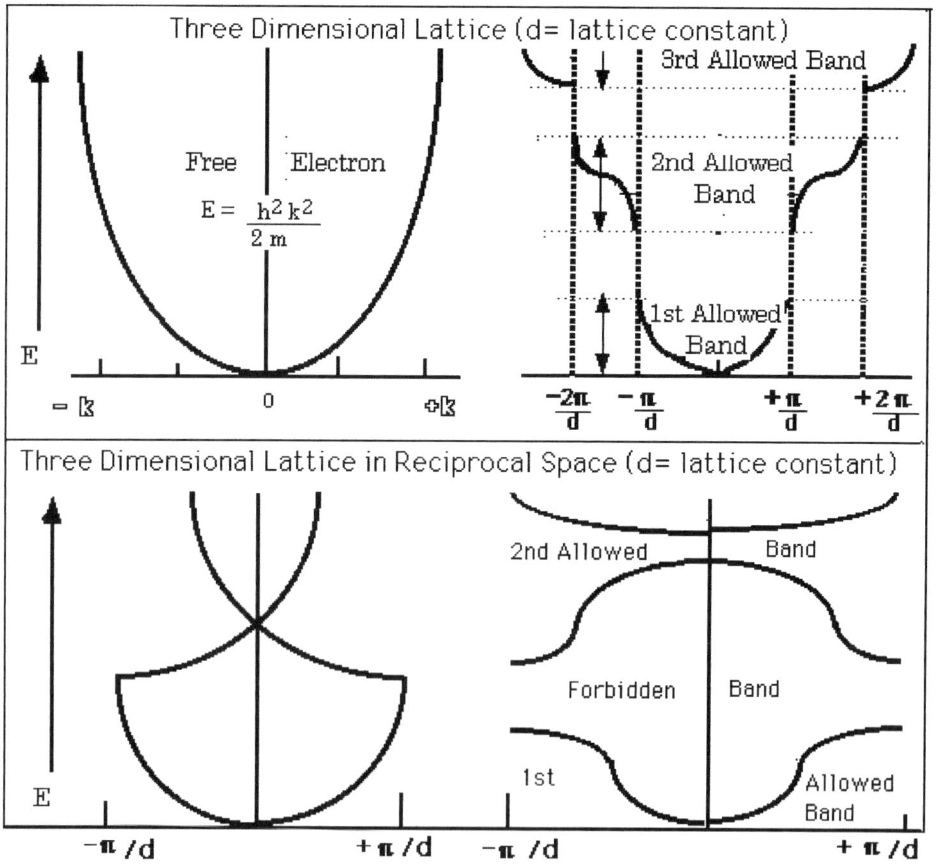

We have shown the least complicated one which turns out to be the simple cubic lattice. Such bands are called "Brilluoin" zones and , as we have said, are the allowed energy bands of electrons in any given crystalline latttice. A number of metals and simple compounds have been studied and their Brilluoin structure determined. However, when one gives a representation of the energy bands in a solid, a "band-model" is usually presented. The following diagram shows three band models:

2.1.13.- Energy Band Models

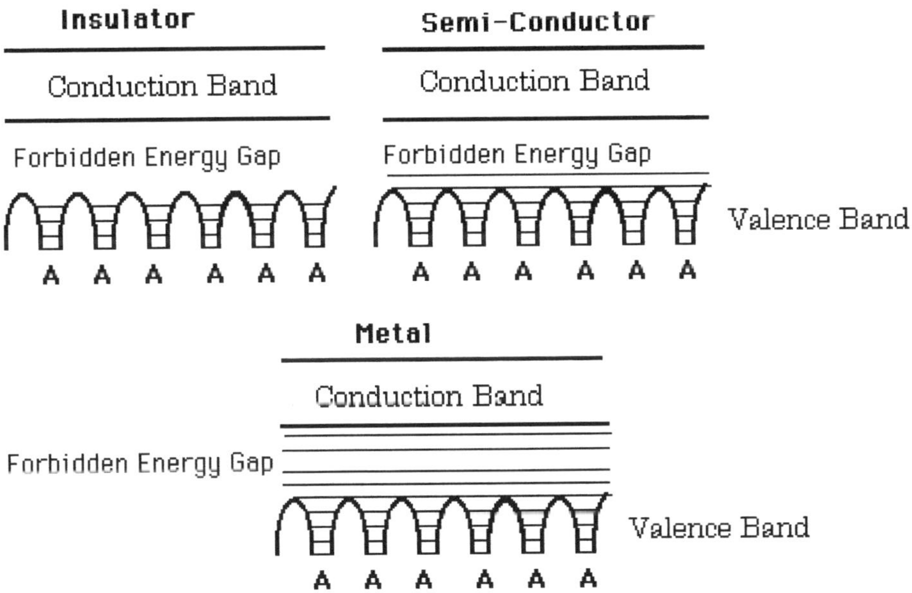


In the solid, electrons reside in the valence band but can be excited into the conduction band by absorption of energy. The energy gap of various solids depends upon the nature of the atoms comprising the solid. Semi-conductors have a rather narrow energy gap (forbidden zone) whereas that of insulators is wide (metals have little or no gap). Note that energy levels of the atoms "A" are shown in the valence band. These will vary depending upon the nature atoms present. We will not delve further into this aspect here since it is the subject of more advanced studies of electronic and optical materials.

Returning to the unit-cell, we can also utilize the vector method to derive the origin of the Miller Indices. The general equation for a plane in the lattice is:

2.1.14.- $\mathbb{n}_{x,y,z} \circ \mathbb{r} = d$

where $\mathbb{n}_{x,y,z}$ is the lattice unit vector in any of the three dimensions, $\mathbb{r}$ is the general vector of the lattice from the origin and d is, of course, the distance between planes of the lattice from the origin. A better perspective is shown in the following diagram:

2.1.15.-

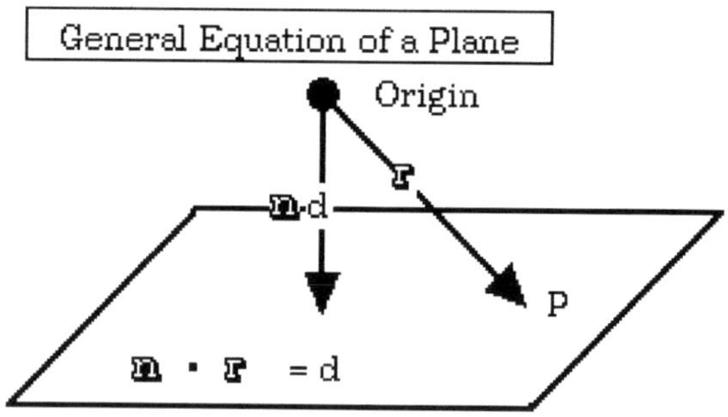


If $\mathbb{n}_{x,y,z}$ is normal to the real lattice plane (as in the reciprocal Lattice), we can define the x, y and z directions of the unit cell in terms of the unit cell vectors and the corresponding intercepts of the reciprocal lattice, i.e.- h, k, and l:

2.1.16.- $x = \mathbb{x} / h \qquad y = \mathbb{y} / k \qquad z = \mathbb{z} / l$

Then, the lattice unit vector will be:

2.1.17.- $\mathbb{n}_{h,k,l} = 1 / \{h^2 + k^2 + l^2\}^{1/2} \{h\,\mathbb{a} + k\,\mathbb{b} + l\,\mathbb{c}\}$

where we have converted the x, y and z directions into intercepts of the reciprocal lattice cell. As we shall see, this equation is also useful in deriving equations describing symmetry operations and plane spacings of the real lattice.

To show that the above discussion is true, consider the following. Let us set up a plane of a cubic lattice as shown in the following diagram:

2.1.18.-

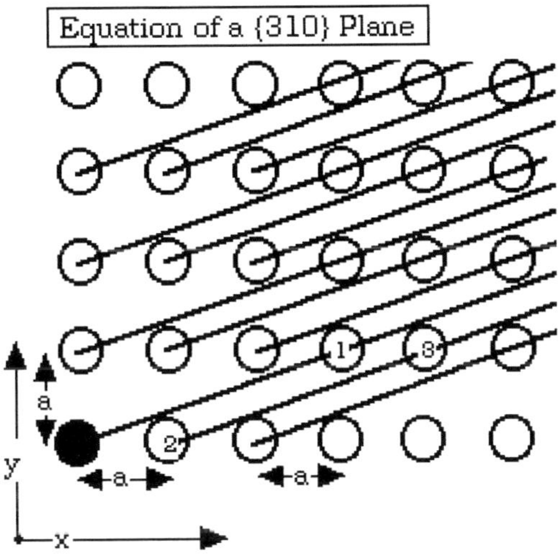


In this case, we have shown a {310} plane of a cubic lattice. Note that we move 3 steps of "a" in the x-direction and one in the "y" direction to define this plane. Since a plane is two-dimensional, this representation is accurate. The equation of any one of the elements of this plane is thus:

2.1.19.- $$a = 3x + y$$

and if we had b or c (both of which equal "a" in a cubic lattice), we could then use a , b, & c to represent the lengths of the sides of the unit-cell which in turn define the planes **within the unit cell**. For the cubic lattice, then we have:

2.1.20.- $$d = 1/(h^2 + k^2 + l^2)^{1/2} \cdot [h\,\mathbb{x} + h\,\mathbb{y} + h\,\mathbb{z}]$$

Note that just three points have been used to define our plane (see 2.1.6.). The original definition was given by Whewell in 1825 and by Grassman in 1829, but was popularized by Miller in 1839. Since three points of a lattice can be used to define a plane, it is obvious that such planes can be defined by using only three points in space, i.e.- a , b & c, to characterize the set of planes within the lattice. Miller Indices are the reciprocals of the intercepts, a , b & c, of the chosen plane on the x , y , z - directions in the lattice (and relate to the $n_1$, $n_2$ and $n_3$ intercepts which we used in our definition of the unit-cell). How they are used is shown in the following table:

2.1.21.- Points on the Lattice and Corresponding Miller Indices

| a, b, c | MILLER INDICES |
|---|---|
| (1, 0, 0) | {100} |
| (1/2, 0, 0 ) | {200} |
| (0 , 1/2, 0) | {020} |
| (0, 0, 1/2) | {002} |

Here, we have defined the sum total of the intercepts in terms of the unit-cell distance being set equal to 1.00, i.e.-

2.1.22.- $$a/(1/h) + b/(1/k) + c/(1/k) = 1$$

Thus, these intercepts are given in terms of the actual unit-cell length found for the specific structure, and not the lattice itself. The Miller Indices are thus the indices of a *stack of planes* within the lattice. Planes are important in solids because, as we will see, they are used to locate atom positions within the lattice structure.

The final factor to consider in our study of how to define the structure of solids is that of the **angle** between the x, y, and z directions in the lattice. In our example so far, all of the angles were 90° in all directions (also called "orthogonal"). If the angles are not 90°, then we have additional lattices to define. For a given unit-cell defined by the unit-cell lengths, a, b & c, the corresponding angles have been defined as:

2.1.23. - Notation Used for Defining Angles in the Solid Lattice

$$\alpha = 2\pi / n_1$$
$$\beta = 2\pi / n_2$$
$$\gamma = 2\pi / n_3$$

where $\alpha$ is the angle in the x-direction, $\beta$ is in the "y" direction, etc. The intercepts are those of the lattice and not of the unit-cell. Note that $\alpha$ , $\beta$, $\gamma = 90\ °$ for the cubic lattice as given above as 2.1.1.

This completes our discussion of the basis and factors developed by past investigators to describe and conceptulize the structure of solids. You will note that we have not yet fully described the symmetry factor of solids. The reason for this is that we use symmetry factors to characterize solid structure without resorting to the theoretical basis of structure determination. That is, we have a standard method for categorizing solid structures. We say that salt, NaCl, is cubic. That is, the $Na^+$ ion and the $Cl^-$ ion are alternately arranged in a close-packed cubic structure. The next section now investigates these structure protocols.

**2.2.**- Solid State Structure Conventions and Protocols

We have already disussed structure factors and symmetry as they relate to the problem of defining a cubic unit-cell and find that still another factor exists if one is to completely define crystal structure of solids. This turns out to be that of the individual arrangement of atoms within the unit-cell. This then gives us a total of three (3) factors are needed to define a given lattice. These can be stipulated as follows:

2.2.1.- <u>Factors Needed to Fully Define a Crystal Lattice</u>

I - unit-cell axes , intercepts and angles
II - rotational symmetry
III - localized space group symmetry

In the first Factor , if the cell lengths are not equal, i.e. $a \neq b \neq c$, we no longer have a cubic lattice, but some other type. And if the angles are no longer 90 °, i.e.- $\alpha \neq b \neq g$ , then we have a change in the lattice type as well. Additionally, we can have the situation where $\alpha = \beta \neq \gamma$ , $\alpha \neq \beta = \gamma$ and $\alpha \neq \beta \neq \gamma$, or alternatively, $\alpha \neq \beta = \gamma$, $\alpha = \beta \neq \gamma$ and $\alpha \neq \beta \neq \gamma$ . The number of combinations that we can make from these 3-lengths and 3-angles is seven (7) and these define the 7 unique lattice structures which are called BRAVAIS LATTICES after the originator of the concept. These have been given specific names:

| 2.2.2.- | CUBIC | ORTHORHOMBIC |
|---|---|---|
| | HEXAGONAL | MONOCLINIC |
| | TETRAGONAL | TRICLINIC |
| | TRIGONAL | |

We can therefore arrange the seven (7) systems into a hierarchy based on symmetry, as shown in the following diagram

2.2.3.-

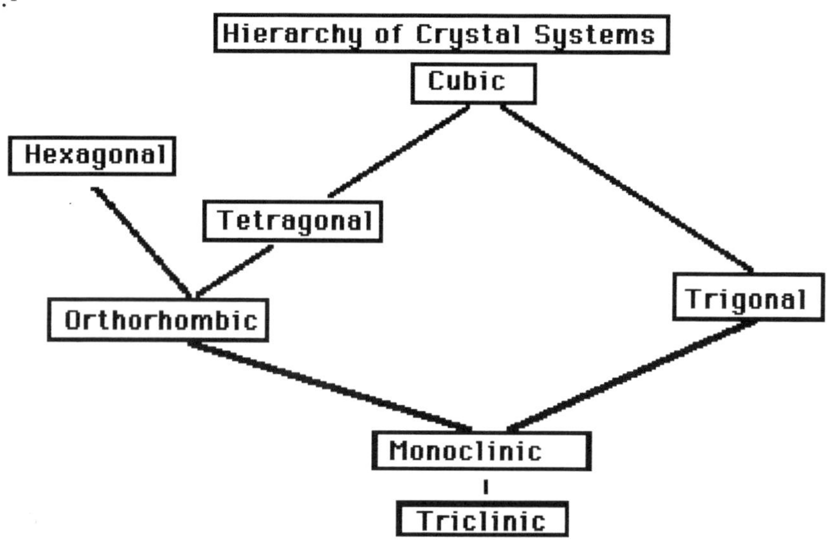


Each System is a Special Case of Those Above It

The lattice with the highest symmetry is listed at the top, with the least symmetrical lattice being shown at the bottom. Each of these seven lattices may have sublattices, the total being **14**.

The following diagram, given on the next page as 2.2.4., shows the symmetry of these 14 Bravais space lattices in terms of the seven crystal systems given in 2.2.3. The specific restrictions are listed in Table 2-1 on the following page along with the relation between axes and angles

2.2.4.-

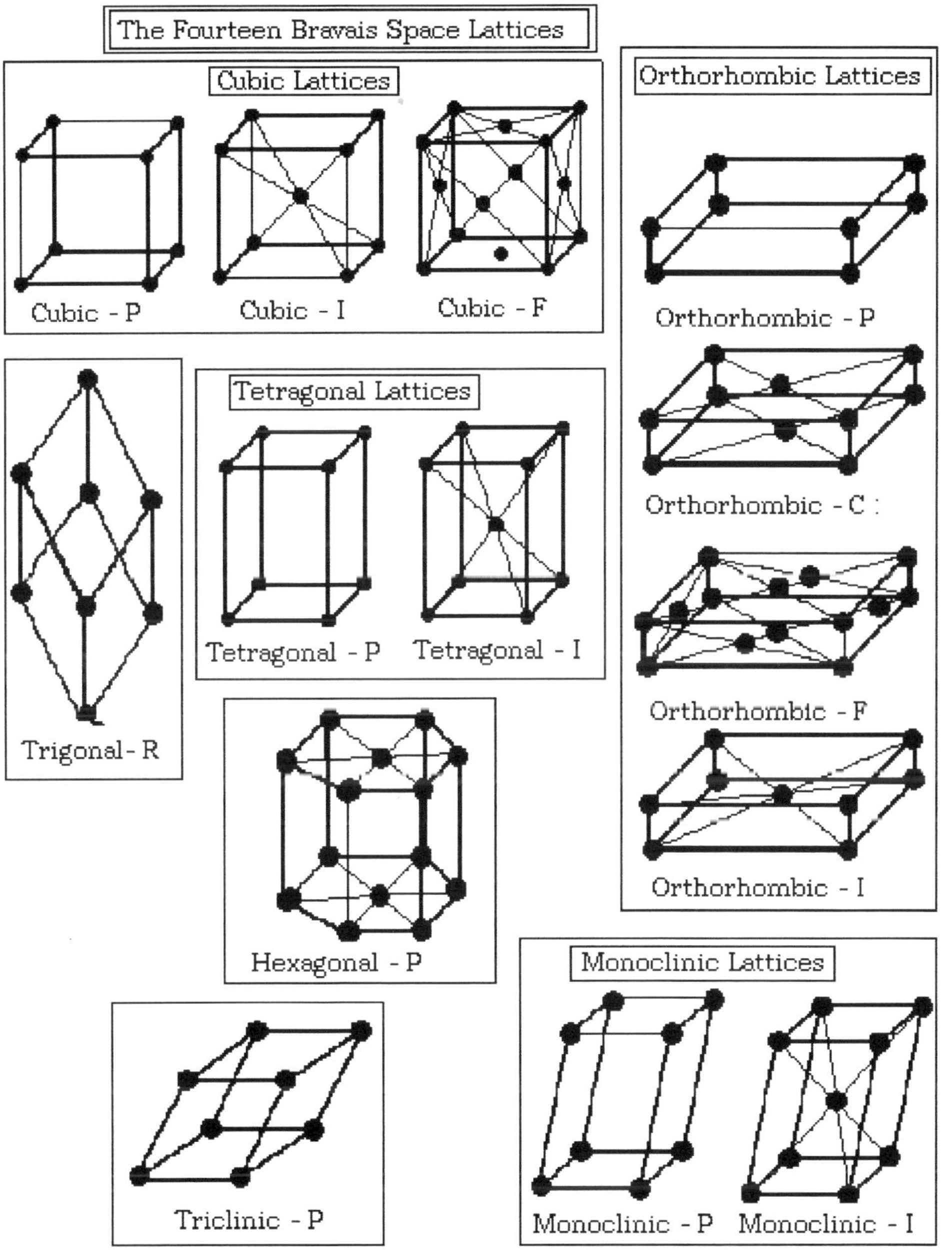
The Fourteen Bravais Space Lattices
Cubic Lattices
Cubic - P
Cubic - I
Cubic - F
Orthorhombic Lattices
Orthorhombic - P
Orthorhombic - C
Orthorhombic - F
Orthorhombic - I
Trigonal - R
Tetragonal Lattices
Tetragonal - P
Tetragonal - I
Hexagonal - P
Monoclinic Lattices
Monoclinic - P
Monoclinic - I
Triclinic - P

2.2.6.-

Inversion Mirror Symmetry

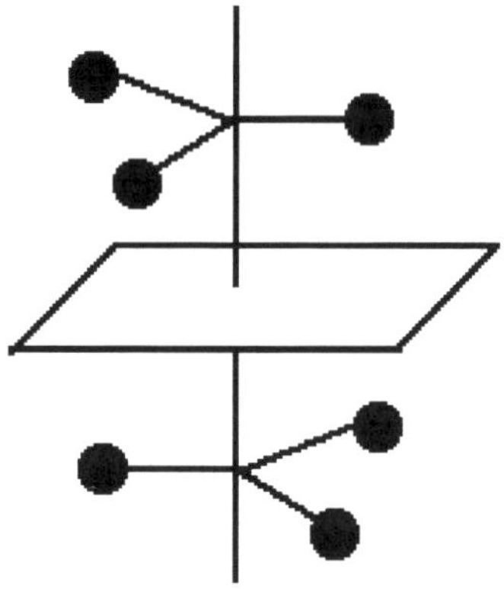

inverted. It is this kind of rotation and inversion that will be seen to be vital in fully describing the structure of solids.

We have already described crystal structures in terms of propagation units, i.e.- translation, in which the crystal is composed of a small number of individual atoms arranged in a unit which is **proliferated** to form the solid. Each assembly of atoms, i.e.- propagation units, is related to any other by one of four simple operations. These are:

2.2.7.- The Four Symmetry Operations

> Translation, i.e.- from one plane to another
> Rotation about an axis of the crystal
> Reflection across a plane
> Inversion through a point

These are the four main operations required to define the symmetry of a crystal structure. The most important is that of translation since each of the other procedures, called symmetry operations, must be consistant with the translation operation in the crystal structure. Thus, the rotation operation must be through an angle of $2\pi$ / n, where n = 1, 2, 3, 4 or 6.

Note that we have already shown that the cubic lattice can be rotated only in this manner, excepting six. We find that the number 6 comes from the hexagonal lattice itself where the arrangement of atoms is obvious (see 2.2.4.). Each of the operations involving crystal assembly units creates a special group:

**Translational** symmetry operations generate the 14 Bravais lattices.

The **rotational** operations generate a total of 32 Point Groups derived from these symmetry operations on the 14 Bravais lattices.

The **reflections and inversions** symmetry operations generate a total of 231 groups, called **Space Groups** which include the 32 Point Groups.

In Point Groups, one point of the lattice remains invarient under symmetry operations, i.e.- there is no translation involved. Space Groups are so-named because in each group all three- dimensional space remains invarient under operations of the group. That is, they contain translation components as well as the three symmetry operations. We will not dwell upon the 231 Space Groups since these relate to determining the exact structure of the solid. However, we will show how the 32 Point Groups relate to crystal structure of solids.

In 1890, Schœnflies formulated a group of symbols describing the various rotations possible for Point Groups. These were replaced in 1936 by the International Symbols or Hermann-Mauguin Symbols, since the point group system could not be extended to describe Space Groups. A comparison of the Point Group Symbols are given in the following Table, given on the next page as Table 2.2. The arrangement shown for Schœnflies symbols in 2.2.8 on the next page is inverted. It is this kind of rotation and inversion that will be seen to be vital in fully describing the structure of solids. Subscripts in Table 2-2 are used to indicate the degree of the rotational axes present, i.e.- $C_3$ = a three(3) fold rotational axis. Horizontal symmetry is indicated by "h" while vertical symmetry = v.

Table 2.2. - Comparison of Schœnflies and International Symbols

| Crystal System | | Point Group System | | Bravais |
|---|---|---|---|---|
| Name | Geometry | Schœnflies | International | Lattice |
| Cubic | a = b = c<br>α = β = γ = 90° | T<br>O<br>Th<br>Td<br>Oh | 23<br>432<br>m 3<br>$\bar{4}$ 3 m<br>m 3 m | P<br>I<br>F |
| Hexagonal | a = β ≠ c<br>α = β = γ<br>α = β = 90 °<br>(γ = 120 °) | C6<br>D6<br>C2h<br>C6v<br>C3h<br>D3h<br>D6h | 6<br>622<br>6/m<br>6mm<br>$\bar{6}$<br>$\bar{6}$ m2 or $\bar{6}$ 2m<br>6/m | P |
| Tetragonal | a = b ≠ c<br>a = β = γ = 90° | C4<br>D4<br>C4h<br>C4v<br>S4<br>D2d<br>D4h | 4<br>422<br>4/m<br>4mm<br>$\bar{4}$<br>$\bar{4}$ m2 or $\bar{4}$ 2m<br>4/mmm | P<br>I |
| Trigonal | a = b = c<br>α = β = 90 °<br>90 ° > γ < 120° | C3<br>D3<br>C3v<br>C3i &<br>D3d | 3<br>32<br>3m<br>$\bar{3}$ & $\bar{3}$ 2/m | |
| Orthorhombic | a ≠ b ≠ c<br>α= β = γ = 90° | D2<br>C2v<br>D2h | 222<br>mm2<br>mmm | P<br>C<br>F |
| Monoclinic | a ≠ b ≠ c<br>α = β = 90° ≠ γ | C2<br>C3<br>C2h | 2<br>m<br>2/m | P<br>C |
| Triclinic | a ≠ b ≠ c<br>a ≠ β ≠ γ | C1<br>Ci | 1<br>$\bar{1}$ | P |

In the Schœnflies notation, the elements defined as :

2.2.8.- C = rotation axis only
D = dihedral (rotation plus dihedral rotation axes)
I = inversion symmetry
T = tetrahedral symmetry
O = octahedral symmetry

In the Hermann-Mauguin Symbols, the same rotational axes are indicated, plus any inversion symmetry that may be present. The numbers indicate the number of rotations present, m shows that a mirror symmetry is present and the inversion symmetry is indicated by a bar over the number, i.e.- $\bar{6}$.

At this point, you may find that the subject of symmetry in a crystal structure to be confusing. However, by studying the terminology carefully in Tablc 2-2, one can begin to sort out the various lattice structures and the symbols used to delineate them. All of the crystal systems can be described by use of either Schœnflies or Hermann-Mauguin symbols, coupled with the use of the proper geometrical symbols.

In 1965, specific diagrams were developed by Wcinrcich to illustrate the 32 point-group symmetries. Appropriate Schœnflies symbols were also given. These drawings are given on the following page as 2.2.10. If they are examined closely, it becomes easier to assess and compare the differences in symmetry operations involved for each type of lattice structure. As an example, consider the following.

In the $C_{1h}$ point group, there is a **one-fold rotation axis** plus horizontal symmetry (horizontal mirror). Note also the difference between $C_{1h}$ and $C_{2h}$.

It is easy to see that these point groups are related to one another, but that they specify quite different rotational symmetries. In contrast, $C_2$ has a 2-fold rotation axis but no horizontal symmetry (but $C_{2h}$ has horizontal symmetry). Contrast the $C_{2v}$ point group with vertical symmetry to the $C_2$

2.2.10. **The 32 Point Group Symmetries**

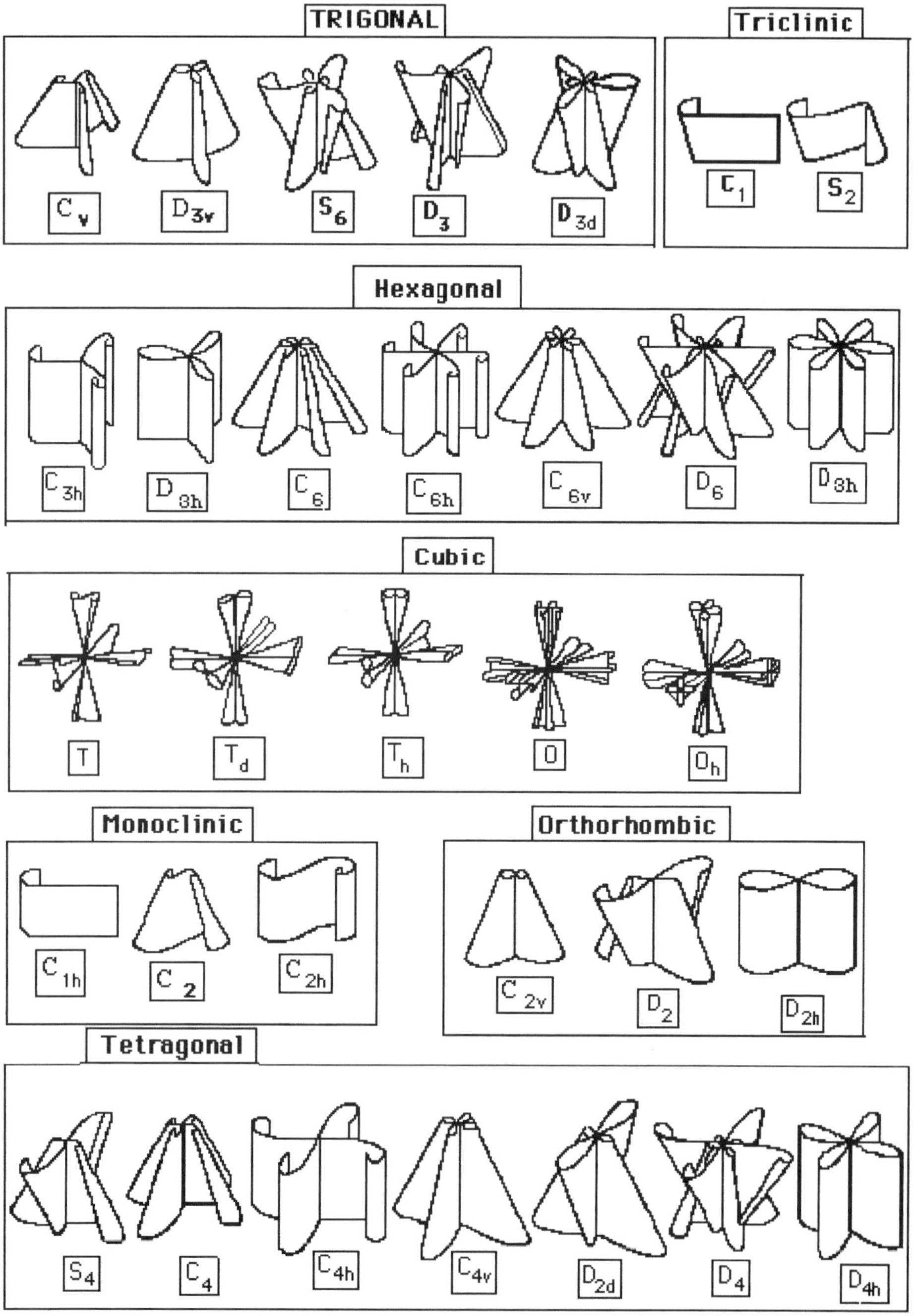

and $C_{2h}$ point groups. The $D_{2h}$ point group has a two-fold rotation axis, a horizontal mirror, and 2 diad rotation axes. Compare also the $C_v$ and $D_{3v}$ point groups and the $C_{3h}$ and $D_{3h}$ groups.

Now examine the symmetry elements for the cubic lattice. It is easy to seethat the number of rotation elements, plus horizontal and vertical symmetry elements is quite high. This is the reason why the Cubic Structure is placed at the top of 2.2.3. Even though the lattice points of 2.2.1. are deceptively simple for the cubic structure, the symmetry elements are not

There is one other factor contributing to the overall symmetries of the lattice structure. This factor involves the local symmetry of the atomic groups which actually form the structure. Examples are the "solid-state building blocks" given above, e.g.- the tetrahedon - the group, $PO_4^{3-}$, and the octahedron - like group, $NbO_6^-$. It is easy to see that if a structure is composed of such building blocks, they will impose a **local structural** symmetry on the lattice, in addition to the other symmetries already present.

The result is that Factor III of 2.2.6. given above imposes further symmetry restrictions on the 32 point groups and we obtain a total of 231 **space groups.** We do not intend to delve further into this aspect of lattice contributions to crystal structure of solids, and the factors which cause them to vary in form. It is sufficient to know that they exist. Having covered the essential parts of lattice structure, we will elucidate how one goes about determining the structure for a given solid.

## **2-3**: HOW TO DETERMINE THE STRUCTURE OF COMPOUNDS

We want to review how one goes about actually determining the structure of a given solid. There are two factors we need to consider:

1) what are the steps required in actually determining a structure?

2) what kind of information is obtained?

2.3.5.-

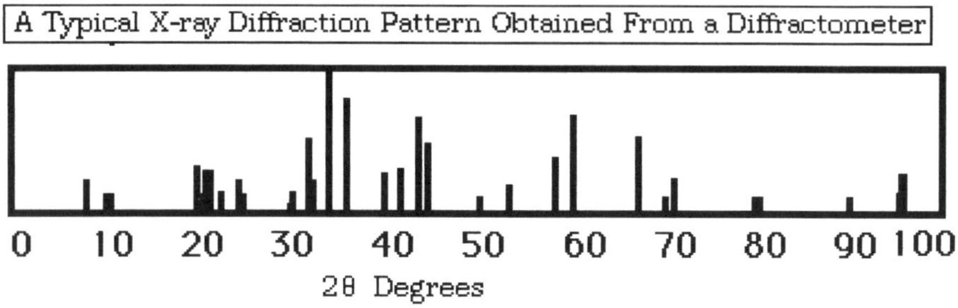


line. If we do not know what the nature of the material is, then we can use the obtained pattern to determine its probable composition.Using the set of diffraction lines, we pick the **three most intense lines in the pattern.**

By referring to the "POWDER DIFFRACTION FILE", published by the AMERICAN SOCIETY FOR TESTING AND MATERIALS, 1916 Race St., Philadelphia, Penna. 19103, we can look up the most probable composition. Usually, we will know how the material was made and the components used to make it. If not, we can analyze for constituents. In this case, we would find La and Al, and would surmize that we have an oxidic compound, since we made the compound by firing the two oxides together.

There are two compounds possible, namely - $LaAlO_3$ and $LaAl_{11}O_{18}$. (How do we know?- Consult the volumes published by the American Ceramic Society entitled "Phase Diagrams for Ceramicists" for $La_2O_3$ and $Al_2O_3$ which show that only two compounds form when these two oxides are reacted).

In the Diffraction File, the 3 **most intense lines** of $LaAlO_3$ are d = 2.66 Å (100), 3.80 Å (80) and 2.19 Å (80). We find that this set of lines does not fit our pattern. We do find that the pattern given for $LaAl_{11}O_{18}$ is identical to the x-ray pattern we obtained. Thus, we have characterized our material. Then, the {h,k,l} values can be calculated from special formulas developed for this purpose. These are given in the following Table 2-3. These equations allow us to calculate {h,k,l} values, once we know the d values and structure, as calculated from the 2θ values found from the diffractometer, using the Bragg equation.

Table 2- 3

Plane Spacings for Various Lattice Geometries

| CUBIC | TETRAGONAL |
|---|---|
| $1/d^2 = h^2 + k^2 + l^2 / a^2$ | $1/d^2 = h^2 + k^2 / a^2 + l^2 / c^2$ |
| **HEXAGONAL** | **ORTHORHOMBIC** |
| $1/d^2 = {}^4/_3\ [(h^2 + k^2 + l^2)/ a^2] + l^2/c^2$ | $1/d^2 = (h^2/a^2) + (k^2/b^2) + (l^2/c^2)$ |
| **RHOMBOHEDRAL** | |
| $1/d^2 = (h^2+k^2+l^2)\sin^2\alpha + 2(hk+kl+hl)(\cos^2\alpha \cos\alpha)/\alpha^2 (1 - 3\cos^2\alpha + 2\cos^3\alpha)$ | |
| **MONOCLINIC** | |
| $1/d^2 = \{1/\sin^2\beta\}\{h^2/a^2 + (k^2 \sin^2\beta) / [b^2 + l^2 / c^2 - 2hl \cos\beta / a c]$ | |
| **TRICLINIC** | |
| $1/d^2 = 1/V^2 \{ S_{11} h^2 + S_{22} k^2 + S_{33} l^2 + 2 S_{12} hk + 2 S_{23} kl + 2 S_{13} hl$ | |
| where: V = volume of unit cell, $S_{11} = b^2 c^2 \sin^2\alpha$, $S_{22} = a^2 c^2 \sin^2\beta$, $S_{22} = a^2 c^2 \sin^2\beta$, $S_{33} = a^2 b^2 \sin^2\gamma$, $S_{12} = abc^2 (\cos\alpha \cos\beta - \cos\gamma)$, $S_{23} = a^2 bc (\cos\beta \cos\gamma - \cos\alpha)$ and $S_{13} = ab^2 c (\cos\gamma \cos\alpha - \cos\beta)$ | |

For the hexagonal composition given above, i.e.- the compound $LaAl_{11}O_{18}$, the {hkl} values were obtained by trying certain values in the hexagonal formula and seeing if the results give valid numbers, consistant with the numbers used. That is, a trial and error method was used to obtain the correct results. For example, one would start with {100} and determine what value of d conforms to this plane. Then, {200} and {300} would be used, etc. If we do this, we obtain the values shown in Table 2-4 as:

TABLE 2-4

Conversion of 2 θ Values of the Diffraction Pattern to {hkl} Values

| 2 θ | I/I$_0$ | d | {hkl} | 2 θ | I/I$_0$ | d | {hkl} |
|---|---|---|---|---|---|---|---|
| 8.02 | 16 | 11.02 | 002 | 36.18 | 74 | 2.48 | 114 |
| 16.1 | 6 | 4.81 | 004 | 39.39 | 27 | 2.29 | 023 |
| 18.86 | 32 | 4.71 | 001 | 40.94 | 29 | 2.20 | 0010 |
| 20.12 | 28 | 4.41 | 012 | 42.79 | 66 | 2.11 | 025 |
| 22.07 | 10 | 4.03 | 013 | 45.01 | 46 | 2.01 | 026 |
| 24.23 | 21 | 3.67 | 001 | 53.36 | 15 | 1.72 | 029 |
| 24.55 | 11 | 3.63 | 014 | 58.57 | 36 | 1.58 | 127 |
| 32.19 | 44 | 2.78 | 110 | 60.07 | 60 | 1.54 | 0211 |
| 32.50 | 15 | 2.76 | 008 | 67.35 | 48 | 1.39 | 220 |
| 33.23 | 15 | 2.70 | 112 | 71.61 | 18 | 1.32 | 0214 |
| 34.02 | 100 | 2.64 | 017 | 95.42 | 14 | 1.04 | 2214 |

This allows us to determine the unit cell lengths for our compound as:

Hexagonal: $a_0 = 5.56.$ Å ; $b_0 = 22.04$ Å

Additionally, we can list the x-ray parameters and convert them to structural factors as shown: These values are averaged over all of the reflections used for calculation. Note that this pattern has several planes where the "d" value is more than ten.

Let us consider one other example. Suppose we obtained the following set of 2θ values and intensities for a compound. These are given in the following Table 2-5:

TABLE 2-5

DIFFRACTION LINES AND INTENSITIES OBTAINED

| $I/I_0$ | 2 θ in degrees | $I/I_0$ | 2 θ in degrees |
|---|---|---|---|
| 96 | 29.09 | 10 | 90.55 |
| 100 | 48.40 | 17 | 97.80 |
| 56 | 48.40 | 19 | 118.26 |
| 50 | 57.48 | 11 | 121.00 |
| 14 | 60.28 | 7 | 144.34 |
| 18 | 78.38 | 9 | 148.49 |
| 14 | 80.78 | | |

In looking over the Table, we find that the first three values are the strongest diffraction lines. After calculating "d" values and looking up the set of strong lines which correspond to our set, we find that the probable compound is $CdO_2$ , or cadmium peroxide. This compound turns out to be cubic in structure, with $a_o$ = 5.313 Å . When we calculate the {h,k,l} values of the diffracting planes, the strongest line is found to be {200}.

We can then make the determination that since $Cd^{2+}$ is a strongly diffracting atom (it has high atomic weight, which is one way of stating that it has many electron shells, i.e.-$1s^2$ $2s^2$ $2p^6$ $3s^2$ $3p^6$ $3d^{10}$ $4s^2$ $4p^6$ $4d^{10}$), the structure is probably face-centered cubic. Indeed, this turns out to be the case. In the unit cell, Cd atoms are in the special positions of: {0,0,0}, {1/2,1/2,1/2}; 0,1/2.1/2}; {1/2,1/2,0}. There are four molecules per unit cell. We could continue further so as to calculate intensities, $I_c$ , using atomic scattering factors already present in prior literature. We would scale our calulated intensities to our observed intensities by $\Sigma I_c = \Sigma I_o$ . We then calculate a reliability factor, called R, from $R = \Sigma (I_o - I_c) / \Sigma I_o$ . A low value indicates that our selection of lattice parameter was correct. If not, we choose a slightly different value and apply it. The details of the procedure for determining exact structure and atomic positions in the lattice are well known, but are beyond the scope of this Chapter.

## 2.4. - SYMMETRY DISTRIBUTION OF CRYSTALS

Inorganic materials do not often possess large unit-cells because their chemical formulae are much simpler than organic molecules. Unit-cell volumes for organic crystals can be as large as 10,000 $Å^3$ and proteins tend to be at least 10 to 100 times larger than that. Most inorganic cell volumes are in the range of 10-1000 $Å^3$ but SiC can have cell volumes up to 1500 $Å^3$.

Certain viruses (which are close to proteins in composition) can have cell dimensions up to 3600 $Å^3$ and cell volumes of 46 billion $Å^3$. Proteins (which are actually polypeptide chains consisting of amino acids linked by peptide bonds) can have molecular weights in the millions. The human protein, hemoglobulin, is composed of four polypeptides, two identical side-chains called "alpha" and two differing side chains of similar composition called "beta". Each alpha chain has 141 amino acids and each beta chain has 146 amino acids, giving a total molecular weight of about 64,500. It is one of the simpler biological molecules. When crystalline materials are surveyed as to the type of symmetry they exhibit in the solid state, we find that only those compounds with simple compositions have high symmetry, i.e.- they crystallize in the cubic form. Those that do not, including most organics, tend to form crystals of low symmetry as shown in the following table, given on the next page.

The percentages are based upon 5572 inorganic, 3217 organic and 224 protein compounds. Two-thirds (66%) of the inorganic compounds have symmetries higher than orthorhombic whereas 85% of the organics do not.

Table 2-6
Distribution of Crystalline Materials Among the Seven Crystal Systems*

| System | Inorganic | Organic | Proteins |
|---|---|---|---|
| Triclinic | 2% | 6% | 2% |
| Monoclinic | 14 | 49 | 35 |
| Orthorhombic | 18 | 30 | 43 |
| Trigonal | 12 | 4 | 6 |
| Tetragonal | 14 | 6 | 6 |
| Hexagonal | 11 | 2 | 2 |
| Cubic | 30 | 4 | 5 |

* This data was taken from "Crystal Data"- Am. Crystallographic Assoc. Monograph - W. Nowacki- Ed. (1967)

For inorganics, the leading space groups are: Fm3m, Pnma, $P6_3/mmc$, Fd3m, $P\ 2_1/c$, Pm3m, C2/c and R3m. Since complicated formulae tend to produce low symmetry, many of the organic crystals appear in the lowest symmetry catagories like monclinic or triclinic.

About 80% of inorganic and 60% of organic compound structures are centric, that is- they have a center of symmetry. Most protein structures are not centric since nearly all living systems, including humans, have a "handedness". The handedness arises from the replication process in which helices (e.g.- proteins, RNA and DNA) wind and unwind, thereby transmitting genetic code. Helices possess a screw-axis symmetry so that most crystals of biological origin belong to space groups having screw axes. Among the proteins, the most populous space groups are: $P\ 2_1\ 2_1\ 2_1$ (34%), $P2_1$,(23%), C2 (11%), and $P\ 2_1\ 2_1\ 2$ (7%). In biological molecules, inversion symmetry and mirror planes are virtually non-existant.

Screw axes are also common among crystals of the simpler organic compounds. But, many of these have mirror and inversion symmetry as well. The most common space groups for organic compounds are: $P\ 2_1/c$ (26%), $P\ 2_1\ 2_1\ 2_1$ (13%), $P\ 2_1$ (8%), and C2/c (7%).

Most of the differences between inorganic, organic and protein crystals lies in the fact that ionic forces (found mostly in inorganic structures) depend upon interatomic distances but not on angle. In contrast, covalent bonds (such as those which predominate in organic and protein compounds) depend upon angle and therefore the crystal(s) is usually acentric, i.e.- without a center of symmetry.

The human body is composed of **eukaryotic** cells. Such cells have an assortment of intracellular membrane-bound *organelles* which carry out the functions of that particular cell. Examples might be liver cells vs: brain cells. Every cell in the human body contains chromosomes. Among the many organisms that have separate sexes, there are two basic types of chromosomes: autosomes and sex chromosomes. Autosomes control the inheritance of all the characteristics and functions of the cell except the sex-linked ones, which are controlled by the sex chromosomes. Humans have 22 pairs of autosomes and one pair of sex chromosomes. For any living cell to persist from one generation to another, it must store information and use it to pass to the next generation of cells. All cells have solved this problem by coding information in certain large molecules, i.e.- deoxyribonuclecic acid= DNA) and ribonucleic acid= RNA. The compositions of DNA and RNA have been shown to consist solely of the 4 nucleic acids, Adenine (A), Thymine (T), Guanine (G) and Cytosine (C). The overall code of the chromosomes is called the "human genome". In 1953, Crick and Watson solved the structure of DNA. With the help of Rosalin Franklin, a co-worker, they showed that pairs of nucleic acids (G+C and A+T) formed a spiral molecule of high molecular weight in which a double strand was twisted into a helix.

In 1990, work was started to characterize the human genome which had been shown to consist of about 3 billion base pairs. The final result was announced in the year 2000. All of the chromosomes have been characterized. The human genome has been shown to contain some 30,000 genes (which are sections of the chromosome which code for specific proteins). Each cell produces the type of proteins needed for it to function. The function of mRNA is to transfer information from the DNA, so as is to fix the limits of the protein needed. The vast majority of the proteins found in living organisms are composed of only 20 different kinds of amino acids, repeated many times and strung together in a particular order. Each type of protein has its own unique sequence of amino acids. This sequence, known as its primary structure, actually determines the shape and function of the protein. The secondary and tertiary structure refers to the looping and folding of the protein chain back upon itself. This overall form determines the actual chemical function of the individual protein. Some proteins are enzymes as well.

Although there are but 30,000 different genes in the human body, it has been shown that as many as 200,000 proteins are produced for use in the individual human cells. Thus, it is clear that the data given in Table 2-6 may be regarded as preliminary. Time will show whether the observations made in Table 2-6 regarding protein structure are correct.

## 2.5.- Phase Relationships Among Two or More Solids

Up to this point, we have considered only one solid at a time. However, when two (2) or more solids are present, they can form quite complicated systems which depend upon the nature of each of the solids involved. To differentiate and to be able to determine the differences between the phases that may arise when two compounds are present (or are made to react together), we use what are termed "phase-diagrams" to illustrate the nature of the interactions between two solid phase compositions. You will note that some of this material was presented earlier in Chapter 1. It is presented here again to further emphasize the importance of phase diagrams.

Consider the following. Suppose we have two solids, "A" and "B". It does not matter what the exact composition of each may be. A will have a specifc melting point (if it is stable and does not decompose at the M.P.) and likewise for the compound, B. We further suppose that the M.P. of B is higher than that of A. Furthermore, we suppose that A and B form a **solid solution** at all variations of composition. What this means is that from 100% A-0% B to 50% A-50% B to 0% A-100% B, the two solids dissolve in one another to form a completely homogenous single phase. If we plot the M.P. of the system, we find that three different curves could result, as shown in the following diagram, given as 2.5.1. on the next page.

In this case, the melting point of the ideal solid solution should increase linearly as the ration of B/A increases. However, it usually does not. Either a negative deviation or a positive deviation is regularly observed.

2.5.1.-

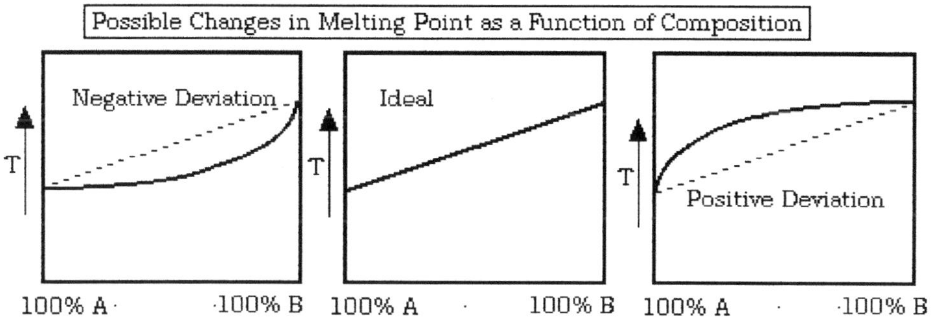


In any phase diagram, composition is plotted against temperature. In this way, we can see how the interactions between phases change as the temperature changes and the behavior as each solid phase then melts. Either two-phase or three phase systems can be illustrated. This is shown in the following:

2.5.2.-

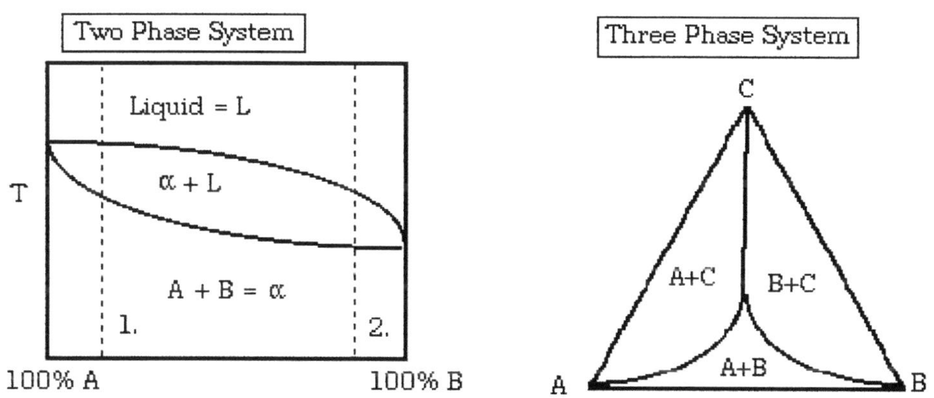

Here, the two-phase diagram is simplified to show a hypothetical phase involving "A" and "B" compounds which form a solid solution from 100% A to 100% B. The solid solution is labelled as "α".

The melting temperature of A is higher than that of B. Therefore, the melting temperature of α drops as the composition becomes richer in B. At specific temperatures on the diagram (see 1. & 2.), a two-phase system appears, that of a liquid plus that of α. Finally, as the temperature rises, the melt is homogenous and the solid, α, has melted. In the three-phase system, only the relationship between A, B and C can be illustrated on a two-dimensional drawing. A three-dimensional diagram would be required to show the effect of temperature as well.

The phase diagrams we have shown are based upon the fact that A and B form solid mutually soluble solid state solutions. If they do not, i.e.- they are not mutually soluble in the solid state, then the phase diagram becomes more complicated. As an example, consider the following, which is the case of limited solid solubility between A and B:

2.5.3.-

A Simple Eutectic Phase Diagram

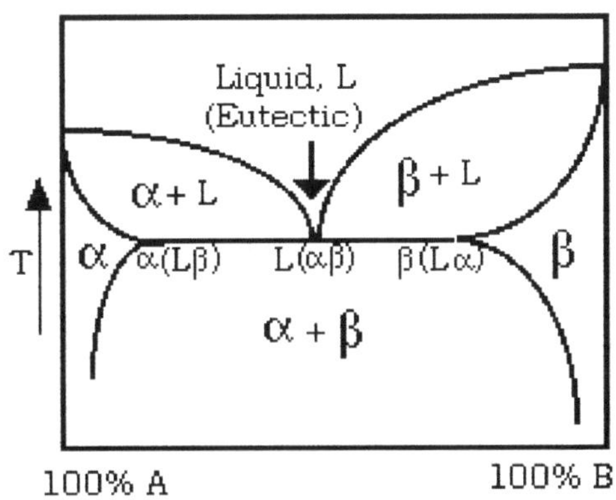


Here, we have the case where A & B form two slightly different componds, α and β. The composition of α is $A_xB_y$ while that of β is $A_uB_v$ , where the subscripts indicate the ratio of A to B (these numbers may be whole numbers or they may be fractional numbers). At low B concentrations, $\alpha$ exists as a solid (the left side of the diagram). As the temperature increases, a melt is obtained and $\alpha$ remains as a solid in the melt (L) {This is indicated by the details given just below the straight line in the diagram}. As the temperature is increased, then $\alpha$ melts to form a uniform liquid. In the middle concentrations, $\alpha$ and $\beta$ exist as two separate phases. At 75% A-25% B, $\beta$ melts to form a liquid plus solid $\alpha$ whereas just the opposite occurs at 25% A and 75% B. When A equals B, then both $\alpha$ and $\beta$ melt together to form the liquid melt, i.e.- the "eutectic point". In two similar cases but differing cases, i.e.- a system with a melting point minimum and another with a different type of limited solid solubility, the behavior differs as shown in the following diagram:

2.5.4.-

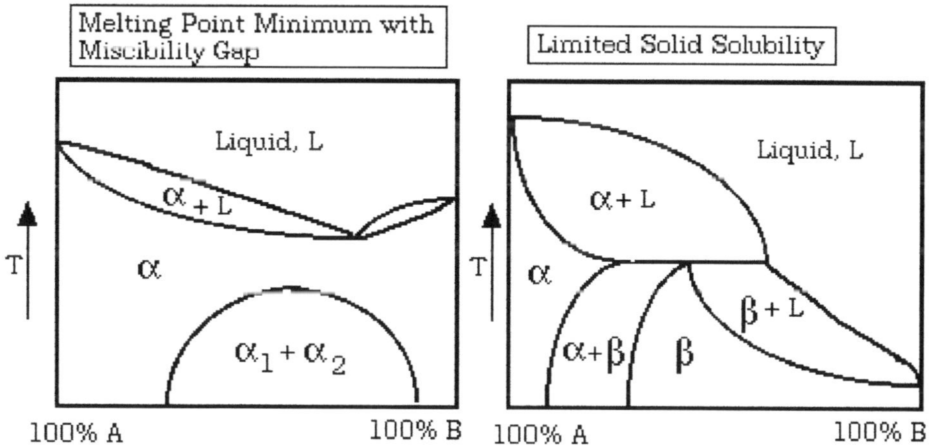


In the case on the left, a composition exists which in which a minimum melting point exists. Again, A and B form $\alpha$ which partially melts to form $\alpha$ + L. However, two forms of $\alpha$ also exist, a compositional area known as a "miscibility gap", i.e.- " $\alpha_1$ and $\alpha_2$". But at higher temperatures, both of these meld into the single phase, $\alpha$. Finally, we obtain the melt + $\alpha$. Note that at about 80% A-20% B, a melting point minimum is seen where melts directly instead of forming the two-phase system, $\alpha$ + L.

In the case of limited solid solubility, the phase behavior becomes more complicated. Here, both α and β form (where the actual composition of α is $A_xB_y$ while that of β is $A_uB_v$ , as given before. Note that the values of x and y change, but that we still have the α phase. The same holds for u and v of the β phase). At low A concentrations, α exists alone while β exists alone at higher B concentrations. A region exists where the two phase system, + exists. We note that as the temperature is raised, a two phase system is also seen consisting of the melt liquid phus a solid phase. The phase-behavior shown on the right side of the diagram arises because the two phases, A and B, have limited solubility in each other.

Now, let us consider the case where three (3) separate phases appear in the phase diagram. In this case, we have three (3) separate phases that appear in the phase diagram. These phases are α , β , and σ , whose compositions are: α = $A_xB_y$ , β = $A_uB_v$ and σ = $A_cB_d$ , respectively (the values of x, y, u, v, c, and d all differ from each other so that $A_xB_y$ is a specific compound as are the others). This shown as follows:

2.5.5.-

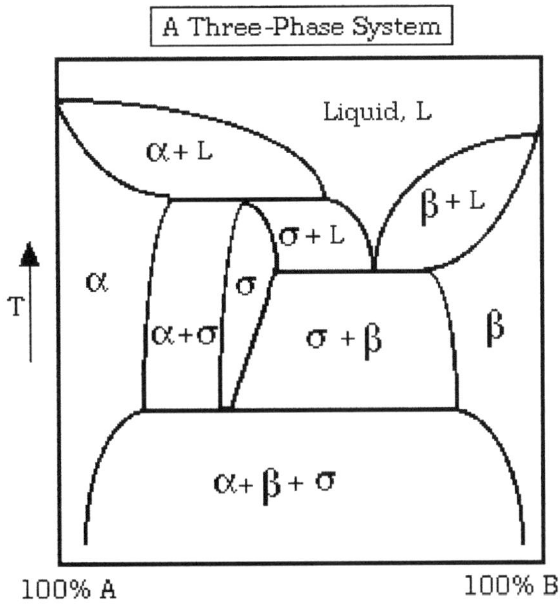

By studing this phase diagram carefully, you can see how the individual phases relate to each other. As you can see, a phase diagram can become quite complicated. However, in most cases involving real compounds, the phase diagrams are usually simple. Those involving compounds like silicates can be complex, but those involving alloys of metals show simple behavior like limited solubility.

Now, let us return to a further discussion of the properties of solids. Summarizing to this point, we have shown that only certain propagation units can be stacked to infinity to form **close-packed solids**. We have also shown how the units fit together to form specific solids with specific symmetries. The structure of solids has also been reviewed in some detail, including differences in crystal symmetry of inorganic and organic compounds. What we find is that the solid contains stacking defects. That is, we find that we cannot form the solid without encountering some sort of defect as we stack molecules to near infinity to form our "perfect" solid. Obviously, this arises because of the application of the 2nd Law of Thermodynamics which involves entropy. The next chapter analyzes the defect solid in some detail.

## SUGGESTED READING

1. "Solids-Elementary Theory for Advanced Students"- G. Weinreich, J. Wiley & Sons, Inc., New York (1965).

2. "Solid State Physics" - A.J. Dekker, Prentice-Hall, Inc., Englewood Cliffs, NJ (1958). (See Chaps. 1,2 &3 in particular).

3. "Stereographic Projections of the Colored Crystallographic Point Groups" , J.A. McMillan, *Am. J. Phys.*, **35**, 1049 (1967).
4. "Imperfections in Crystal", J.M. Honig, *J. Chem. Ed.*, **34**, 224 (1957).
5. "The Physical Chemistry of Solids", R.J. Borg and G.J. Dienes, Academic Press, NY (1992).

6. "Solid State Chemistry", Prentice-Hall ,N. B. Hannay, Englewood Cliffs, NJ (1967)

## Problems for Chapter 2

1. Redraw the Figure given in 2.1.15. and 2.1.18 so as determine the equation for the {420} plane.

2. Define the symmetry elements for: $C_{2h}$ $C_{6v}$ $D_{3h}$ $D_4$ & $D_{2d}$

3. Given the following "d-spacings" for the diffraction lines of a given cubic powder, calculate the {h,k,l} values of the plane spacings.

| Intensity | 2Θ in Degrees |
|---|---|
| 96 | 29.09 |
| 100 | 33.70 |
| 56 | 48.4 |
| 50 | 57.48 |
| 18 | 78.38 |
| 19 | 118.26 |

4. Iron (Fe) has the body-centered structure at 298 °K and a density of 7.86 gm with a lattice spacing of $a_0$ = 2.876 Å. Calculate Avogaddro's number from these data.

# Chapter 3

## Defects in Solids

In the first chapter, we defined the nature of a solid in terms of its building blocks plus its structure and symmetry. In the second chapter, we defined how structures of solids are determined. In this chapter, we will examine how the solid actually occurs in Nature. Consider that a solid is made up of atoms or ions that are held together by covalent/ionic forces. It is axiomatic that atoms cannot be piled together and forced to form a periodic structure without mistakes being made. The 2nd Law of Thermodynamics demands this. Such mistakes seriously affect the overall properties of the solid. Thus, defects in the lattice are probably the most important aspect of the solid state since it is impossible to avoid defects at the atomistic level. Two factors are involved:

1) The entropy effect
2) The presence of impurities in any given solid.

What we mean by this is that the chemical properties of the solid are **not** determined solely by its structure but also by the nature of its chemical composition and in particular by any **defects** that may be present. We will find that both physical and chemical properties of a solid are largely determined by the type and nature of defects present. It is axiomatic in inorganic chemistry that:

> **"The perfect solid does not exist in Nature and its reactive properties are determined, to a great extent, by the defects present in its structure".**

We have already stated that some defects are related to the **entropy of the solid**, and that a perfect solid would violate the second law of thermodynamics. The 2nd law states that **zero entropy** is only possible at absolute zero temperature. However, most solids exist at temperatures far above absolute zero. Thus, most of the solids that we encounter **are** defect-solids. The defects are usually "point defects", which are atomistic

dimensions. Indeed, this is exactly the case found for defects in solids. Specific names have been given to each of these three types of defects. Thus a one-dimensional defect is called a "point" defect, a two-dimensional defect a "line" or "edge" defect and a three-dimensional defect is called a "plane" or "volume" defect.

Point defects are changes at atomistic levels, while line and volume defects are changes in stacking of planes or groups of atoms (molecules) in the structure. Note that the arrangement (structure) of the individual atoms (ions) are not affected, only the method in which the structure units are assembled. Let us now examine each of these three types of defects in more detail, starting with the one-dimensional lattice defect and then with the multi-dimensional defects. We will find that specific types have been found to be associated with each type of dimensional defect which have specific effects upon the stability of the solid structure.

a. <u>The Point Defect in Homogeneous Solids</u>

Point defects were mentioned in a prior chapter. We now need to determine how they affect the structure and chemical reactivity of the solid state. We will begin by identifying the various defects which can arise in solids and later will show how they can be manipulated to obtain desirable properties not found in naturally formed solids. Since we have already defined solids as either homogeneous and heterogeneous, let us look first at the homogeneous type of solid. We will first restrict our discussion to solids which are stoichiometric, and later will examine solids which can be classified as "non-stoichiometric", or having an excess of one or another of one of the building blocks of the solid. These occur in semi-conductors as well as other types of electronically or optically active solids.

Suppose you were given the problem of identifying defects in a homogeneous solid. Since all of the atoms in this type of solid are the same, the problem is somewhat simplified over that of the heterogeneous solid (that is- a solid containing more than one type of atom or ion). After some introspection, you could speculate that the homogeneous solid could have the following types of point defects:

3.1.1.- Types of Point Defects Expected in a Homogeneous Solid

| | | | |
|---|---|---|---|
| * | Vacancies | * | Substitutional Impurities |
| * | Self-interstitial | * | Interstitial Impurities |

On the left, in 3.1.1., are the two types of point defects which involve the lattice itself, while the others involve impurity atoms. Indeed, there do not seem to any more than these four, and indubitably, no others have been observed. Note that we are limiting our defect family to point defects in the lattice and are ignoring line and volume defects of the lattice. These four point defects, given above, are illustrated in the following diagram, given as 3.1.2. on the next page.

Note that what we mean by an "interstial" is an atom that can fit into the spaces between the main atoms in the crystalline array. In this case, we have shown a hexagonal lattice and have labelled each type of point defect. Note that we have shown a vacancy in our hexagonal lattice, as well as a foreign interstitial atom which is small enough to fit into the **interstice** between the atoms of the structure. Also shown are two types of substitutional atoms, one larger and the other smaller than the atoms composing the principal hexagonal lattice. In both cases, the hexagonal packing is disrupted due to a "non-fit" of these atoms in the structure.

Additionally, we have illustrated another type of defect that can arise within the homogeneous lattice of 3.1.2. (in addition to the vacancy and substitutional impurities that are bound to arise). This is called the "self-interstitial". Note that it has a decisive effect on the structure at the defect. Since the atoms are all the same size, the self-interstitial introduces a **line-defect** in the overall structure. It should be evident that the line-defect introduces a difference in packing order since the close packing **at the arrows** has changed to cubic and then reverts to hexagonal in both lower and upper rows of atoms.

3.1.2.-

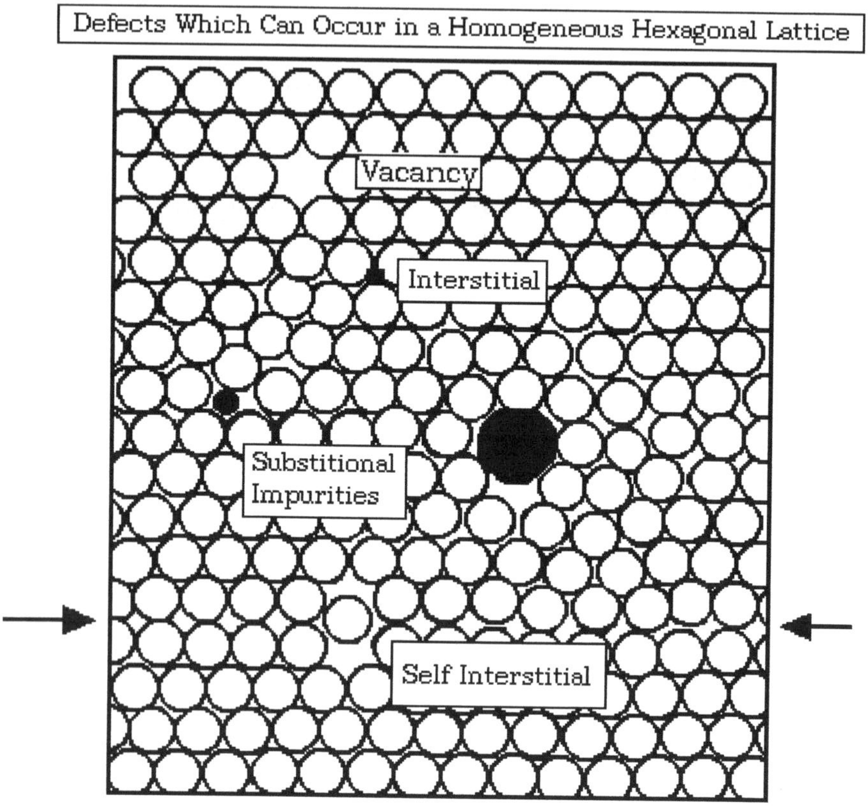


It may be that this type of defect is a major cause of the line or edge type of defects that appear in most homogeneous solids. In contrast, the other defects produce only a disruption in the **localized** packing order of the hexagonal lattice, i.e.- the defect does not extend throughout the lattice, but only close to the specific defect.

Now, suppose that we have a **solid solution** of two (2) elemental solids. Would the point defects be the same, or not? An easy way to visualize such point defects is shown in the following diagram, given as 3.1.3. on the next page. It is well to note here that homogeneous lattices usually involve metals or solid solutions of metals (alloys) in contrast to heterogeneous lattices which involve compounds such as ZnS.

3.1.3.- Defects in the Homogeneous Solid Containing 2 Solids in Solution

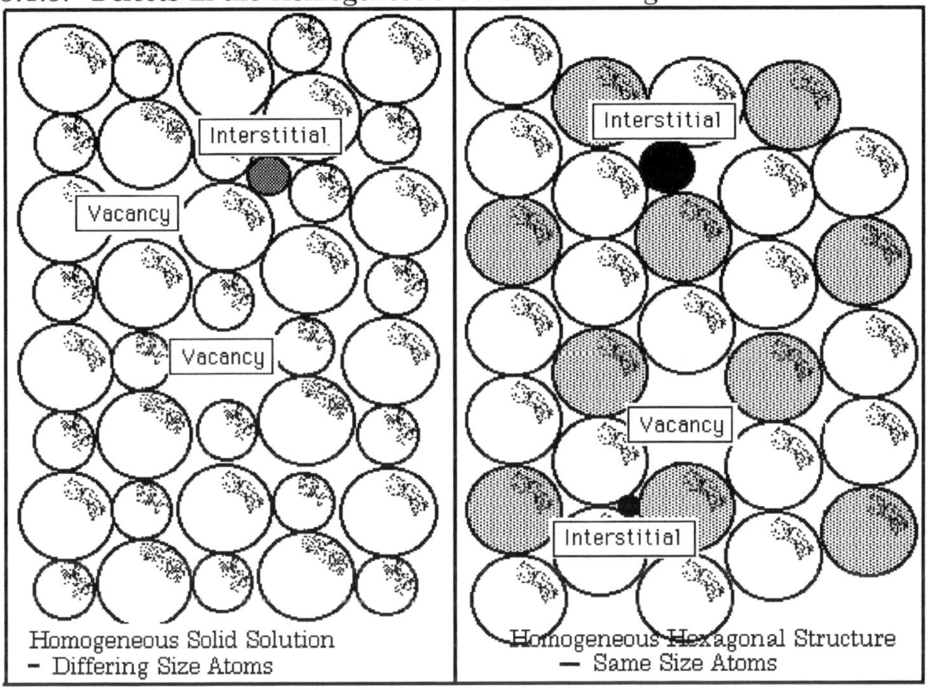


Here, we use a hexagonally-packed representation of atoms to depict the close-packed solid. In this case, we have shown both types of homogeneous solids. That is, one solid is composed of the same sized atoms while the other is composed of two different sized atoms.

On the right are the types of point defects that could occur for the same sized atoms in the lattice. That is, given an array of atoms in a three dimensional lattice, only these two types of lattice point defects could occur where the size of the atoms are the same. The term "vacancy" is self-explanatory but "self-interstitial" means that one atom has slipped into a space between the rows of atoms (ions). In a lattice where the atoms are all of the same size, such behavior is energetically very difficult unless a severe disruption of the lattice occurs (usually a "line-defect" results. This behavior is quite common in certain types of homogeneous solids. In a like manner, if the metal-atom were to have become misplaced in the lattice and were to have occupied one of the interstitial positions, as shown in the different sized atom solid, then the lattice is disrupted by its presence at the interstitial position. This type of defect has been observed.

Summarizing, three types of point defects are evident in a homogeneous lattice. In addition to the Vacancy, two types of substitutional defects can also be delineated. Both are direct substitutions in the "lattice", or arrangement of the atoms. One is a smaller atom, while the other is larger than the atoms comprising the lattice. In both cases, the lattice arrangement affects the hexagonal ordering of the lattice atoms around it. The lattice is seen to be affected for many lattice distances.

It is for this reason that compounds containing impurities sometimes have quite different chemical reactivities than the purest ones. That also has an effect upon the chemical reactivity of the solid. However, the interstitial impurity does not affect the lattice ordering at all. Now, let us look at another type of defect in the solid. Let us consider the heterogeneous lattice

b. The Point Defect in Heterogeneous Solids

The situation concerning defects in heterogeneous inorganic solids is similar to that given above, except for one very important factor, that of **charge** on the atoms. Covalent inorganic solids are a rarity while ionicity or partial ionicity seems to be the norm. Thus, heterogeneous solids are usually composed of charged moieties, half of which are positive (cations) and half negative (anions). In general, the total charge of the cations will equal that of the anions (Even in the case of semi-conductors, where the total of the charges is not zero, the excess charge (n- or p- type) is spread over the whole lattice so that no single atom, or group of atoms, ever has a charge different from its neighbors). In a given structure, cations are usually surrounded by anions, and vice-versa. Because of this, we can regard the lattice as being composed of a **cation sub-lattice** and an **anion sub-lattice**. (Remember what was stated in Chapter 1 concerning the fact that most structures are oxygen-dominated). What we mean by a "sub-lattice" is illustrated in the following diagram:

3.1.4.-

A Cubic Lattice Showing the Cation and Anion Sub-Lattices

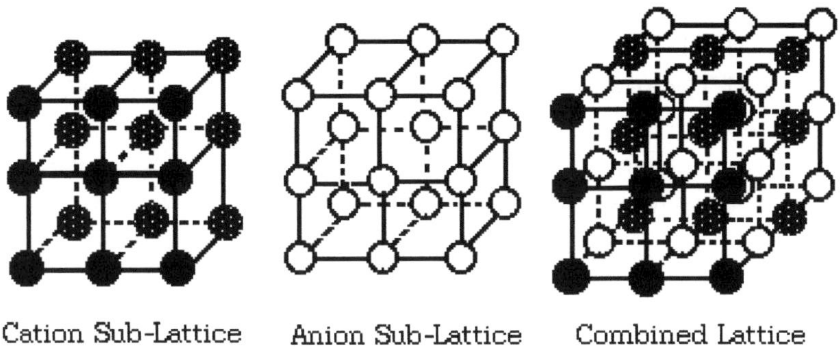


In this case, we have shown both the cation and anion "sub-lattices separately, and then the combination. It should be clear that all positive charges in the cation sub-lattice will be balanced by a like number of negative charges in the anion sub-lattice, even if excess charge exists in onc or the other of the sub-lattices. However, if an atom is missing, the overall lattice readjusts to compensate for this loss of charge. If there is a different atom present, having a differing charge, the charge-compensation mechanism again manifests itself. Thus, a cation with an extra charge needs to be compensated by a like anion, or by a nearest neighbor cation with a lesser charge. An example of this type of charge-compensation mechanism for a divalent cation sub-lattice would be the following **defect equation:** :

3.1.5.- $2\ Ca^{2+} \leftrightharpoons Li^{+} + Sb^{3+}$

where the $Sb^{3+}$ and $Li^{+}$ are situated on nearest neighbor cation sub-lattice sites, in the divalent $Ca^{2+}$ sub-lattice. Note that a total charge of 4+ exists on both sides of the above equation.

> **Thus, the charge compensation mechanism represents the single most important mechanism which operates within the *defect ionic* solid.**

We find that the number and types of defects, which can appear in the heterogeneous solid, are limited because of two factors:

1) The charge-compensation factor
2) The presence of two sub-lattices in the ionic solid.

These factors restrict the number of point defect types we need to consider in ionic heterogeneous lattices (having both cations and anions present). For ionic solids, the following types of defects have been found to exist:

* Schottky defects (absence of both cation and anion)
* Cation or anion vacancies
* Frenkel defects (Cation vacancy plus same cation as interstitial)
* Interstitial impurity atoms (both cation and anion)
* Substitutional impurity atoms(both cation and anion)

These defects are illustrated in the following diagram, given as 3.1.6. on the next page.

Note that, in general, anions are larger in size than cations due to the extra electrons present in the former. A hexagonal lattice is shown in 3.1.6. with both Frenkel and Schottky defects, as well as substitutional defects. Thus, if a cation is missing (cation vacancy) in the cation sub-lattice, a like anion will be missing in the anion sub-lattice. This is known as a Schottky defect (after the first investigator to note its existence).

In the case of the Frenkel defect, the "square" represents where the cation was supposed to reside in the lattice before it moved to its interstitial position in the cation sub-lattice. Additionally, "Anti-Frenkel" defects can exist in the anion sub-lattice. The substitutional defects are shown as the same size as the cation or anion it displaced. Note that if they were not, the lattice structure would be disrupted from regularity at the points of insertion of the foreign ion.

3.1.6.-

Point Defects Which Can Occur in the Heterogeneous Ionic Solid

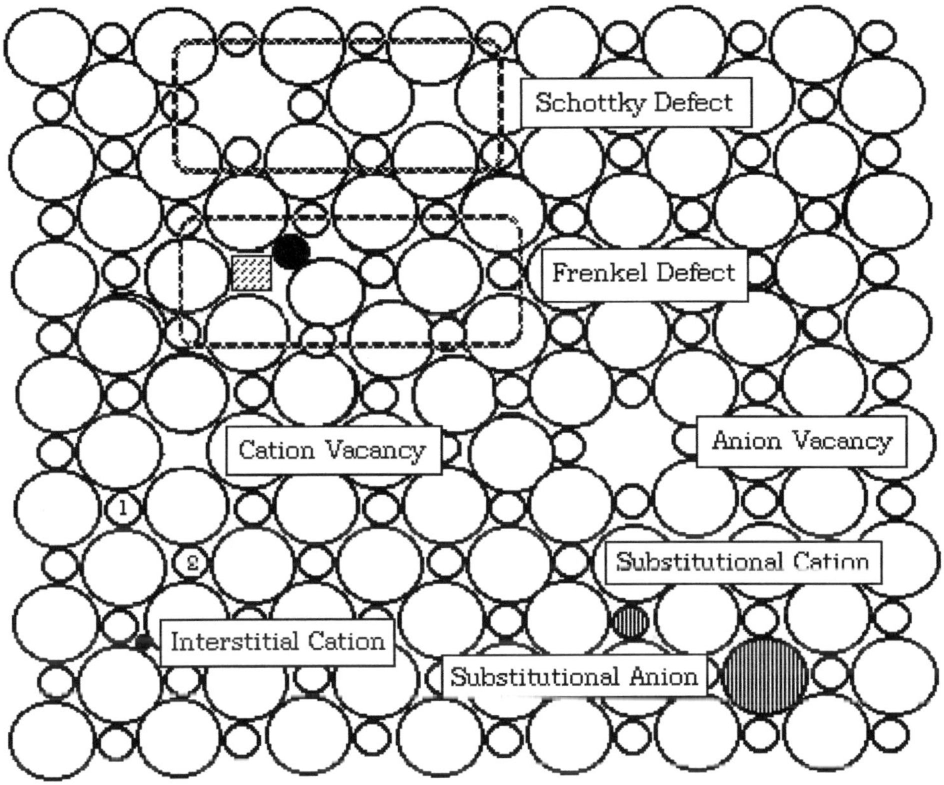


All of these point defects are intrinsic to the heterogeneous solid, and arise due to the presence of both cation and anion sub-lattices. The factors responsible for their formation are entropy effects (stacking faults) and impurity effects. At the present time, the highest-purity materials available still contain about 0.1 part per billion of various impurities, yet are 99.9999999 % pure. Such a solid will still contain about $10^{14}$ impurity atoms per mole. So it is safe to say that all solids contain impurity atoms, and that it is unlikely that we shall ever be able to obtain a solid which is **completely pure and does not contain defects.**

c. The Line Defect

In a three-dimensional lattice, we have observed planes of atoms (or ions) composing the lattice. Up to now, we have assumed that these planes maintain a certain relation to one another. That is, we have shown that there are a set of planes as defined by the {hkl} values, which in turn depends upon the type of Bravais lattice that is present. However, we find that it is possible for these rows of atoms to "slip" from their equilibrium positions. This gives rise to another type of lattice defect called "line defects". In the following diagram, we present a hexagonal lattice in which a line defect is present:

3.1.7.-

Effect of a Line Defect upon Hexagonal Packing

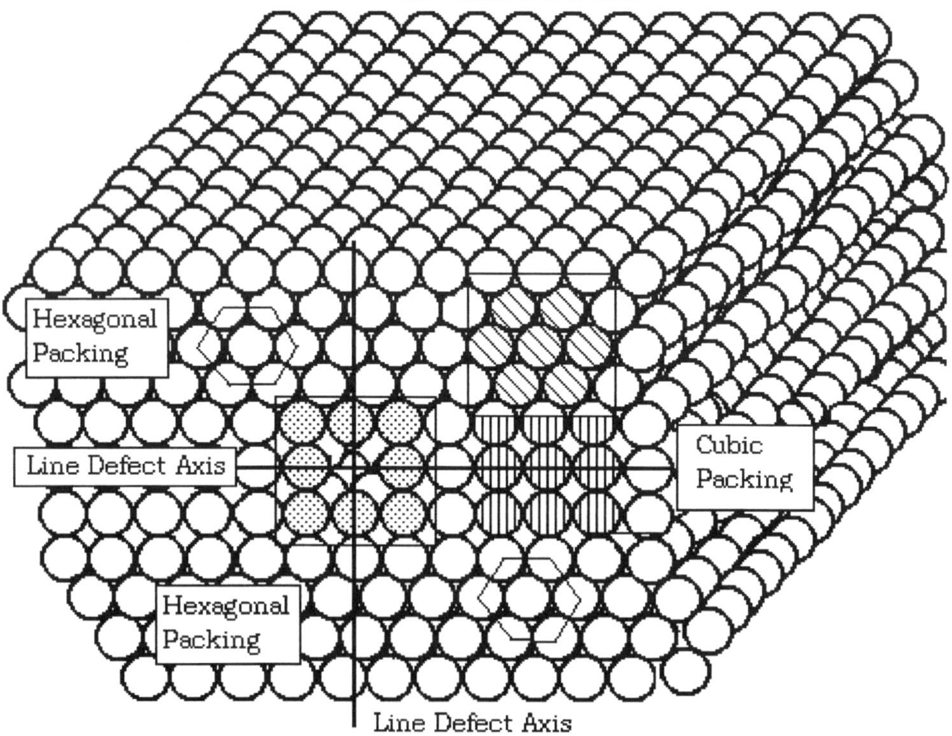


Here, rows of atoms comprising a hexagonal lattice are shown. The first four layers are composed of close-packed hexagonal layers of atoms. However, at the point shown where the line-defect axes cross in the diagram, a row of atoms is missing. This has changed the packing of the two atom layers closest to the missing row of atoms to that of cubic close-packing. It is this lack of continuity which causes the line defect because one layer has slipped from its equilibrium position. While it might not seem that this situation is serious, it causes a strain to appear in the lattice structure in which a compression is present on one side of the line defect and a strain on the other. A representation of this is shown in the following diagram:

3.1.8.-

The Effects of Line Defects Upon a Lattice

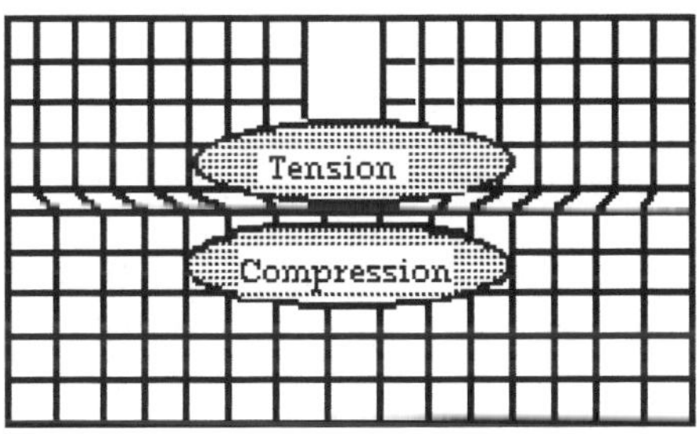


A line or edge dislocation in the Solid. On one side of the Lattice, where a line is missing, the lattice is under tension while the other side of the lattice is under compression.

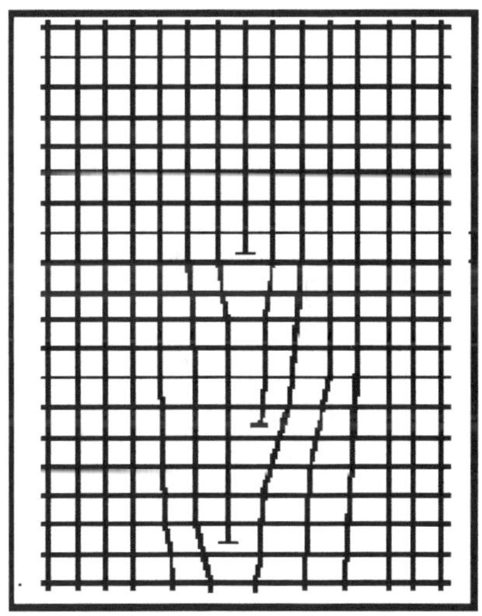

Another Perspective of Line Defects

Note that a row of atoms is missing. This has the effect of introducing a tension on the lattice since the underneath rows of atoms have to compensate for the lack of the one row. Therefore, a tension is placed on the upper rows of atoms because of the mandated change in lattice spacings in the lower rows, where a compressive force is present.

There have been several cases where it has been possible to directly observe line imperfections by suitable preparation and microscopic examination of the surface in reflected light. One example is the MgO crystal. MgO is a cubic crystal and it is possible to etch it along the {100} direction (this is the direction along the x-axis). What is observed is a series of surface lines of specific length. The reason that these defect lines show up is that they are more easily etched by acid along the direction of the edge dislocation. Note that in its simplest form, an edge dislocation is an omission of a line of atoms composing the lattice. The area within the lattice around the line defect is under both compression and tension due to the difference in atom-density as one passes through it in a direction perpendicular to the line defect. Another view of the same type of defect is shown in the following diagram:

3.1.9.-

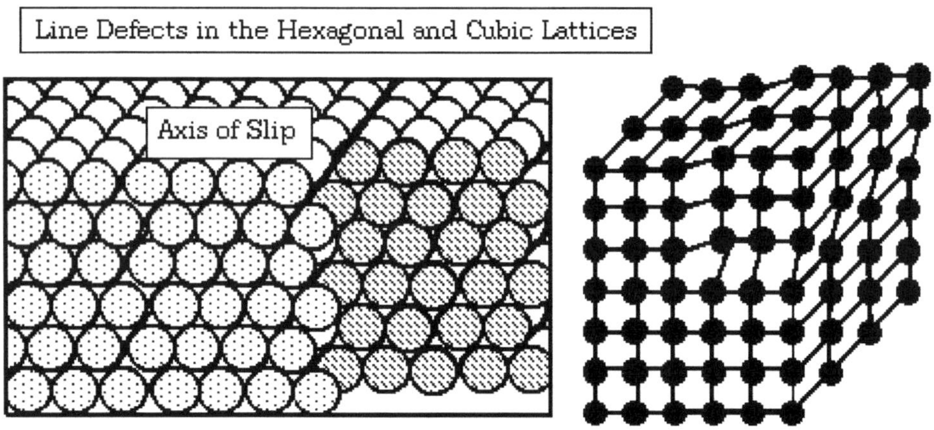


Here, the axis of slip is shown in the hexagonal lattice as being on, or near the surface of the array of atoms. In the cubic lattice, a slip plane is depicted where a line of atoms is missing and the lattice has moved to accommodate this type of lattice defect.

It should be clear that the presence of line defects in a crystal lattice leads to a disruption of the continuity of the lattice just as the presence of point defects affects the packing of a given lattice. The line defect, however, is a physical change in the lattice whereas the point defect can be classified as a chemical change. That is, the point defect causes a change in the electronic structural arrangemant of the atoms while the line defect has no effect on the electronic properties.

d. <u>The Volume Defect</u>

The volume defect is somewhat more difficult to visualize in two dimensions. Let us suppose that a line defect has appeared while the crystal structure was forming. This would be a situation similar to that already shown in 3.1.3. where a line defect was shown. The compression-tension area of the defect has a definitive effect upon the growing crystal and causes it to deform around the line defect. This is shown in the following diagram:

3.1.10.- The Volume Defect - Beginnings of a Screw Dislocation

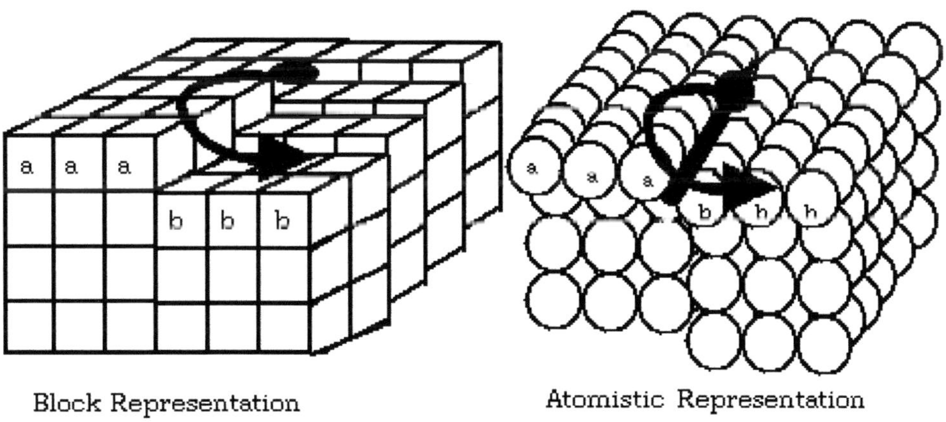


In one case, we have shown a block representation of the growth pattern of the crystal and in the other an atomistic representation. What happens is that a series of "steps" appear around the line defect and the crystal begins to assume a spiral growth pattern. That is, the points shown as "a" , being at least one atom unit higher in the crystal, begin to grow over the atom units at "b". This results in a growth situation similar to that shown in the following diagram. Here, the spiral pattern of growth is evident, and proceeds from the line defect dislocation. Both a top view and a side view are given. This type of crystal growth arises from the point of the line defect which has introduced the compression-tension factor into the growing lattice.

3.1.11.-

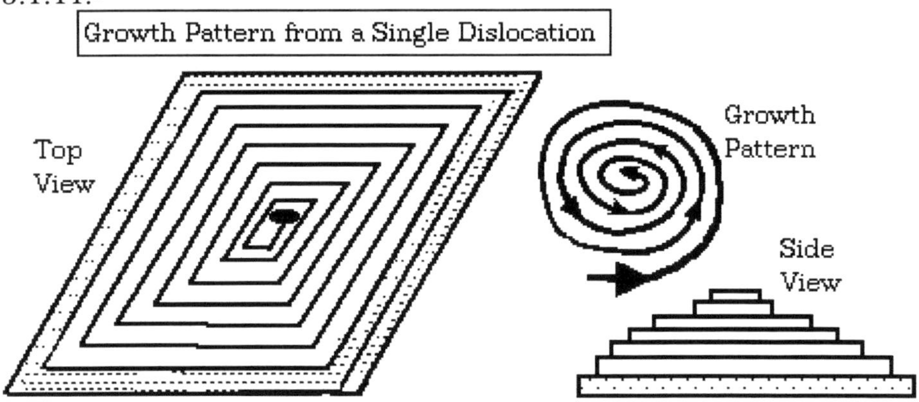


This type of volume defect in the crystal is known as a "screw dislocation", so-called because of its topography. Note that the spiral dislocation of the growing lattice deposits around the line defect at right angles to the line defect.

In the following diagram, given as 3.1.12. on the next page, another representation is shown, detailing how the dislocation line (line defect) becomes a screw-dislocation.

Note that "b" in this diagram is the same as that in 3.1.8. Because edge and volume defects propagate throughout the lattice, they affect the physical properties of the solid, whereas it is the point defects that affect the chemical properties of the solid. These latter properties include electrical and resistive, optical and reactivity properties of solids. Thus, we can now classify ***defects in solids*** as:

EXTRINSIC - Edge & Volume (physical)
INTRINSIC - Point (chemical)

3.1.12.-

Spiral Growth Around a Dislocation Line

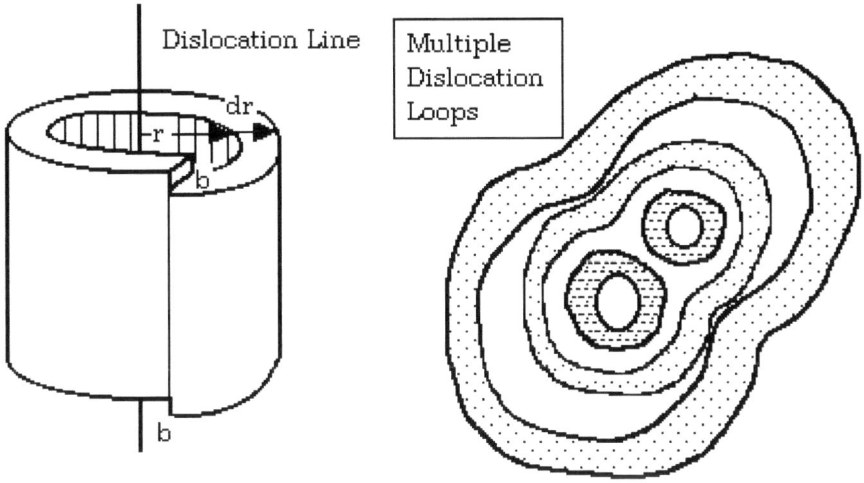


It is the intrinsic defects that have the most interest for us, since they affect the chemical properties of the solid while extrinsic defects havelittle effect. Extrinsic defects are the proper study for those interested in the mechanics of solids, particularly metals.

A broad variety of commercial products are based upon controlled point-defects. These include transistors, integrated circuits, photosensors, color-television, fluorescent lamps, just to name a few. None of these would be possible without control of point defects. As an example, consider single crystal silicon. A boule is grown about 200 to 300 mm. in diameter (8 to 12 inches). These weigh about 200 lbs before they are sliced into flat discs. The native defects in silicon grown from the melt are vacancies and interstials, impurities and oxygen entering the crystal from the growth conditions (from the $SiO_2$ crucible used to hold the liquid mass and from the atmosphere [here, the inert gas used to blanket the growth chamber does not eliminate all of the surface-adsorbed gaseous molecules]). Si vacancies and interstituals tend to agglomerate to form "surface "pits" which interfere with the formation of the integrated circuits on the surface of the silicon disc as they are being manufactured.

These, then, are the set of possible defects for the Plane Net, and the following summarizes the types of intrinsic defects expected. Note that we have used the labelling: V = vacancy; i = interstitial ; M = cation site; X = anion site and s = surface site. We have already stated that surface sites are special. Hence, they are included in our listing of intrinsic defects.

| 3.2.2- | VACANCIES | CHARGED PARTICLES |
|---|---|---|
| | $V_M$ , $V_X$ , $V^+_M$ , $V^-_X$ | $e^-$ , $\pi^+$ |
| | INTERSTITIAL SITES | SURFACE SITES |
| | $M_i$ , $X_i$ , $M^+_i$ , $X^-_i$ | $M_s$ , $X_s$ , $M^+_s$ , $X^-_s$ |

Two items of interest can be noted:

1) We can have pure vacancies in the lattice
2) These vacancies can be charged.

The same can be said for interstitial sites and surface sites as well. What this means is that the point defect can acquire a charge.

For example, let us consider a cation vacancy in ZnS. The vacancy can be represented as: $V_{Zn^{2+}}$. However, with a $V_{Zn^{2+}}$ present, a lack of charge of 2+ exists in the lattice. This deficit of charge must be compensated in the lattice by a like amount of negative charge such as a $V_{S^{2-}}$ . However, we have already stated that the $V_{Zn^{2+}}$ point defect can also acquire a charge such as $V^+_{Zn^{2+}}$ . In this case we have a total deficit in charge of 3+. Obviously, the presence of the anion vacancy, $V_{S^{2-}}$ , is not sufficient to maintain charge compensation in the lattice. Therefore some other mode of charge compensation is required. One possibility is, of course, the charged anion vacancy, $V^-_{S^{2-}}$ . There are others as well. The charging of surface sites and interstitials is not as clear as that of vacancies, and we will not dwell on them further at this point.

One might think that perhaps $V^-_M$ and $V^+_X$ ought to be included in our list of vacancies. Howcvcr, a negatively-charged cation vacancy **alone,** particularly when it is surrounded by negative anions, would not be very stable. Neither should a positively-charged anion vacancy be any more stable. Either arrangement would require high energy stabilization to exist. Therefore, we do not include them in our listing. However, a $V_M$ could capture a positive charge to become $V^+_M$ and likewise for the $V_X$ which then becomes a $V^-_X$ . Both of these ought to be stable when surrounded by the oppositely charged sites of the other sub-lattice (and indeed they are as we have indicated above).

One should note that when a **charged-interstitial** is present in one sub-lattice, the other sub-lattice will contain either a like-charged interstitial or a like-charged substitutional ion which exactly balances the total charge present in the lattice. Such equations might include one or both of the following equations:

3.2.3.- $[\, M_i^{3+} \,]_{M_i^+} \leftrightharpoons [\, X_i^{3-} \,]_{X_i^-}$

$[\, M_i^{3+} \,]_{M_i^+} \leftrightharpoons X^{3-}{}_{X^-}$

The first equation describes the equilibrium between two interstitial sites with excess charge, one on each sub-lattice, while the other equation shows the equilibria between the interstitial cation and that of the substitutional anion, both having a like excess charge. Obviously, various charge-compensating combinations of all of the 6 basic lattice defects are possible, i.e.- the $V_M$ ,$V_X$ ,$M_i$ , $X_i$ , $M_s$ , and $X_s$ defects, as well as between the substitutional cations or anions having a charge differing from those comprising the lattice. The number of combinations is thus:

$$C = 6!/(6\text{-}2)! \text{ , or } C = 30$$

Additionally, we can have at least four (4) types of charged vacancies and interstitial sites as given in 3.2.2. (Charged surface sites are not common but are included here for the sake of completeness). This gives rise to 12 more charge defect equations or a total of 42 defect equations that we can write!

The positive hole requires further explanation. $\pi^+$ is the positive electronic equivalent of the electron, $e^-$ , in solids and they annihilate each other upon reaction:

3.2.4- $e^- + \pi^+ = \text{energy}$

The following diagram shows how the positive hole can exist in the solid.

3.2.5.-

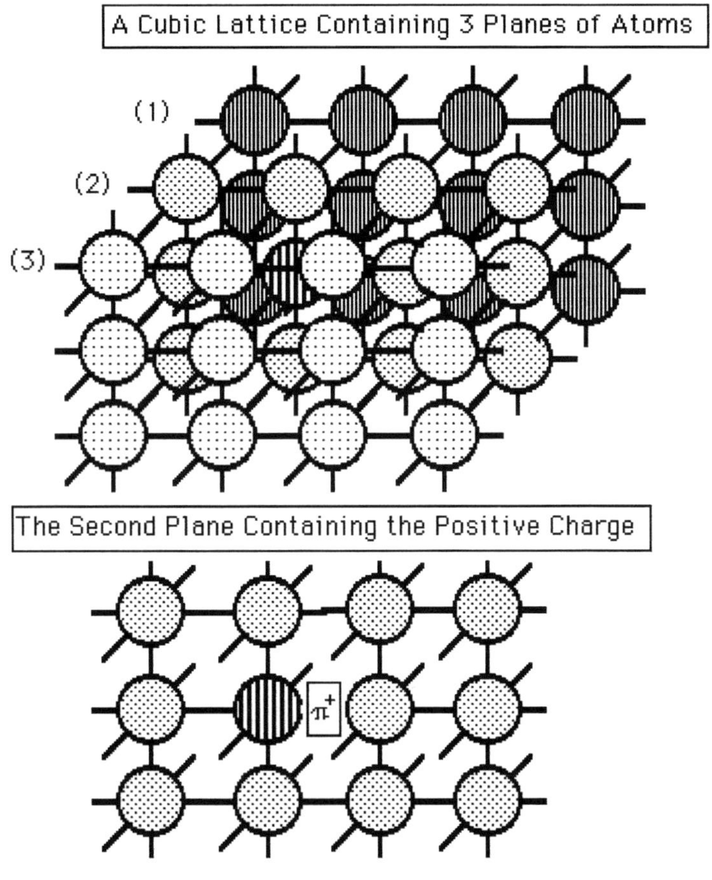


Note that we have labelled all three planes and have removed Plane "2" so as to show the presence of $^{70}Ga_{31}$ in the lattice. This type of charge-substitutional defect is used extensively in the Semiconductor Industry to produce integrated circuits by deliberate "doping" of atoms which have a valency differing from that of the major structural element, i.e. Si as $Si^{4+}$.

A second kind of electronic defect involves the electron. Let us suppose that the second plane of the cubic lattice has a **vacancy** instead of a substitutional impurity of differing valency. This makes it possible for the lattice to capture and localize an extraneous electron at the vacancy site. This is shown in the following diagram. The captured electron then endows the solid structure with special optical properties since it can absorb photon energy. The structure thus becomes optically active. That is, it absorbs light within a well-defined band and is called a "color-center" since it imparts a specific color to the crystal.

3.2.6.-

The Second Plane Containing a Vacancy Which Has Captured an Electron

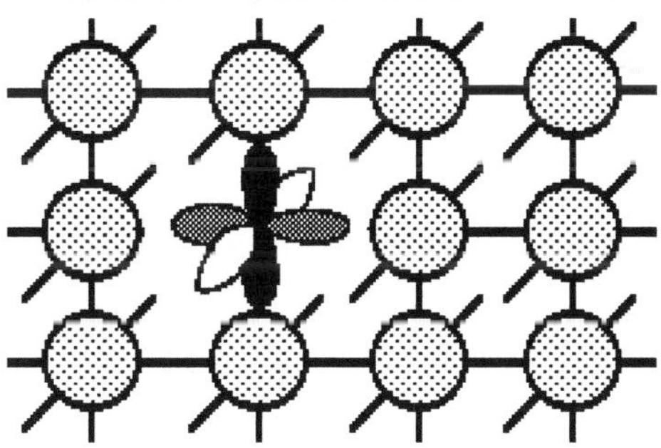

The alkali halides are noted for their propensity to form color-centers. It has been found that the peak of the band changes as the size of the cation in the alkali halides increases. There appears to be an inverse relation between the size of the cation (actually, the polarizability of the cation) and the peak energy of the absorption band. These are the two types of electronic defects that are found in crystals, namely positive "holes" and negative "electrons", and their presence in the structure is related to the fact that the lattice tends to become charge-compensated, depending upon the type of defect present.

Returning to the subject of lattice defect formation, we can now proceed to write a series of defect reactions for the defects which we found for our **plane net:**

3.2.7.- SCHOTTKY : $MX \leftrightharpoons M_{1-\delta}\, X_{1-\delta} + \delta V_m + \delta V_x$

FRENKEL : $MX \leftrightharpoons M_{1-\delta}\, X + \delta M_i + \delta V_m$

ANTI-FRENKEL: $MX \leftrightharpoons M\, X_{1-\delta} + \delta V_x + \delta X_i$

F - CENTER: $MX \leftrightharpoons M\, X_{1-\delta} + \{V_x/e^-\} + \delta/2\ X^2 \Uparrow$

In this case, we use δ as a small fraction since the actual number of defects is small in relation to the overall number of ions actually present. For the F-Center, the brackets enclose the complex consisting of an electron captured at an anion vacancy. Note that these equations encompass **all** of the mechanisms that we have postulated for each of the individual reactions. That is, we show the presence of vacancies in the Schottky case and interstitial cations for the Frenkel case involving either the cation or anion. The latter, involving an interstitial **anion** is called, by convention, the "Anti-Frenkel" case. The defect reaction involving the "F-Center" is also given.

Actually, the formation of an F-center is more complicated than this. Although F-centers can be formed by several methods, the best way to do so is by exposing a sodium chloride crystal to sodium metal vapors. In that case, the following defect reactions have been observed to take place:

3.2.8.- Defect Reactions for the NaCl Crystal

$$Na^{\circ} \leftrightharpoons Na^{+} + e^{-}$$

$$Na^{+} + \{ Na^{+}, Cl^{-}\} \leftrightharpoons 2\,\{Na^{+}\,[\,Cl^{-}\,,\,V_{Cl}\,]\}$$

$$V_{Cl} + e^{-} \leftrightharpoons [\,V_{Cl}\,/\,e^{-}]$$

These equations illustrate how the crystal responds to the presence of sodium vapor, i.e.- excess $Na^{+}$, by forming anion vacancies, to form the F-center.

Up to this point, we have only considered stoichiometric lattices. Let us now consider non-stoichiometric lattices. What we mean by this is that, in non-stoichiometric crystals, the heterogeneous lattice can contain an excess of either the cation or the anion in contrast to the homogeneous lattice which may only contain an excess of charge. What happens then is that the heterogeneous lattice responds by creation of even more lattice defects.

## **3-3** : NON-STOICHIOMETRIC SOLIDS

In a non-stoichiometric crystal, the lattice may have either excess charge or excess cations and/or anions situated in the lattice. Consider the semi-conductor, Ge. It is a homogeneous solid and is expected to contain excess charge. The defect reactions associated with the formation of p-type and n-type lattices are:

3.3.1- **n-type:** $Ge + \delta As \leftrightharpoons [Ge/As_\delta] + \delta e^-$

**p-type:** $Ge + \delta Ga \leftrightharpoons [Ge/Ga_\delta] + \delta \pi^+$

where δ is a small excess over stoichiometry. The excess charges shown are spread over the entire lattice, as stated before.

In the heterogeneous solid, a different mechanism concerning charge dominates. If there are associated vacancies, a different type of electronic defect, called the "M-center", prevails. In this case, a mechanism similar to that already given for F-centers operates, except that two (2) electrons occupy neighboring sites in the crystal. The defect equation for formation of the M-center is:

3.3.2.- M-center: $KCl \leftrightharpoons KCl_{1-\delta} + \delta/2\, Cl_2 \Uparrow + \delta/4\, [V^-_{Cl} \mid V^-_{Cl}]$

The following diagram, given as 3.3.3. on the next page, illustrates the "M-Center":

This type of electronic defect, found only in heterogeneous solids, consists of two associated, negatively-charged anion vacancies, one on each plane as shown.

3.3.3.-

An "M-Center" Shown on Two Planes of a Cubic Lattice

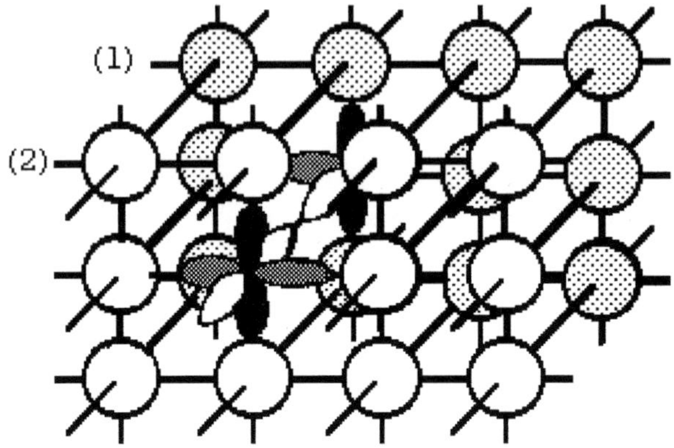


They are usually joined along the {110} plane of the lattice of the face-centered salt crystal, although we have not shown them this way (The {100} plane is illustrated in the diagram). Note that each vacancy has captured an electron in response to the charge-compensation mechanism which is operative for all defect reactions. In this case, it is the anion which is affected whereas in the "F-center", it was the cation which was affected (see equation 3.2.8. given above). These associated, negatively-charged, vacancies have quite different absorption properties than that of the F-center.

Additionally, the heterogeneous solid will contain defects in its lattice, as noted above. This involves heterogeneous solids containing impurities and we use the same type of notation given above. For the case of an AgCl crystal containing the $Cd^{2+}$ cation as an impurity, we have:

3.3.4- $$2\ Ag^+ \leftrightharpoons [\ Cd^{2+}{}_{Ag^+}\ ,\ V_{Ag^+}]$$

where the $Cd^{2+}$ cation occupies one $Ag^+$ site and the vacancy the other, as nearest neighbors. This is an example of a **heterotype impurity** system since different cations are involved. Although the vacancy occupies one site, it is not charged since the total charge differential lies on the $Cd^{2+}{}_{Ag}$ site. Note that the symbol that we have used denotes a doubly-charged cadmium cation on a monovalent **silver** cation site in the lattice. This illustrates the fact that the normal lattice of an insulator compound tends to charge compensate itself fully, but excess charge does not appear. The same is not true with semi-conducting compounds, which may be heterogeneous or homogeneous in nature.

Another heterotype example is $CdCl_2$ containing $Sb^{3+}$ . Here, we can write at least three different equations involving defect equilibria:

3.3.5.- Defect Reactions Involving the $CdCl_2{:}Sb^{3+}$ System

$$2\ Cd^{2+} \leftrightharpoons Sb^{3+} + \pi^{+} + V_{Cd}$$
$$2\ Cd^{2+} \leftrightharpoons Sb^{3+} + V^{+}_{Cd}$$
$$2\ Cd^{2+} \leftrightharpoons Sb^{3+} + Li^{+}$$

In the last equation, we have the instance where charge-compensation has occurred due to inclusion of a monovalent cation. A vacancy does not form in this case. All of these equations are cases of impurity substitutions in the normal lattice..

Another genre is the so-called **homotype impurity** system. One example is the substance, nickelous oxide, which is a pale-green insulator when prepared in an inert atmosphere. If it is reheated in air, or if a mixture of NiO and $Li^{+}$ is reheated in an inert atmosphere, the NiO becomes a black semi-conductor. This is a classical example of the effect of defect reactions upon the intrinsic properties of a solid. The defect reactions involved are:

3.3.6.-

$$2\ Ni^{2+} \leftrightharpoons [\ Ni^{3+} / V_{Ni}\ ] + \pi^{+}$$
$$2\ Ni^{2+} \leftrightharpoons Ni^{3+} + Li^{+}$$

This behavior is typical for transition metals which easily undergo changes in valency in the solid state. That is, homotype impurity systems usually involve cations which can undergo valence changes easily, i.e.- transition metal cations. In the NiO case, we have gone from an insulator to a semi-conductor just by introducing lattice defects through a thermal effect.

Having introduced these examples, let us now examine a method of symbolism useful in characterizing defect reactions in solids.

## 3.4.- DEFECT EQUATION SYMBOLISM

Whether you realize it or not, we have already developed our own symbolism for defects and defect reactions based on the Plane Net. It might be well to compare our system to those of other authors, who have also considered the same problem in the past. It was Rees (1930) who wrote the first monograph on defects in solids. Rees used a box to represent the cation vacancy, as did Libowitz (1974). This has certain advantages since we can write equation 3.3.5. as shown in the following:

3.4.1- Symbolisizm Used for Defect Equations

$$2\ Cd^{2+} \leftrightharpoons \boxed{Sb}^{+}_{Cd} + \square_{Cd} + \pi^{+}$$

$$2\ Cd^{2+} \leftrightharpoons \boxed{Sb}^{+}_{Cd} + \boxed{\pi}^{+}_{Cd}$$

$$2\ Cd^{2+} \leftrightharpoons \boxed{Sb}^{+}_{Cd} + \boxed{Li}^{+}_{Ni} + \pi^{+}$$

Likewise, the other equations of 3.3.6. would become:

3.4.2-
$$2\ Ni^{2+} \leftrightharpoons Ni^{3+} + \square_{Ni} + \pi^{+}$$

$$2\ Ni^{2+} \leftrightharpoons Ni^{3+} + \boxed{Li}^{+}_{Ni}$$

Although the results are equal as far as utility is concerned, we shall continue to use our symbolism, for reasons which will become clearer later on. The following Table is a comparison of defect symbolism, as used by prior Authors. Note that our symbolism most resembles that of Krœger, but not in all aspects. These prior authors also considered other intrinsic defects that we have not touched, namely interstices and the so-called "anti-structure" occupation.

Table 3-1

Comparison of Point Defect Notation by Various Authors

| | Rees(1930) | Krœger (1954) | Libowitz (1974) |
|---|---|---|---|
| Cation Site Vacancy: | $\square_M$ | $V_M$ | $\square_M$ |
| Anion Site Vacancy: | $\square_X$ | $V_X$ | $\square_X$ |
| Cation Interstitial | $\Delta_M$ | $M_i$ , $M^+_i$ | $M_i$ |
| Anion Interstitial | $\Delta_X$ | $X_i$ , $X^-_i$ | $X_i$ |
| Negative Free Charge | e | $e^-$ | $e^-$ |
| Positive Free Charge | p | $h^+$ | $h^+$ |
| Interstices | --- | $\alpha V_i$ | $\alpha V_i$ |
| Unoccupied Interstitial | --- | $V_i$ | $\Delta$ |
| Anti-structure Occupation | --- | --- | $M_M$ , $X_X$ , $M_X$ , $X_M$ |

The latter deals with an impurity **anion** on a cation site coupled with an impurity **cation** on an anion site, both with the proper charge. We have mentioned interstices but not in detail. They appear as a function of structure (Refer back to Chapter 1 - Diagram 1.3.2.). There is one in a tetrahedron, four in a body centered cube, and six in a simple cube. Thus, $\alpha$ in $\{\alpha V_i\}$ is 1, 4 or 6, respectively. We shall need this symbol later, as well as $V_i$ , the unoccupied interstitial.

Before we proceed to analyze defect reactions by a mathematical approach, let us consider an applications of solid state chemistry. In this example, the effect of a defect on the properties of the solid is described.

## **3-5** - SOME APPLICATIONS FOR DEFECT CHEMISTRY

In order to understand some of the aspects of these examples, we need to discuss a certain amount of background for each area of the solid state being considered.

a. <u>Phosphors:</u>

Phosphors are inorganic materials which convert incident radiant energy to visible light within a device. The device chosen can be a cathode-ray tube, i.e.- a television tube, or a fluorescent lamp. A phosphor consists of a matrix modified by an additive chosen so that it becomes optically active within the matrix, or compound. This is an example of a substitutional impurity in a lattice wherein the additive, usually called an "activator", introduces a lattice defect that is optically active. However, the added impurity still follows **all** of the rules found for defects in a lattice, as shown by the following example.

In the prior literature, it was found (Kinney- 1955) that $Ca_2P_2O_7$ (i.e.- calcium pyrophosphate) could be activated by $Sb^{3+}$ to form a phosphor: $Ca_2P_2O_7$: $Sb_{.02}$. This formalism actually indicates the composition:

$$[(Ca_{0.99}\,,\,Sb_{0.01})_2\,P_2\,O_7)]\,.$$

where a solid-state solution of calcium pyrophosphate and antimony pyrophosphate is indicated. The brightness response of this phosphor was moderate when excited by ultraviolet radiation but was improved four times by the addition of $Li^+$. The optimum amount of $Li^+$ proved to be that exactly equal to the amount of $Sb^{3+}$ present in the phosphor, i.e.-

$$2\ Ca^{2+} \leftrightharpoons Sb^{3+} + Li^+.$$

The explanation lies in the defect reactions controlling the formation of the phosphor itself. The defect reactions occurring were found to be the substitution of a trivalent cation on a divalent site and the defects reactions thereby associated. This is shown in the following table which compares these two methods of preparing such materials. In this case, the increase in brightness was found to be related to the amount of activator actually being incorporated into the lattice. It is well known that phosphor brightness is proportional to the numbers of $Sb^{3+}$ ions (activators) actually incorporated into cation sites of the pyrophosphate structure.

Table 3-2

Defect Reactions Which Occur in the Phosphor: $(Ca_{.99}, Sb_{.01})_2P_2O_7$.

| | PHOSPHOR BRIGHTNESS |
|---|---|
| $2\ Ca^{2+} \leftrightharpoons Sb^{3+}{}_{Ca} + V_{Ca} + \pi^+$ | 25 % |
| or $2\ Ca^{2+} \leftrightharpoons Sb^{3+}{}_{Ca} + V^+{}_{Ca}$ | |
| $2\ Ca^{2+} \leftrightharpoons Sb^{3+}{}_{Ca} + Li^+{}_{Ca}$ | 100 % |

Phosphors are prepared by heating the ingredients at high temperature (> 1000 °C.) to obtain a compound having high crystallinity. The sintering process decreases entropy and is **counterproductive** to the formation of vacancies in the pyrophosphate lattice. In the absence of $Li^+$, vacancy-formation was retarded with the result that the amount of $Sb^{3+}$ which could be incorporated into activator sites actually decreased. In contrast, four times as many activator ions were incorporated into the lattice when the charge-compensating $Li^+$ ions were present on nearest neighbor sites. Since it was necessary to have a vacancy which could form the charge-compensated site, i.e.- $V^+{}_{Ca}$, required for the incorporation of the impurity, $Sb^{3+}{}_{Ca^{2+}}$, these lattice sites did not form, due to the sintering process mandatory for forming the phosphor.

Note that we have written two defect reactions for the case of vacancy formation in Table 3-2. Pyrophosphate is an insulator and the formation of a positively-charged vacancy is much less likely than the vacancy plus a free positive charge. This brings us to a rule found in defect chemistry that seems to be universal, namely:

ALTHOUGH MORE THAN ONE DEFECT REACTION MAY BE APPLICABLE TO A GIVEN SITUATION, ONLY ONE IS USUALLY FAVORED BY THE PREVAILING THERMODYNAMIC AND ELECTRON-COMPENSATION CONDITIONS.

## 3.6.- DEFECT EQUILIBRIA AND THEIR ENERGY

Just as chemical reactions can be described and calculated in terms of thermodynamic constants and chemical equilibria, so can we also describe defect formation in terms of equilibria. Using the Law of Mass Action, viz-

3.6.1- Law of Mass Action:

$$bB + cC \leftrightharpoons dD + eE \;;\; K = a^d{}_D \; a^e{}_E \; / \; a^b{}_B \; a^c{}_C$$

where the equilibrium constant, K, is written in terms of the activities and coefficients of the equation given on the left side of 3.6.1. We can also write an equation for any given defect in the lattice. For the intrinsic defects in an MX lattice, we would have:

3.6.2.- <u>Frenkel Defects (for the MX crystal)</u>

$$M_X \leftrightharpoons M_i + V_M \;;\; K_F = a_{M_i} \; a_{V_M} \; / \; a_M$$

3.6.3.- <u>Schottky Defects (for the MX crystal)</u>

$$MX \leftrightharpoons V_M + V_X \;;\; K_{Sh} = a_{V_M} \; a_{V_M} \; / \; a_{M_X}$$

Note that we have also specified these equilibrium constants in terms of the activity of the **associated defects.** We can also write thermodynamic equations for these defects:

3.6.4.- Chemical Thermodynamics

$$\Delta G = \Delta H - T\,\Delta S = -\,RT \ln K \qquad K = \exp \Delta S/R \cdot \exp - \Delta H/RT$$

3.6.5.- Defect Thermodynamics

$$K_d = \exp \Delta S_d / R \cdot \exp - \Delta H_d / RT$$

where d refers to the specific defect. Note that we must know either $\Delta H_d$ or $\Delta S_d$ in order to calculate the equilibrium constant. Alternately, if we know the equilibrium constant (or can estimate it), we need to know only one of the others to calculate the energy associated with the defect. Thc following summarizes the knowledge we have developed above for the Schottky and Frenkel defects:

1. There is an Activation Energy for defect formation. In many cases, this energy is low enough that defect formation occurs at, or slightly above, room temperature.

2. Defects may be described in terms of thermodynamic constants and equilibria. The presence of defects changes both the local vibrational frequencies in the vicinity of the defect and the local lattice configuration around the defect.

One question we may logically ask is how are we to know what types of defects will appear in a given solid? The answer to this question is as follows:

IT HAS BEEN FOUND: **"There are two associated effects on a given solid which havc opposite effects on stoichiometry. Usually, one involves the cation site and the other the anion site. Because of the differences in defect-formation-energies, the concentration of other defects is usually negligible".**

Thus, if Frenkel Defects predominate in a given solid, other defects are usually not present. Likewise, for the Schottky Defect. Note that this applies for **associated defects**. If these are not present, there will still be **2 types** of defects present, each having an opposite effect upon stoichiometry.

## **3.7-** DEFECT EQUILIBRIA IN VARIOUS TYPES OF COMPOUNDS

Up to now, we have been concerned with the MX compound as a hypothetical example of the solid state. We will now extend these concepts to more complicated ones such as binary and ternary compounds. We will use the precepts already developed for the simple MX compound. For the sake of simplicity, we restrict ourselves to binary compounds, that is- one cation and one anion. An example of a ternary compound is $ABX_S$ , where A and B are different cations, and S is a small whole number. Our example of a binary compound will be:

$$MX_S$$

In this case, we may have: $MX_2$ , $MX_3$ or even $MX_4$. We can distinguish between four states for these hypothetical compounds, including specific combinations of these states, to wit:

stoichiometric vs: non-stoichiometric
non-ionized vs: ionized

This gives us four combinations:

1) Non-ionized Stoichiometric
2) Ionized Stoichiometric
3) Non-ionized Non-stoichiometric
4) Ionized Non-stoichiometric

We will deal with the first two cases, namely 1) defects in the stoichiometric case and 2) ionization of defects in stoichiometric compounds. Since the case of non-stoichiometric compounds can be very complicated, we will present treatment of the last two in the Appendix at the end of this Chapter.

a. STOICHIOMETRIC BINARY COMPOUNDS OF $MX_S$

In the real world of defect chemistry, we find that in addition to the simple defects, other types of defects can also appear, depending upon the type of crystal we are dealing with. These may be summarized as shown in the following, given as 3.7.1. on the next page.

According to our nomenclature, as used in the table, $V_M$ is a vacancy at an M cation site, etc. The first five pairs of defects given above have been observed experimentally in solids, whereas the last four have not. This answers the question posed above, namely that defects in solids occur in pairs.

### 3.7.1- Defects in the $MX_S$ Compound

| | PAIRS OF DEFECTS |
|---|---|
| Schottky | $V_M + V_X$ |
| Frenkel | $V_M + M_i$ |
| Anti-Frenkel | $X_i + V_X$ |
| Anti-Structure | $X_M + M_X$ |
| Vacancy- AntiStructure | $V_M + M_X$ |
| | |
| Interstitial | $M_i + X_i$ |
| Anti-Structure Interstitial | $M_X + X_i$ |
| Anti-Structure Interstitial | $X_M + M_i$ |
| Anti-Structure Vacancy | $X_M + V_X$ |

We have now introduced into our nomenclature a distinction between structure and anti-structure defects. What this means is that stacking faults can sometimes result in the formation of $X_M$ and $M_X$ defects. These have been recognized as high-energy defects since they exist as cations (anions) surrounded by a complete positive (negative) charge. In fact, it is difficult to see just how they could exist. However, there have been some cases studied where they do exist, and so we have included them in our listing of defects although we originally stated that such defects are not likely to be very prevalent. However, note that these defects generally exist as a vacancy plus the anti-structure defect, not as associated anti-structure defects together, as we showed above for the homogeneous solid.

Table 3-3, given on the next page, summarizes the various pairs of defects possible for binary compounds. Equilibria are given along with the appropriate equilibrium constant. Note that these equations are rather simple and can be used to specify the equilibrium constants for these defects present in the lattice. These types of defects have been observed and studied in the compounds given under "Example" in this Table. These are the major types of defects to be expected in most inorganic compounds, where the number of sites in the lattice is constant.

TABLE 3-3

POSSIBLE DEFECTS IN THE $MX_S$ COMPOUND

(For a Constant Number of Sites)

| Type of Defect | Defect Pair | Example | Defect Equation | Equilibrium Constant |
|---|---|---|---|---|
| Schottky | $V_M + V_X$ | TiO | $0 \leftrightharpoons V_M + S\, V_X + \alpha V_i$ | $K_{Sh} = V_M\, V_X{}^S\; V_i{}^{\alpha}$ |
| Frenkel | $V_M + M_i$ | ZnO | $M_M + \alpha V_i \leftrightharpoons V_M + M_i$ | $K_F = V_M\; M_i \;/\; M_M\; V_i{}^{\alpha}$ |
| Anti-Frenkel | $X_i + V_X$ | $LaH_2$ | $X_X + \beta\, V_i \leftrightharpoons X_i + V_X$ | $K_{AF} = X_i\, V_X \;/\; X_X\, V_i{}^{\beta}$ |
| Anti-Struct. | XM+MX | AuZn | $M_M + X_X \leftrightharpoons M_X + X_M$ | $K_{AS} = M_X\, X_M \,/\, M_M\, X_X$ |

If there is an excess of sites present in the lattice, the situation is somewhat different (and much more complicated). In at least one case, one type **has** been observed and so we cannot completely ignore them, even though they are fairly rare.

Table 3-4 gives a complete listing of these types of defects and the names given to them.

TABLE 3-4

(Excess Number of Sites)

| Name of Defect | Defect Pair | Type | Defect Equation | Equilibrium Constant |
|---|---|---|---|---|
| Vac.-Struc. | $V_M + M_X$ | NiAl | $\delta M_M \leftrightharpoons \delta M_X + (1-\delta)\, V_M$ | $K_{VS} = M^{\delta}{}_X V_M{}^{1+\delta} V_i{}^{\alpha} /$ $M^{\delta} + \alpha\, ]V_i$ |
| Struct.Vac. | $V_X + X_M$ | | $\delta X_X \leftrightharpoons X_M (1+\delta)\, V_X + \alpha V_i$ | $K_{SV} = X_M V_X{}^{1+\delta}\, V_i{}^{\alpha} / X^{\delta}{}_X$ |
| Interstitial | $M_i + X_i$ | | $\delta( M_M + SX_X) \leftrightharpoons M_i + S\, X_i$ $- \; (1+\delta+\alpha)\, V_i$ | $K_i = M_i\, X_i\, S\, V_i{}^{1+\delta+\alpha} \;/$ $[M_M{}^{\delta}\; X_X{}^{S\delta}]$ |
| Interstitial Struct. | $M_X + X_i$ | | $M_M + (1+\delta) X_X \leftrightharpoons M_X$ $+ (1+\delta) X_i (1+\delta+\alpha)\, V_i$ | $K_{IS} = M_X\, X_i{}^{1+\delta} V_i{}^{(1+\delta+\alpha)}$ $/ \; M_M\, X_X{}^{1+\delta}$ |
| Struct-Interstial | $M_i + X_M$ | | $(1+\delta) M_M + SX_X \leftrightharpoons$ $(1+\delta) M_i + \delta\, X_M$ | $K_{SI} = M_i\, X_M{}^{\delta} \;/$ $M_M{}^{1+\delta}\, X_X{}^S$ |

These five defects are based upon an excess in the number of sites available. This excess we call "$\delta$". Note that we are not speaking of the ratio of cations to anions, i.e.- stoichiometry, but of the total number of sites. To see how this is possible, consider the Vacancy-Structure type.

We have a $V_M$ (a cation vacancy) associated with an $M_X$ , an M atom on an anion site.The total number of atoms remains constant, but there is an excess of cations, notably $M_X$ . Fortunately, we do not have to deal with these equations very much but include them for the sake of completeness. Note that we have used a hypothetical compound to represent all of the possible compounds that we might encounter.

In order to relate these equation to the real world, the following table lists the equations given in Table 3-3 in terms of a $CaCl_2$ crystal.

TABLE 3-5
POSSIBLE DEFECTS IN THE $CaCl_2$ COMPOUND
(For a Constant Number of Sites)

| Type of Defect | Defect Pair | Defect Equation | Equilibrium Constant |
|---|---|---|---|
| Schottky | $V_{Ca}+ V_{Cl}$ | $0 \leftrightharpoons V_{Ca} + 2V_{Cl} + \alpha V_i$ | $K_{Sh} = V_{Ca} V^2_{Cl} \; V_i{}^{\alpha}$ |
| Frenkel | $V_{Ca}+ Ca_i$ | $M_M + \alpha V_i \leftrightharpoons V_M + M_i$ | $K_F - V_{Ca} Ca_i / Ca_{Ca} V_i{}^{\alpha}$ |
| Anti-Frenkel | $Cl_i + V_{Cl}$ | $Cl_{Cl} + \beta V_i \leftrightharpoons Cl_i + V_{Cl}$ | $K_{AF} = Cl_i V_{Cl} / Cl_{Cl} V_i{}^{\beta}$ |
| Anti-Struct. | $Cl_{Ca}+Ca_{Cl}$ | $Ca_{Ca} + Cl_{Cl} \leftrightharpoons Cl_{Ca}+Ca_{Cl}$ | $K_{AS} = \dfrac{Cl_{Ca}+Ca_{Cl}}{Ca_{Ca}+ Cl_{Cl}}$ |

We have used $CaCl_2$ because it is a familiar material and the equations which result are rather simple. Note that in most cases, the equilibrium constant is given in terms of the **ratio** of defects to the number of ions in their **correct** position in the lattice.

Let us now consider the concentration of defects that might appear in the $MX_S$ crystal.

## b. DEFECT CONCENTRATIONS IN $MX_S$ COMPOUNDS

To see how these equations might be used, consider the following. First, suppose we want to be able to determine the number of intrinsic defects in a given solid. Since pairs of defects predominate in a given solid, the number of each type of intrinsic defect, $N_i$ (M) or $N_i$ (X), will equal each other, For Schottky defects in the $MX_S$ crystal, we have:

3.7.2- $$N_i (V_M) = N_i (S V_X)$$

This makes our mathematics simpler since we can rewrite the Schottky equation of Table 3-3 as:

3.7.3- $$0 \leftrightharpoons N_{i (V_M)} + S N_{i (V_M)} + \alpha V_i$$

Here, we have expressed the concentration as the ratio of defects to the number of M- atom sites (this has certain advantages as we will see). We can than rewrite the defect equilibria equations of Table 3-3 and 3-4 in terms of numbers of **intrinsic** defect concentrations, shown as follows:

Some of these equations are complicated and we need to examine them in more detail so as to determine how they are to be used.

3.7.4.- EQUILIBRIUM CONSTANTS and INTRINSIC DEFECTS

| | |
|---|---|
| SCHOTTKY: | $k_{Sh} = N_i \ (S N_i)^2 = S^S N_i^{S+1}$ |
| FRENKEL: | $k_F = N_i^2 / (1 - N_i)(\alpha - N_i)$ |
| ANTI-FRENKEL: | $k_{AF} = N_i^2 / (S - N_i)(\alpha - N_i)$ |
| ANTI-STRUCT: | $k_{AS} = N_i^2 / (1 - N_i)(S - N_i)^S$ |
| VAC.-STRUCT: | $k_{VS} = S^S (S + 1)^{S+1} \cdot N_i^{2S+1} / (S - N_i - S N_i)^S$ |
| STRUCT-VAC: | $k_{SV} = (S+1)^{S+1} N_i^{S+2} / (S - N_i - S N_i)$ |
| INTERSTIAL: | $k_I = S^S N_i^{S+1} / (\alpha - N_i - S N_i)^{\alpha+S+1}$ |

Equation 3.6.10. given above shows that intrinsic defect concentrations will increase with increasing temperature and that they will be low for high Enthalpy of defect formation. This arises because the entropy effect is a positive exponential while the enthalpy effect is a negative exponential. Consider the following examples of various types of compounds and the natural defects which may occur (depending upon how the compounds were originally formed):

**TiO** is cubic with the NaCl structure. A sample was annealed at 1300 °C. Density and X-ray measurements revealed that the intrinsic defects were Schottky in nature ($V_{Ti}$ + $V_O$) and that their concentration was 0.140. In this case, S = 1 so that:

3.7.5.- $K_{Sh} = 0.0196 = 2 \times 10^{-2}$

This crystal is quite defective since 1 out of 7 Ti-atom-sites ($0.14^{-1}$) is a vacancy, and likewise for the oxygen-atom-sites.

Another example is:

**$CeH_2$** . From thermodynamic measurements, it was found that the intrinsic defects were Anti-Frenkel in nature, i.e.- ($H_i$ + $V_H$). An equilibrium constant was calculated as:

3.7.6.- $k_{AF} - 3.0 \times 10^{-4}$ at a temperature of 600 °C.

This compound has the cubic fluorite structure with **one** octahedral interstice per Ce atom. Therefore, $\alpha = 1$, and S = 2 for $CeH_2$ . We can therefore write:

3.7.7.- $k_{AF} = N_i^2 / (2 - N_i)(1 - N_i) = 3.0 \times 10^{-4}$

**or**

$N_i = 2.4 \times 10^{-2}$ (600 °C.)

This means that 1 out of 42 hydride atoms is interstitial, and 1 out of 84 hydride-atom-sites is vacant.

Let us now review what we have covered concerning stoichiometric binary compounds:

1. We have shown that defects occur in pairs. The reason for this lies in the charge-compensation principle which occurs in all ionic solids.

2. Of the nine defect-pairs possible, only 5 have actually been experimentally observed in solids. These are:

Schottky
Frenkel
Anti-Frenkel
Anti-Structure
Vacancy-Structure.

3. We have given defect-equations for all nine types of defects, and the Equilibrium Constants thereby associated. However, calculation of these equilibria would require values in terms of **energy** at each site, values which are difficult to determine. A better method is to convert these EC equations to those involving **numbers of each type of intrinsic defect,** as a ratio to an intrinsic cation or anion. This would allow us to calculate the actual number of intrinsic defects present in the crystal, at a specified temperature.

Let us now summarize what we have covered:

1. We have shown that defect equations and equilibria can be written for the $MX_S$ compound, both for stoichiometric and non-stoichiometric cases.
2. In addition, we have shown that further defect formation can be induced by external reacting species, and that these act to form specific types of defects, depending upon the chemical nature of the crystal lattice.
3. We have also shown that the intrinsic defects can become ionized to form species not found in nature.

**3-8.** - THE EFFECTS OF PURITY (AND IMPURITIES)

Our study has led us to the point where we can realize that the primary effect of impurities in a solid is the formation of defects, particularly the Frenkel and Schottky types of associated defects. Thus, the primary effect obtained in a high purity solid is the **minimization** of defects. Impurities, particularly those of differing valencies than those of the lattice, cause charged vacancies and/or interstitial sites. We can also increase the reactivity of a solid to a certain extent by making it more of a defect crystal by the addition of selected impurities. It is not so apparent as to what happens to a solid as we continue to purify it. To understand this, we need to examine the various grades of purity as we normally encounter them. Although we have emphasized inorganic compounds thus far (and will continue to do so), the same principles apply to organic crystals as well. COMMERCIAL GRADE is usually about 95% purity (to orient ourselves, what we mean is that 95% of the material is that specified, with 5% being different (unwanted-?) material. Laboratory or "ACS-REAGENT GRADE" averages about 99.8% in purity.

The GRADES listed are named for the usage to which they are applied, and are usually minimum purities required for the particular application. Fiber optic materials are currently prepared by chemical vapor deposition techniques because any handling of materials introduces impurities.

3.8.1.- GRADES OF PURITY FOR COMMON CHEMICALS

| GRADE | % | ppm | Impurity Atoms per Mole of Compound |
|---|---|---|---|
| Commercial | 95 | 50,000 | $3.0 \times 10^{22}$ |
| Laboratory | 99.8 | 2000 | $1.2 \times 10^{21}$ |
| Luminescent | 99.99 | 100 | $6 \times 10^{19}$ |
| Semi-conductor | 99.999 | 10 | $6 \times 10^{18}$ |
| Crystal Growth | 99.9999 | 1 | $6 \times 10^{17}$ |
| Fiber-Optics | 99.999999 | 0.01 | $6 \times 10^{15}$ |

Furthermore, this is the only way found to date to prepare the required materials at this level of purity.

The frontiers of purity achievement presently lie at the fraction of parts per quadrillion level. However, because of Environmental Demands, analytical methodology presently available far exceeds this. We can now analyze metals and anions at the femto level (parts per quadrillion = $10^{-15}$) if we wish to do so.

**Nevertheless, it is becoming apparent that as high purity inorganic solids are being obtained, we observe that their physical properties may be somewhat different than those usually accepted for the same compound of lower purity** .

The higher-purity compound may undergo solid state reactions somewhat different than those considered "normal" for the compound. If we reflect but a moment, we realize that this is what we might expect to occur as we obtain compounds (crystals) containing far fewer intrinsic defects.

It is probably true that many of the descriptions of physical and solid state reaction mechanisms now existing in the literature are only partially correct. It would appear that part of the frontier of knowledge for Chemistry of The Solid State will lie in measurement of physical and chemical properties of inorganic compounds as a function of purity.

## Suggested Reading


1. A.C. Damask and G.J. Dienes, *Point Defects in Metals*, Gordon & Breach, New York (1972).
2. G.G. Libowitz, "Defect Equilibria in Solids", *Treatice on Solid State Chem.*- (N.B. Hannay- Ed.), **1**, 335-385, (1973).
3. F.A. Krœger, *The Chemistry of Imperfect Crystals*, North-Holland, Amsterdam (1964).
4. F.A. Krœger & H.J. Vink in *Solid State Physics, Advances in Research and Applications* (F. Seitz & D. Turnbull-Eds.), pp. 307-435 (1956).
5. J.S. Anderson in *Problems of Non-Stoichiometry* (A. Rabenau-Ed.), pp.1-76, N. Holland, Amsterdam (1970).
6. W. Van Gool, *Principles of Defect chemistry of Crystalline Solids*, Academic Press, New York (1964).
7. G. Brouwer, "A General Asymmetric Solution of Reaction Equations Common in Solid State Chemistry", Philips Res. Rept., **9** , 366-376 (1954)

8. A. B. Lidiard, "Vacancy Pairs in Ionic Crystals", *Phys. Rev.*, **112**, 54-55 (1958).
9. J.S. Anderson, "The Conditions of Equilibrium of Nonstoichiometric Chemical Compounds, *Proc. Roy. Soc. (London)* , **A185**, 69-89 (1946).
10. N.N. Greenwood, *Ionic Crystals, Lattice Defects & Non-Stoichiometry,* Butterworths, London (1968).

## PROBLEMS FOR CHAPTER 3

1. Identify the types of point defects likely to appear in a homogeneous solid. Contrast those to the typical defects which may appear in a heterogeneous solid.

2. Given the strontium chloride crystal, write the defect reaction(s) expected if lithium chloride is present as an impurity. Do likewise for the antimony chloride impurity. Also, write the defect reactions expected if **both** impurities are present in equal quantities.

3. Draw one or more "plane-nets" for the "P" cation combined with a "U" anion. Indicate all of the possible defects that can appear. Write the symbol of each as you proceed. Include pairs of defects as needed.

4. Write equations for as many of the thirty (30) defects reactions of your "PU" plane-net as you can. Do not forget the defect-pairs.

5. For a divalent lattice containing both $Ca^{2+}$ and $S^{=}$ , write all 30 of the equilibrium equations if both $As^{3+}$ and $Cl^{-}$ are substitutional impurities in the lattice. (Include all of the possible vacancies and interstitial sites). Be sure to include the site where each ion is situated.

6. Calculate the activation energy at room temperature for a cubic crystal containing $3 \times 10^{-3}$ vacancies.

7. Rewrite all of the equations in Table 3-2 in terms of the compound, $CaCl_2$ , and compare them to those given in Table 3-3. What effect does non-stoichiometry have upon the numbers and types of defects that may be prevalent? Do the same for the compound $AsCl_3$.

8. Write a series of equations for the Frenkel defect, similar to those given for the Schottky defect, i.e.- Equations 3.7.2 to 3.7.3.

9. Using a "MX" plane net, illustrate the defects shown in 3.7.1. for the hypothetical compound, MX. Do the same for the compound, $MX_2$ , using the configuration: X-M-X.

10. Given: $k_{Sh} = 3 \times 10^{-3}$ for $CaCl_2$ , calculate the number of intrinsic defects present in this crystal. If $CaCl_2$ is face-centered cubic, use the same equilibrium constant to calculate the intrinsic Frenkel, Anti-Frenkel and Interstial defects expected in this crystal.

11. The native defects in ZnO are: $V^{=}_{Zn}$ , $Zn_i$ , $V_O$ , $Zn^{+}_i$ , $V_O^{+}$ , & $V^{-}_{Zn}$.

a. Write the defect reactions leading to their formation.

b. Write the defect reactions for ZnO that has been subjected to a reducing atmospher, i.e.- a reaction with $H_2$. Predict the intrinsic defects which will be present.

12. Draw a heterogeneous lattice, using circles and squares to indicate atom positions in a simple cubic lattice. Indicate both Schottky and Frenkel defects, plus the simple lattice defects. Hint- use both cation and anion sub-lattices.

13. The following vacancy parameters have been measured for a solid as:

$$\Delta H_v = 2200 \text{ cal/mol}$$
$$\Delta S_v = 2.0 \text{ k}$$

Calculate the vacancy concentration at 160 °K and at 298 °K. (Hint: see Appendix III)

14. The following vacancy parameters have been measured for gold as:

$$\Delta H_V = 21.62 \text{ kcal/mol}$$
$$\Delta S_V = 2.0 \text{ K}$$

Calculate the vacancy concentration at the melting point of 1063 °K and at 298 °K.

**Appendix I.**

Concentrations of Defects in Non-Ionized Non-Stoichiometric Compounds

This presentation is presented for those who wish to examine the mathematics of both non-stoichiometric intrinsic- defect compounds and the ionization of defects in both stoichiometric and nonstoichiometric compounds as represented by:

$$MX_{S\pm\delta}$$

A. DEFECTS IN NON-STOICHIOMETRIC $MX_{S \pm \delta}$ COMPOUNDS

We now proceed as we did for the stoichiometric case, namely to develop defect- concentration equations for the non-stoichiometric case. Consider the effect of Anti-Frenkel defect production. From Table 2-1, we get $K_{AF}$, with its associated equation, $k_{AF}$. In Table 2-2, we use $K_{Xi}$ for X-interstitial sites. Combining these, we get:

App 1.1.- $K_{AF} = K_{V_X} \cdot K_{Xi} = N_i^2 / (S-N_i)(\alpha - N_i)$

When both $V_X$ and $X_i$ coexist in the lattice, the deviation from stoichiometry (from 3.7.10.) becomes:

App1.2.- $\delta = [X_i] - [V_X]$

Using the equilibrium constant of 3.7.9., i.e.-

App 1.3.- $K_{V_X} = p_{X_2}^{1/2} [V_X] / X_X = p_{X_2}^{1/2} [V_X] / S - [V_X]$

and the appropriate one from Table 3-2 (i.e.- $K_{Xi}$ ), we get (assume for simplicity that $S = \alpha = 1$) :

App 1.4.- $\delta = \alpha\, p_{X_2}^{1/2} K_{Xi} / p_{X_2} K_{Xi} + 1 - S K_{V_X} / K_{V_X} + p_{X_2}^{1/2}$

We can rearrange terms in App 1.4. to obtain:

App 1.5.- $K_{Xi} (1- \delta)\, p_{X_2} - \delta( K_{V_X} K_{Xi} + 1)\, p_{X_2}^{1/2} - K_{V_X}^{2} (1- \delta) = 0$

Using App 1.1. and $N_i << 1$, we now obtain:

App 1.6.- $N_i^2 \; (1- \delta) p_{X_2} - \delta K_{V_X} p_{X_2}^{1/2} - K_{V_X}^{2} (1- \delta) = 0$

Solving for $p_{X_2}$ yields :

App 1.7.-

$$p_{X_2} = K_{V_X}^{2} ( \delta^2 + 2 N_i (1- \delta^2) \pm \delta [\delta^2 + 4 N_i (1 - \delta^2)]^{1/2} / 2 N_i^4 (1 - \delta)^2$$

Since at stoichiometric composition, $\delta$ must equal zero, this rather formidable equation reduces to:

App 1.8.- $p_{X_2}^{o} = K_{V_X}^{2} / N_i^2$

where $p_{X_2}^{o}$ is the pressure of $X_2$ gas in equilibrium with the $MX_S$ crystal at the stoichiometric composition. This gives us the opportunity to divide App 1.7. by App 1.8. to obtain:

App 1.9.- $p_{X_2} / p_{X_2}^{o} = \delta^2 + 2 N_i (1- \delta^2) \pm \delta [\delta^2 + 4 N_i (1 - \delta^2) / 2N_i^2 (1 - \delta)^2$

We can therefore calculate $\delta$ in terms of the ratio of $p_{X_2}$ to $p_{X_2}^{o}$ and $N_i$. This result is shown in the following diagram;

App 1.10.-

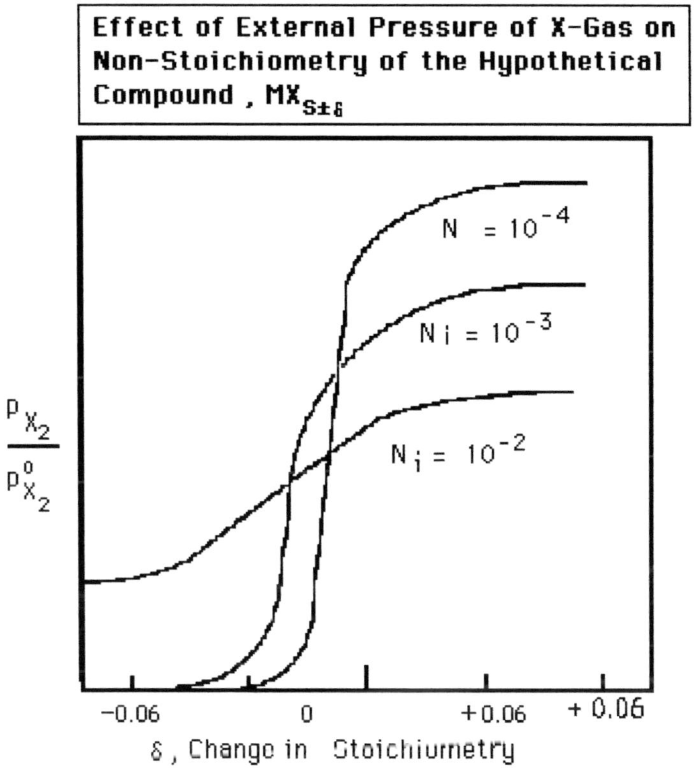


Herein is shown how δ changes from negative to positive at the higher pressure ratios. For a hypothetical $MX_S$ compound (S = 1), which contains Anti-Frenkel defects, to obtain a 0.7% deviation from the stoichiometric composition requires a pressure increase of some 5000 fold, when the original intrinsic defect concentration is $10^{-4}$. However, if it is $10^{-2}$, only a two-fold increase in pressure is needed to cause the same effect on a deviation of 0.7%, i.e.- $\delta = \pm\ 0.007$ in $MX_{S\pm\delta}$.

**This means that a highly defective structure is much more prone to cause the formation of many more defects than one containing only a few defects**

Although we will not treat the other types of pairs of defects, it is well to note that similar equations can also be derived for the other intrinsic defects. What we have really shown is that external reactants can cause further changes in the non-stoichiometry of the solid. Let us now consider ionized defects. It should be clear that an external gaseous factor has a major effect upon defect formation. The equations given above are very complicated and represent more closely what actually happens in the real world of defect formation in crystals.

## APPENDIX II
## ANALYSIS OF A REAL CRYSTAL USING A THERMODYNAMIC METHOD

In the non-stoichiometric case where ionization of defects is the norm, the mathematics become too complicated so that the equations are not solvable. However, we can use a thermodynamic method to obtain the results we want. We will present here the case of silver bromide whose use in photographic film highlights the use of defect chemistry for practical purposes.

### I. THE AgBr CRYSTAL WITH A DIVALENT IMPURITY, $Cd^{2+}$

Consider the crystal, AgBr. Both cation and anion are monovalent, i.e.- $Ag^+$ and $Br^-$. The addition of a divalent cation such as $Cd^{2+}$ should introduce vacancies, $V_{Ag}$, into the crystal, because of the charge-compensation mechanism. To maintain electro-neutrality, we prefer to define the system as:

App. 2-1.- $(1-\delta)\ Ag^+ Br^- : \delta\ Cd^{2+}\ S^=$

Fortunately, AgBr is easy to grow as a single crystal, using Stockbarger Techniques. Possession and measurement of a single crystal greatly facilitates our measurement of defects.

The defects we expect to find are:

App. 2-2.- $V_{Ag}$ , $Ag_i$ , $e^-$ , $\pi^+$, $Cd_{Ag}$ , and $[Cd_{Ag} , V_{Ag}]$

The last defect is one involving two nearest neighbor cation lattice-sites The following Table presents the defect reactions governing this case.

TABLE 2-App

DEFECT REACTIONS IN THE AgBr CRYSTAL CONTAINING $Cd^{2+}$

a. $0 \leftrightharpoons e^- + \pi^+$

b. $Ag_{Ag} + \alpha V_i \leftrightharpoons V_{Ag} + Ag_i$

c. $Ag_{Ag} + \alpha V_i \leftrightharpoons V_{Ag}^- + Ag_i^+$

d. $Ag_i \leftrightharpoons Ag_i^+ + e^-$

e. $1/2\ Br_2 \leftrightharpoons Br_{Br} + V_{Br}$

f. $Cd_{Ag} \leftrightharpoons Cd_{Ag}^+ + e^-$

g. $Cd_{Ag} + V_{Ag}^- \leftrightharpoons [Cd_{Ag} , V_{Ag}]$

h. $Cd_{Ag}^+ + V_{Ag}^- \leftrightharpoons [Cd_{Ag} , V_{Ag}]$

i. <u>For Constant Cd</u>

$$Cd_T = Cd_{Ag} + Cd_{Ag}^+ + [Cd_{Ag} , V_{Ag}] = K$$

j. <u>For Electro-neutrality</u>

$$e^- + V_{Ag}^- = \pi^+ + Ag_i^+ + Cd_{Ag}^+$$

Experimentally, we find that if we fix the $Cd^{2+}$ content at some convenient level, it is necessary to anneal the AgBr crystals at a fixed temperature for times long enough to achieve complete equilibrium. If the temperature is changed, then both type and relative numbers of defects may also change. Singly-charged defects predominate, i.e.-

App. 2-3.- $2\ V_{Ag}^- = Ag_i^+ + Cd_{Ag}^+$

These in turn may form the complex:

App. 2-5.- $Cd_{Ag}^+ + V_{Ag}^- \leftrightharpoons [Cd_{Ag} , V_{Ag}]$

App. 2-6.- $V_{Ag} \leftrightharpoons V_{Ag}^- + \pi^+$

c. Non-stoichiometry (excess cation)

These are the eight equations (App. 2-12. to App. 2-18.) required to calculate the defect concentrations arising from the effects of the external factor, $p_{Br_2}$.

From measurements of conductivities, transfer numbers (electro-migration of charged species), lattice constants and experimental densities, it has been shown that Frenkel defects predominate (Lidiard - 1957). This means that:

App. 2-19.- $Ag_i^+ , V_{Ag}^- \gg Br_i^- , V_{Br}^+$

Furthermore, $V_{Ag}^- \cong Ag_i^+$ so that in terms of our equilibrium constants we get:

App. 2-20.- FOR FRENKEL DEFECTS: $K_\varepsilon \gg K_\gamma$ and $K_\varepsilon \gg K_\delta$

Thus, we need only to consider the above two (2) defects, namely - $V_{Ag}^-$ and $Ag_i^+$ , since they are the major contributors to non-stoichiometry. By calculating $p^o_{Br_2}$ as before (when $\delta = 0$ , see 2.7.22 & 2.7.23.), we can express our overall defect equation as:

App. 2-21.- $p^{1/2}_{Br_2} / (p^o_{Br_2})^{1/2} = \{\delta / 2\varepsilon + [(1+ \delta/2\varepsilon)^2]^{1/2}\} \{\delta/2\beta + [1+(\delta/2\beta)^2]^{1/2}\}$

Because of the conditions given in App. 2-14., the first half of the equation can be set equal to one. **Note that we are using** $\delta$ , $\beta$ , **and** $\gamma$ **as the equilibrium constants . i.e. -**

App. 2-22.- $\varepsilon \equiv K_\varepsilon^{1/2}$ $\beta \equiv K_\beta^{1/2}$ $\gamma \equiv K_\gamma^{1/2}$

In the remaining part of the equation, $\delta \gg \beta$ . By taking logarithms, we can then obtain:

App. 2-23.- $$1/2 \ln p_{Br_2} / p^o_{Br_2} = \ln \delta - \ln \beta$$

This result then leads us to a plot of the effect of partial pressure of $Br_2$ on the deviation from stoichiometry , $\delta$ , for the AgBr crystal. (This work is due to Greenwood - 1968).

For $\beta = 0$ , there is a point of inflection where the slope of the line is defined by the equilibrium constant, i.e.- $\beta = K_\beta^{1/2}$ . The larger this value, the flatter is the curve. All relations regarding the defects can now be derived. The major defects turn out to be:

App. 2-24.-

$$V_{Ag}^- \leftrightharpoons Ag_i^+$$
$$\pi^+ \leftrightharpoons K_\delta \, (1/K_\varepsilon)^{1/2} \, p^{1/2}_{Br_2}$$
$$e^- \leftrightharpoons (K_\varepsilon)^{1/2} \, (1/K_\varepsilon) \, K_\beta ( 1/ p_{Br_2})^{1/2}$$

This shows that both $\pi^+$ and $e^-$ are minority defects dependent on $p_{Br_2}$.

The following gives the standard Enthalpy and Entropy of these defect reactions, according to Krœger (1965):

App. 2-25.-

| DEFECT REACTION | ΔS Cal./mol/°K | ΔH Kcal./mol. |
|---|---|---|
| $Ag_{Ag} \leftrightharpoons Ag_i^+ + V_{Ag}^-$ (Frenkel) | 25.6 | 29.3 |
| $Ag_{Ag} + Br_{Br} \leftrightharpoons V_{Ag}^- + V_{Br}^+$ | - 13.3 | 36 |
| $0 \leftrightharpoons e^- + \pi^+$ | 25 | 78 |
| $1/2\, Br_2 + Ag_{Ag} \leftrightharpoons AgBr + V_{Ag}^- + \pi^+$ | 4.9 | 25.4 |

It is apparent that the Frenkel process coupled with the electronic process are the predominating mechanisms in forming defects in AgBr through the agency of external reaction with $Br_2$ gas.

A final comment: we can use these thermodynamic values to calculate the equilibrium constants according to:

App. 2-27.- $K_i = \exp - \Delta G_i^o / RT$

and can also obtain the activity of the silver atom in AgBr from App. 2-9. By using equation App. 2-9., we can show :

App. 2-28. For $a_{Ag} = 1$ , @ T = 277 °C. ; $\delta = + 10^{12}$
For $p_{Br_2} = 1$ atm. @ 277 °C. ; $\delta = - 10^{-7}$

where the plus or minus indicate an excess or deficit of the silver atom in AgBr. This result is due to Wagner (1959).

## Appendix III
### Statistical Mechanics and the Point Defect

The mathematical method, statistical mechanics, was developed by James Clerk Maxwell of Scotland and Ludwig Boltzmann of Germany to predict and explain the measurable properties of macroscopic systems, i.e.- solids, on the basis of the properties and behaviour of the microscopic constituents of those systems, i.e.- atoms. Statistical mechanics, for example, interprets thermal energy as the energy of atomic particles in disordered states and temperature as a quantitative measure of how energy is shared among such particles. Statistical mechanics draws heavily on the laws of probability, so that it does not concentrate on the behaviour of every individual particle in a macroscopic substance but on the average behaviour of a large number of particles of the same kind.

One law of statistical mechanics states that, in a system in thermal equilibrium, on the average, an equal amount of energy will be associated with each independent energy state. This law states specifically that a system of particles in equilibrium at absolute temperature T will have an average energy of 1/2 kT associated with each degree of freedom (where k is the Boltzmann constant). In addition, any degree of freedom contributing potential energy will have another 1/2kT associated with it.

For a system of n degrees of freedom, of which t have associated potential energies, the total average energy of the system is 1/2(n + t)kT.

For an atom in a solid, vibratory motion involves potential energy as well as kinetic energy, and both modes will contribute a term 1/2kT, resulting in an average total energy of 3kT. Thus, it is the entropy of mixing that forces the creation of a certain number of vacant lattice positions above 0.0 °K. Hence, vacancies are the natural resultof thermodynamic equilibrium and not the result of accidental growth or sample preparation.

In order to use this method, one starts by defining the number of states that may be present. Let $N_L$ be the total number of lattice sites and $N_V$ be the number of vacancies. The number of ways that $N_V$ vacancy-sites can be arranged on $N_L$ sites will be a simple combination, defined by a mixing entropy:

App3.1. $$\Omega = \frac{N_L!}{N_V!(N_L - N_V)!}$$

The mixing entropy is defined by:

App3.2.- $$\Delta S_M = k \ln \Omega$$

In order to solve combinatorial equations like App3 .1., a method called Stirling's approximation for large numbers is used. This gives:

App3.3.- $$\Delta S_M \cong k[N_L \ln N_L - N_V \ln N_V - (N_L - N_V) \ln()]$$

where $N_L$ is the number of lattice sites and $N_V$ is the number of vacancies. It is well to note here that $N_L >> N_V$ so that the entropy of mixing will be:

App3.4.- $$\Delta S_M = [N_V \ln N_L - N_V \ln N_V]$$

or $$\Delta S_M = -k\, N_V \ln(N_V / N_L) = -R\, X_V \ln X_V$$

where $X_V$ is the quotient of $N_V/N_L$. This expression is only valid for noninteracting defects. And, it is the same as the entropy of mixing of an ideal solution, vis:

App3.5.- $\Delta S_M = -R\,[X_A \ln X_A + X_V \ln X_V\,]$

Since $X_A$ is any atom in the solid, $X_A \cong 1$, the entropy of mixing becomes:

App3.6.- $\Delta S_M = -\,R\,X_V \ln X_V$

The complete Helmholtz free energy of the system is then written:

App3.7.- $$\Delta F_V = -\,N_V\,[\Delta E_V - T\Delta S^v{}_V] + RT\left[\frac{[N_L - N_V]}{N_L}\ln\frac{N_L - N_V}{N_L} + \frac{N_V}{N_L}\ln\frac{N_V}{N_L}\right]$$

where $\Delta S^v{}_V$ is the change in vibrational entropy arising from a change in vibrational energy around the vacanr lattice site. If we minimize the free energy with respect to the vacancy concentration, we get:

App3.8.- $$\partial F_V = 0 = \Delta E_V - T\Delta S^v{}_V + RT\ln\frac{N_V}{N_L - N_V}$$

Here, $\Delta E_V$ and $\Delta S_V$ as the standard internal energy and standard entropy of the defects, respectively. This gives us a final result of:

App3.9.- $$\ln\frac{N_V}{N_L - N_V} = \ln X_V = -\,\frac{\Delta E_V - T\Delta S^v{}_V}{RT}$$

As we have already stated, $N_L >> N_V$ so that the atomic fraction of vacancies is:

App3.10.- $$X_V = \exp\,(-\Delta F_V\,/RT) \approx \exp\,(-\Delta G_V\,/RT)$$

or: $$X_V = e^{-\Delta E/RT}$$

Since, $\Delta G_V = \Delta F_V + P\Delta V$ (at constant pressure), we can take the derivitive of App3.10. to obtain:

App3.11- $$\frac{\partial(\Delta \underline{G_V}/RT)}{\partial T^{-1}} = \frac{d \ln \Delta \underline{G_V}}{dT^{-1}} = \frac{\Delta \underline{H_V}}{R}$$

which is a form of the Gibbs-Helmholz equation.

Note that we can use the same statistical mechanical approach to calculate Schottsky pairs, Frenkel pairs, divancies (which are associated vacancies), impurity-vacancy complexes, and line dislocation-point defect complexes.

# Chapter 4

## Mechanisms and Reactions in the Solid State

We have investigated the structure of solids in the second chapter and the nature of point defects of the solid in the third chapter. We are now ready to describe how solids react. This will include the mechanisms involved when solids form by reaction from constituent compounds. We will also describe some methods of measurement and how one determines extent and rate of the solid state reaction actually taking place. We will also show how the presence and/or formation of point defects affect reactivity in solid state reactions. They do so, but not in the manner that you might suspect. We will also show how solid state reactions progress, particularly those involving silicates where several different phases appear as a function of both time and relative ratios of reacting components.

Solids are generally considered chemically inert at room temperatures and the most common-place evidence is often overlooked. That is, solids do not appear to be reactive until they are heated. However, the atoms or ions comprising solids are under constant vibratory motion with the lattice and can "diffuse" from site to site. If vacancies are present, they are continually being "filled" and "emptied" even at room temperature. Those solids based upon iron (Fe) undergo continuous oxidation to form a layer of "rust". Thus, solids are not completely stable and are under continuous change over time.

It should be clear by now that inorganic solids (which consist of atoms bound together by both covalent and ionic forces) do not react by either changing the bonding within a molecular structure or by reacting one-on-one in a mobile phase such as a liquid, as do organic compounds. Solids can only react at the **interface** of another solid, or in the case of a liquid-solid reaction, react with the liquid molecule at the solid interface.

For a solid which exists as a powder, a useful mental concept is that of ping-pong balls of several colors. Here, we imagine that reaction results when the layers become mixed. Furthermore, reaction can only take place between neighboring balls of differing color, provided they are nearest neighbors. Thus, it should be clear that if we pour balls of the same color into a jar and then pour another layer of contrasting color on top, reaction can only take place at the **interface** of the layers, between balls of differing colors. One color can only "diffuse" into the other layer by replacing a ball of the other color.

Our first step will be to delineate known solid state reaction mechanisms.

Mechanisms Relating to Solid State Reactions

Phase changes
Formation of phase boundaries
Rate process changes in solids
Nucleation
Diffusion processes
Diffusion-controlled solid state reactions.

We can further separate these reaction mechanisms into two types, those involving a single phase (homogeneous) and those involving two different phases (heterogeneous). We will examine each of these in turn, starting with phase changes. By phase changes, we mean transformations of form or structure within the solid state.

## **4.1.** - PHASE CHANGES

We can distinguish seven (7) types of phase transformations. For the homogeneous solid, these include the types of changes given in 4.1.1. on the next page. These seven transformations, which involve just **one** composition, are familiar to most. We touched upon them briefly in the first chapter when we considered the transformations involving water. Now consider phase changes involving **two solids** and the types of reaction mechanisms which may occur. These are called heterogeneous reactions and are presented in Table 4-1, also given on the next page.

4.1.1.- PHASE TRANSFORMATIONS (SINGLE PHASE)

| CHANGE | DESIGNATION | |
|---|---|---|
| Gas to Liquid | Condensation | $G \Rightarrow L$ |
| Gas to Solid | Condensation | $G \Rightarrow S$ |
| Liquid to Gas | Evaporation | $L \Rightarrow G$ |
| Liquid to Solid | Solidification | $L \Rightarrow S$ |
| Solid to Gas | Sublimation | $S \Rightarrow G$ |
| Solid to Liquid | Melting | $S \Rightarrow L$ |
| Solid to Solid | Polymorphic Transformation | $S \Rightarrow S$ |

TABLE 4 - 1

TYPES OF SOLID STATE HETEROGENEOUS REACTIONS

| Type of Reaction | Final Product | Example of Reaction |
|---|---|---|
| 1. Decomposition $A \Rightarrow B + C$ | S + S<br>S + G<br>G + G | $3\ AuCl \Rightarrow AuCl_3 + 2\ Au$<br>$CaCO_3 \Rightarrow CaO + CO_2 \Uparrow$<br>$NH_4Cl \Rightarrow NH_3 \Uparrow + HCl \Uparrow$ |
| 2. Synthesis $A + B \Rightarrow C$ | S | $2\ AgI + HgI_2 \Rightarrow Ag_2HgI_4$<br>$MgO + Al_2O_3 \rightarrow MgAl_2O_4$<br>$MgO + SiO_2 \Rightarrow MgSiO_3$ |
| 3. Substitution $A + B \Rightarrow C + D$ | S + S<br>S + G | $Cu + AgCl \Rightarrow CuCl + Ag$<br>$BaCO_3 + TiO_2 \Rightarrow BaTiO_3 + CO_2 \Uparrow$ |
| 4. Consecutive $A \Rightarrow B \Rightarrow C$ | S + S | i) $La_2O_3 + 11\ Al_2O_3 \Rightarrow 2\ LaAlO_3 + 10Al_2O_3$<br>ii) $2\ LaAlO_3 + 10\ Al_2O_3 \Rightarrow LaAl_{11}O_{18}$ |

In both 4.1.1. and Table 4-1, S , L & G refer to solid, liquid and gas, respectively. Note that we have also classified these heterogeneous mechanisms in terms of the same PHYSICAL CHANGES given above for homogeneous transformations. For the most part, the initial material will be a solid while the nature of the final product will vary according to the type of material undergoing solid state reaction.

Review these examples carefully. They are typical of those encountered in solid state chemistry. Thus, "A" reacts to form "B +C" (Decomposition) or

"A + B" reacts to form "C" (Synthesis"). etc. These reactions cover most of those normally found in solid state chemistry. Once you have mastered these types of reactions, you will be able to identify most solid state reactions in terms of the reactants and products. This is important, especially when you may be trying to form an new composition not well known in solid state science. Addionally, you may wish to form a specific composition by a new method to see if it has superior physical or chemical properties over that same material formed by a different solid state mechanism or reaction. In many cases, this has been found to be true and this factor has been responsible for several scientific advances in the solid-state. That is- if you can find a different method for making a material, its properties may prove to be superior to that already known.

## **4.2.** - THE ROLE OF PHASE BOUNDARIES IN SOLID STATE REACTIONS

Consider two (2) solids, A and B. It should be obvious that they will react only when in close proximity, and usually when they are intimately mixed and heated (if they are powders). What this means is illustrated by the following diagram:

4.2.1.-

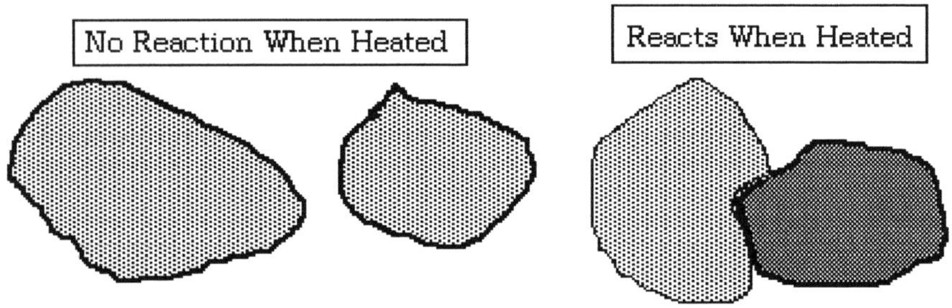


In the case where two particles are involved, those that are not in close proximity will not react, whereas those that are close will undergo solid state reaction with ease. This is due to the fact that cations and/or anions from one structure must be transported, or interchange by some mechanism, to the other structure in order to form a completely new compound. Thus, the degree of dispersion and mixing of one reacting solid with another is important to the overall mechanism of solid state reaction. This cannot be overemphasized since lack of dispersion is the primary cause for production of unwanted phases (or lack of production- that is, when the solid state reaction does not proceed according to expection).

As an example, consider the following. Suppose we have a crucible half-filled with a powder. We now fill the crucible with another powder of different composition and then heat the filled crucible. Any solid state reaction which does occur can only do so at the boundary of the two layers of powders. If the reaction is: A + B ⇒ AB, then we find that the reaction product, which is also a solid, forms as a **phase boundary** between the two layers. The same condition exists in a solid state reaction between two crystalline particles having differing compositions. That is, they can only react at the interface of each particle. This is illustrated in the following diagram, which is a model of how the components react through a phase boundary:

4.2.2.- Formation of a Phase Boundary

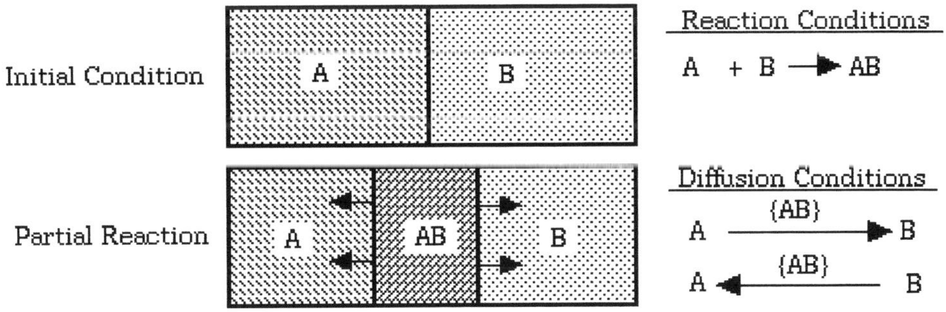


In this case, we have given both the starting conditions and those of the intermediate stage of solid state reaction. It should be clear that A reacts with B, and vice versa. Thus, a phase boundary is formed at the interface of the bulk of each particle, i.e.- between A and AB, and between B and AB. The phase boundary, AB, then grows outward as shown above. Once the phase boundary is established, then each reacting specie must diffuse through the phase {AB} to reach its opposite phase boundary in order to react. That is- "A" must diffuse through "AB" to the phase boundary between “AB” and “B” before it can react with "B" to form “AB” and further increase the dimensions of “AB” at the expense of both “A” and “B”. It is quite important that you master this concept which is fundamental towards understanding both phase boundaries of solid state reaction and diffusion conditions that prevail.

However, this is not how it occurs in Nature. We have presented the above concept because it is easier to understand than the actual conditions which occur. Thus, the overall solid state reaction is dependent upon the rate of diffusion of the two (2) species. These two rates may, or may not, be the same. The reason that A and/or B do not react in the middle, i.e.- the phase {AB}, is that AB has a certain ordered structure which probably differs from either A or B. But there is a more important reason which is not easily illustrated in any diagram.

The actuality is that any given atom of A **does not move** through AB to the phase boundary of A and AB. What really happens is that any given atom displaces **one of the A atoms in {AB}**. The displaced A-atom then causes another displacement by a "hopping" motion. The stated displacement then travels to the right until the other edge of the phase boundary is encountered. There, reaction of the final, displaced A-atom occurs with a B-atom to form AB. Simultaneously, displaced B-atoms are diffusing to the left by the same "hopping" motion. Thus, the walls of {AB} move in a three-dimensional direction at the expense of both of the atomic volumes of A and B. A depiction of this mechanism is given in the following diagram, given as 4.2.3. on the next page.

Here, we have shown an "A" ion (shown as a black sphere) which approaches the surface of "AB" and displaces an "A" ion in this solid phase. A series of "hops", i.e.- from "1" to "7", then occurs in the AB-phase with the final "A" ion ending within the B-phase where a displacement in the normally cubic "B" phase occurs. At the same time, the displaced "B" ion is diffusing in the opposite direction by a series of "hops", i.e.- from "a" to "e" to the interface of "AB" with "A". Note that the "A" phase is shown as a hexagonal phase while "B" is a cubic phase (as is the "AB" phase). It should be clear that the rate of diffusion of "A" will differ from that of "B".

4.2.3.-

A "Hopping" Mechanism in the Reaction: A + B → AB

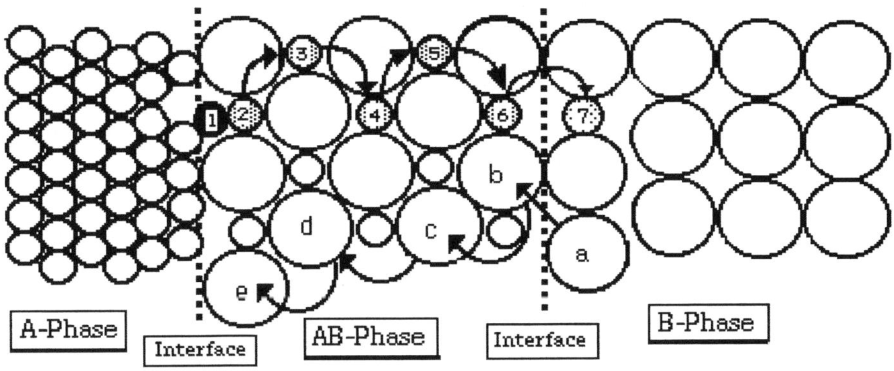


Note also that a change in structure has occurred as the AB arrangement grows.

Let us take another example, using the same type of model, in which A is a gas and B is a solid. A diagram illustrating this mechanism is given as follows:

4.2.4.-

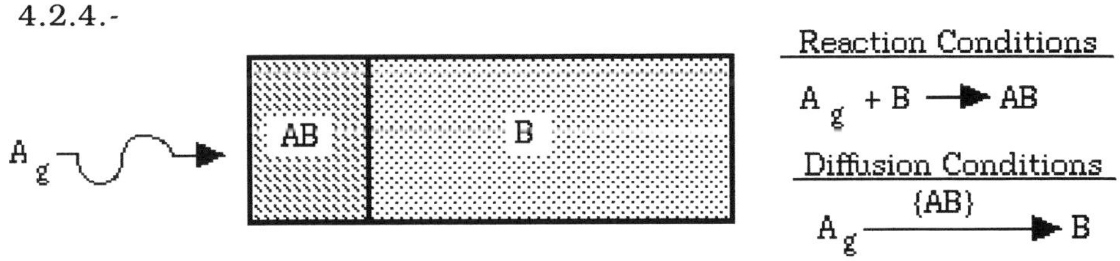


Here, only **one** interface forms as the reaction takes place **initially** at the surface of "B". Once the AB-phase has formed, gaseous A atoms must then diffuse through to the phase boundary in order for the reaction to occur. The wall of {AB} thus moves in two (2) dimensions instead of the three, in contrast to the case of 4.2.2.

Consider another example involving diffusion and reactivity, such as the reaction between particles of:

4.2.5.- $BaO + SiO_2 \Rightarrow BaSiO_3$

In this case, the structure of BaO consists of discrete atoms of $Ba^{2+}$ and $O^{=}$, arranged in a cubic structure. $SiO_2$, on the other hand, consists of $SiO_4$ tetrahedra bound at the apices through mutual oxygen atoms, and arranged in a three-dimensional network. This is shown in the following diagram:

4.2.6.-

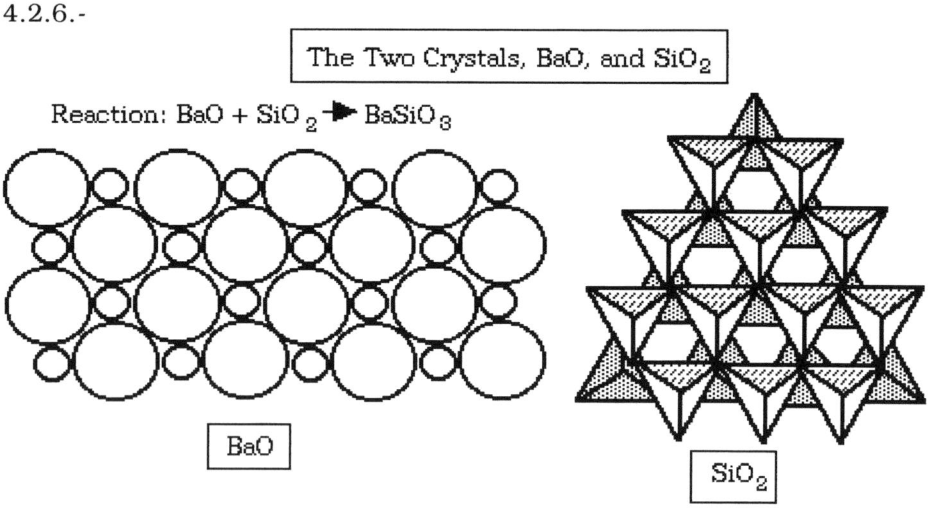


We find that this solid state reaction is **very slow**, even at 800 °C. , and occurs at the interface of the two types of particles. The reaction is slow because it is diffusion-limited. What is happening is that since the silica-network is three-dimensionally bound, **the only reaction that occurs** is caused by the diffusion of $Ba^{2+}$ atoms within the network, as shown in the following:

4.2.7.-

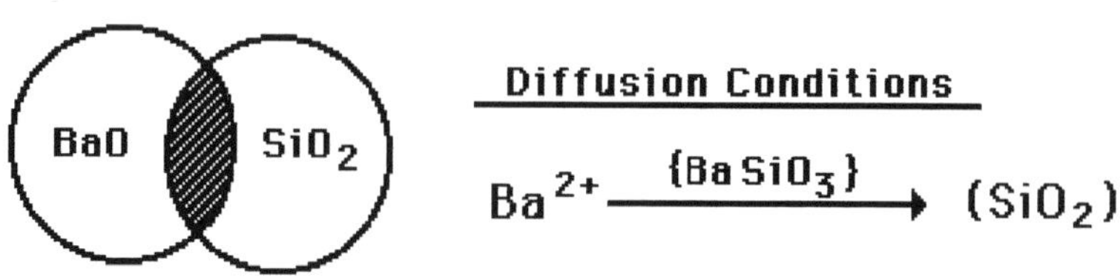

Note that diffusion occurs only in **one direction** because the silica-tetrahedra are **not free to move.** What is actually happening is that the three-dimensional network of tetrahedra is being **rearranged** to form another structure. This illustrates the fact that the actual structure and composition of the two reacting species are the major factor in determining the nature and speed of the solid state reaction.

On the other hand, if we use $BaCO_3$ as a reactant, vis:

4.2.8.- $BaCO_3 + SiO_2 \Rightarrow BaSiO_3 + CO_2 \Uparrow$

we find that the reaction is very **fast.** This has nothing to do with the gaseous $CO_2$ produced, but with the form of the BaO particles. In this case, the initial reaction is:

4.2.9.- $BaCO_3 \Rightarrow BaO + CO_2 \Uparrow$

The BaO is produced in the form of very small particles of nearly atomic proportions which react immediately to form the **silicate.** Actually, the rate of reaction is proportional to the number of **nuclei produced per unit volume.** A nucleus is a point where atoms or ions have reacted and begun the formation of the product structure. In the case of the BaO reaction, the number of nuclei formed per unit of time is small and formation of the structure is diffusion limited. In the case of $BaCO_3$ decomposition, the atomic-proportioned BaO reacts nearly as fast as it is formed so that the number of nuclei per unit volume is enormously increased. It is thus apparent that if we wish to increase solid state reaction rates, one way to do so is to use a decomposition reaction to supply the reacting species. we will further address this type of reaction later on in our discussion.

Note also that we have just introduced the concepts of nuclei and nucleation in our study of solid state reaction processes. Our next step will be to examine some of the mathematics used to define rate processes in solid state reactions. We will not delve into the precise equations here but present them in Appendices at the end of this chapter. But first, we need to examine reaction rate equations as adapted for the solid state.

## 4.3.- REACTION RATE PROCESSES IN SOLIDS

As stated above, we can classify solid state reactions as being either homogeneous or heterogeneous. The former involves reactions by a single compound whereas the latter involves reactions between two different compounds. As stated above in Table 4-1, there are four (4) types of solid state reactions.

4.3.1.- Types of Solid State Reactions

1. Decomposition: $A \Rightarrow B + C$
2. Synthesis : $A + B \Rightarrow C$
4. Substitutional: $A + B \Rightarrow C + D$
4. Consecutive: $A \Rightarrow B \Rightarrow C$

Rate processes for the solid state are generally defined in terms of a rate, ***r***, and a volume, V, usually a molar volume. Thus, we have:

4.3.2.- Homogeneous: $\boldsymbol{r} = dn_i/dt \cdot 1/V_t$

Heterogeneous: $\boldsymbol{r} = dV_t/dt \cdot 1/V_f$

where $n_i$ is initial moles, $V_t$ is volume at time, t, and $V_f$ is final volume. The reason that these equations differ is that a reaction between **two solids** may propagate a **final product** having a larger or smaller volume than that of the original reacting species, while that of a **single compound** remains **stable**. If we define the fraction decomposed at any time as:

4.3.4.- $x \equiv V_t / V_f$

We can express these volumes in terms of x so as to get:

4.3.5.- $dx/dt = k_1 (T) \cdot f(x)$

where $k_1$ is the rate constant for the equation in 4.3.4. and f(x) relates to the definition given in 4.3.2. This equation, i.e.-4.3.5., then has the same general form as those used in the study of kinetics of chemical reactions.

From the Kinetic Theory, when a gaseous system goes from an initial state to a final state, the reacting species must come close enough to react. At the moment of "joining", we have the "activated complex". By using this concept, we can obtain some general equations useful to us. The concept of the "activated complex" is illustrated in the following diagram:

4.3.6.-

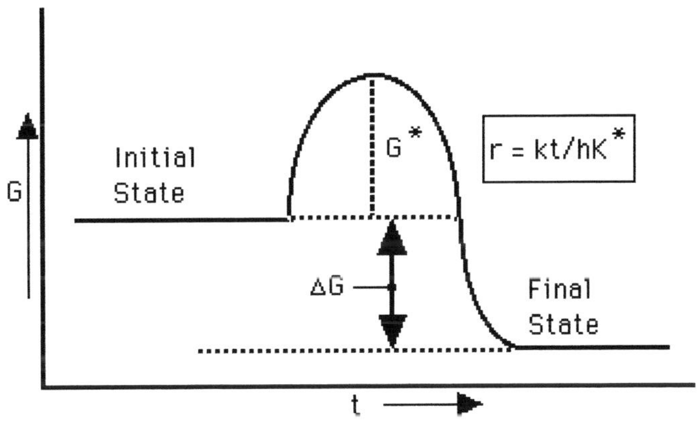


It is easy to see that the free energy of the activated complex, G*, is higher than that of the initial state. K* is the equilibrium constant of the activated complex, and k is the Boltzmann constant. Then, we can write equations describing both ΔG and the activated state as:

4.3.7.- $\Delta G = RT \ln K^*$

$$k_1 = kT/h \exp.(-\Delta G^*/RT) = kT/h \exp.(\Delta S^*/R) \exp.(-\Delta H^*/RT)$$

where the star (*) refers to the activated species. From the Clausius-Clapyeron equation, we have:

4.3.8.- $E_o \equiv RT^2 d(\ln k_1)/dt = \Delta H^* + RT$

where $E_o$ is an internal energy. By defining a frequency factor as:

4.3.9.- $Z = kT/h \exp(-E_o^*/RT)$

We can simply express this result similar to that of 4.3.7. to obtain:

4.3.10.- $k_1 = Z \exp.(-E_o^*/RT)$

which is the ARRHENIUS EQUATION for the rate constant, $k_1$. We can transform the above equations into *general* equations as well, and use them to define the various types of reactions given above, namely:

4.3.11.- RATE EQUATIONS FOR SOLID STATE REACTIONS

Simple nth Order

$$nA \xRightarrow{k_1} B \qquad \text{Rate Equation: } -dx_A/dt = k_1 x^n_A$$

Parallel Reactions

$$\{A \xRightarrow{k_1} B\} \text{ and } \{A \xRightarrow{k_2} C\} \qquad \text{Rate Equation: } -dx_A/dt = k_1 x_A + k_2 x_A$$

Consecutive

$$A \xRightarrow{k_1} B \xRightarrow{k_2} C \qquad \text{Rate Equation: } -dx_A/dt = k_1 x_A - k_2 x_B$$

Three types of rate equations are shown here. These rate equations can be used for quite complicated reactions, but a specific method or measurement approach is needed. How we do this is critical to determining accurate estimation of the progress of a solid state reaction. We will discuss suitable methods in another chapter. We now return to the subject of nucleation so that we can apply the rate equations given above to specific cases. First, we examine heterogeneous processes.

## 4.4.- DEFINING HETEROGENEOUS NUCLEATION PROCESSES

If we are interested in the nucleation of a particle prior to completing the solid state reaction, we need to distinguish between surface and volume nucleation of the particle, since these are the major methods of which we can perceive. Several cases are shown in the following diagram.

4.4.1.-

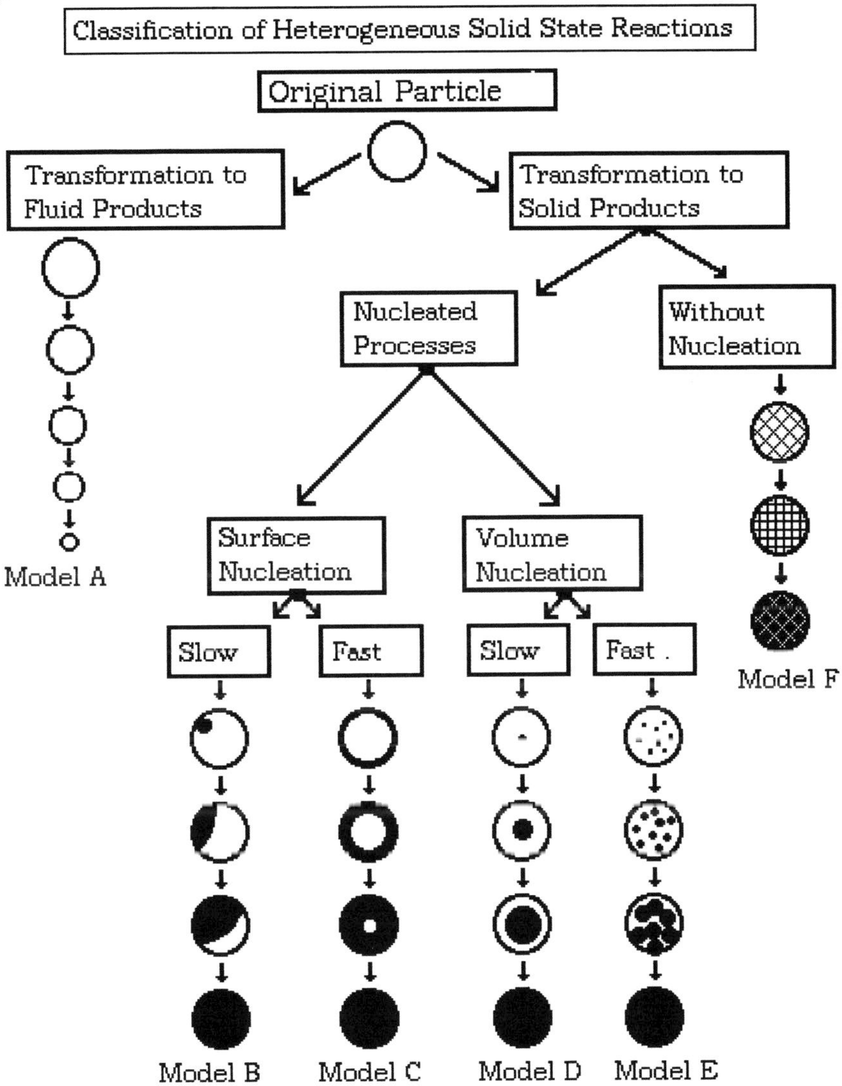


In this diagram, we classify the various types of heterogeneous solid state reactions that can occur. Note that both transformation to solid and fluid products are diagrammed. Also shown are variations where surface and/or volume nucleation is involved.

When solids react and nucleation is the norm, they do so either by surface or volume nucleation. Model B involves **surface nucleation** wherein the surface becomes nucleated in one spot, i.e.- two reacting particles in close proximity. This is shown in 4.4.2. Here, two touching particles have initiated growth of another phase by nucleating at the point where each particle touches the other.

4.4.2.- Model B where Reaction Only Takes Place at the Point Where Two Particles Touch

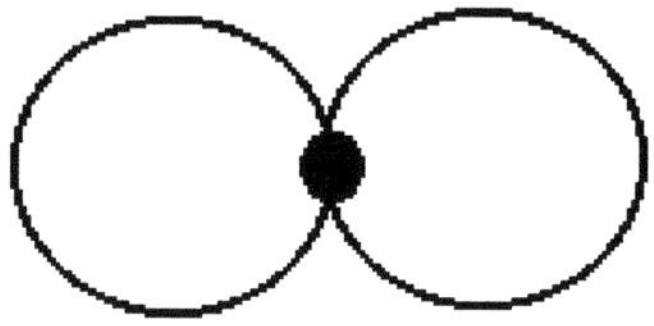

As shown in 4.4.1., two stages usually occur in surface nucleation, nucleation and then nuclei growth. If surface nucleation is fast (Model C) it is likely due to reaction of a **gas** with the solid particle. The reaction of a **liquid** is the other possibility, i.e.-

4.4.3.-
$$A_S + B_G \Rightarrow (AB)_S$$
$$A_S + B_L \Rightarrow (AB)_S$$

If it slow, then nucleation is likely to be due solely to proximity. Model D is an example of **volume nucleation where** decomposition of a solid is involved whereas Model E is that involving gas or liquid nucleation of the solid. Note that if nucleation does not occur, the solid reacts uniformly throughout its whole volume (Model F). However, this mode is rare and the nucleation stages are more likely to occur. We will not dwell upon how these nucleation models were derived and will only present the results here. One is referred to Appendix I wherein one can study the mathematics used to obtain the net-result.

The factors that we must consider include the following:

1) The formation of an incipient embryo, E
2) Numbers of embryos which form
3) The growth of the embryo into a nucleus, N

4) Growth of nuclei
5) Growth and transformation of nuclei into the final product of the solid state reaction by formation of phase boundaries.

What we mean by an embryo is shown in the following diagram wherein the dimensions are distorted from the actual case in order to illustrate our point.

4.4.4.-

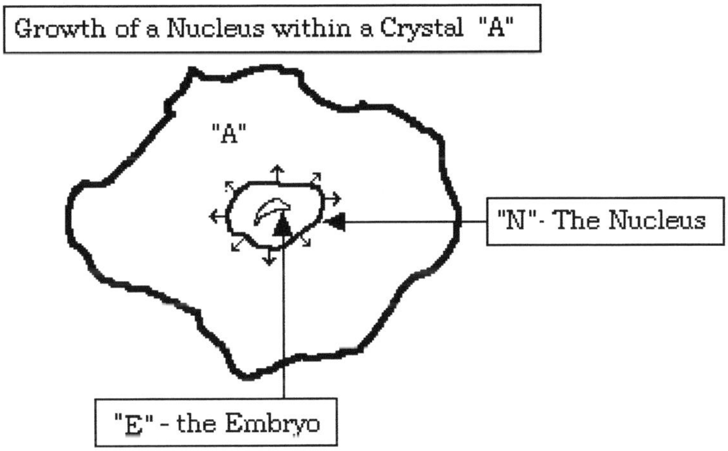


In this case, we must differentiate between the walls of the rapidly forming embryo and those of the incipient nucleus. This drawing shows the formation of an embryo **E** , which then grows into a nucleus, **N**, within the crystallite volume, **A** (none are drawn to scale). The change in free energy in going from B to A will be a function of both volume and surface area (shape). We must discriminate between the interfacial surface energy between B and A, and the internal energy of C. The nucleus, N, then grows by expanding its walls to form a new structure (compound). We note that the growth of nuclei is associated with the motion of a phase boundary, as described above. We will not give the mathematics involved here but reserve them for Appendix II at the end of this chapter.

However, we need to point out that the rate of nucleation, I, is defined as:

4.4.5.- I ≡ Number of nuclei/Unit time/Unit volume

and xdt is the growth rate of the nuclei, as in 4.4.4. The final result involves:

4.4.6.- $r_{crit} = -2\,\gamma\,/\,\Delta G_V$

which is the **Gibbs-Thompson** equation defining the critical radius required for an embryo to form. We also can relate these equations to numbers of nuclei being formed and their growth. This is illustrated as follows in 4.4.7., given on the next page.

In the simple case, B grows at the expense of A. However, in the second case involving parallel reactions, **two embryos** form from A. In the third case, at least two steps are involved in the two consecutive reactions where "A" transforms to "B" which transforms to "C".

In Table 4-2, we show both phase-boundary controlled and diffusion-controlled nucleation reactions as a function of both **constant** and **zero rate of nucleation.**

TABLE 4-2

NUCLEATION AS A FUNCTION OF THE SHAPE FACTOR, θ

| | | PHASE BOUNDARY CONTROLLED $[\beta(t) = k_2\, t^r]$ | | DIFFUSION CONTROLLED $[\beta(t) = D\sqrt{t}]$ | |
|---|---|---|---|---|---|
| | θ | r | E | r | E |
| Constant Rate of Nucleation | spheres | 4 | $3\,E_2 + E_1$ | 2.5 | $3/2\,E_D + E_1$ |
| | plates | 3 | $2\,E_2 + E_1$ | 2.0 | $E_2 + E_D$ |
| | needles | 2 | $E_2 + E_1$ | 1.5 | $1.5\,E_D + E_1$ |
| Zero Rate of Nucleation | spheres | 4 | $3\,E_2$ | 1.5 | $1.5\,E_D$ |
| | plates | 3 | $2\,E_2$ | 1.0 | $E_D$ |
| | needles | 2 | $E_2$ | 0.5 | $0.5\,E_D$ |

## 4.4.7.- EMBRYO GROWTH RATE EQUATIONS FOR VARIOUS CASES

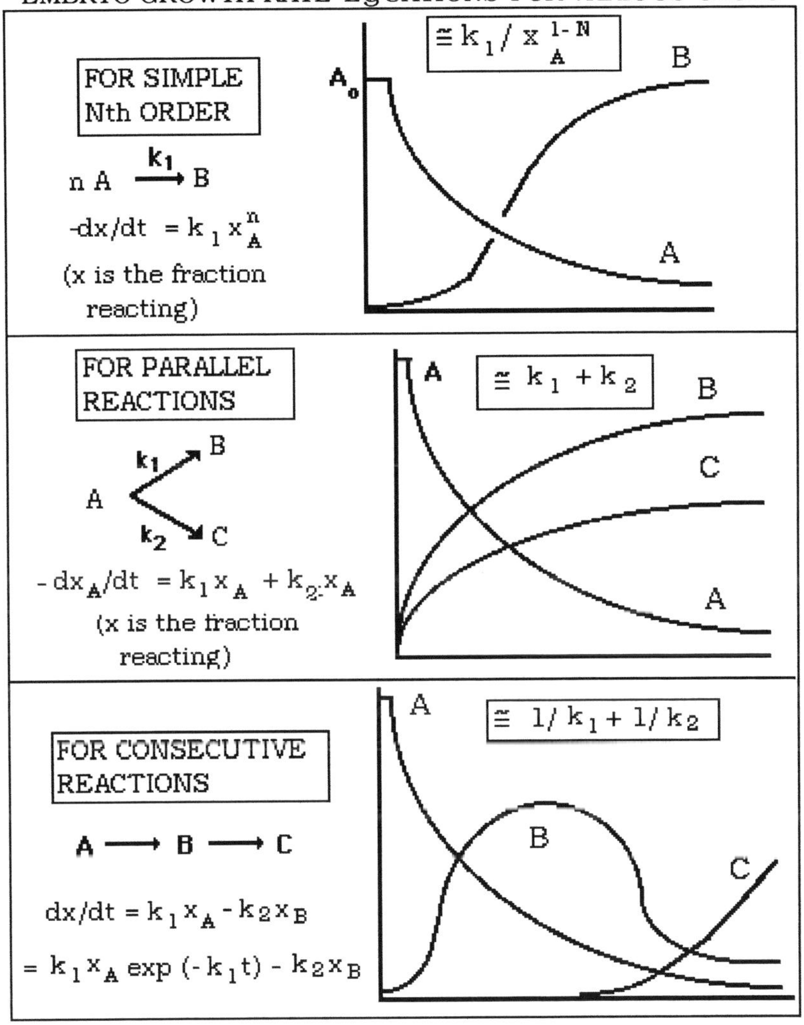


Referring to 4.4.4. and Table 4-2, we see that the shape factor, θ, has a major effect upon r and the exponent of time, t, as well as on the activation energies. $E_1$ is the activation energy for nuclei **formation**, whereas $E_2$ is that for **nuclei growth.** $E_D$ is the activation energy for **diffusion growth.** Note that we have now directly associated diffusion as a mechanism in nuclei formation and that we have already given examples above concerning solid state reaction mechanisms concerning diffusion (see 4.2.2 for example).

Actually, there are three (3) major classifications for nuclei growth in heterogeneous solid state reactions. Restating, these are:

4.4.8.- Phase-boundary Controlled
Random Growth
Diffusion Controlled

We have already dealt with two of these. Section **2** dealt with formation of a phase boundary while we have just completed Section **4** concerning nuclei growth as related to a phase boundary. We will consider diffusion mechanisms in nuclei and diffusion-controlled solid state reactions at a later part of this chapter.

These are areas of solid state chemistry where considerable experimental work has already been accomplished, particularly in the area of tarnishing because of the obvious commercial interest in such reactions.

We are now ready to consider some solid state reactions that relate more directly to the real world. These include the tarnishing reaction and Fick's Laws of Diffusion. Both of these scientific areas have been rigorously studied because of their importance in revealing how diffusion mechanisms are related to everyday solid state reactions which occur on a daily basis.

## **4.5.** THE TARNISHING REACTION

In 1902, Wagner published an analysis, based on diffusion reactions, of the oxidation of the surface of a metal. His interpretation has remained a classic in solid state diffusion analysis. The surface of a metal consists of metal atoms bound to the inner structure by a series of hybrid-bonds. If oxygen gas is present (air), the metal can form an oxide coating:

4.5.1.- $M + 1/2\, O_2 \Rightarrow MO$

In some cases, the oxide-coating protects the surface from further oxide buildup. One example is that of aluminum where an oxide coating appears almost instantaneously once the pristine surface is exposed to air. Yet, there are many cases where the oxide layer continues to buildup until the metal is totally consumed (One example is that of iron and "rust"). How is this possible? Wagner hypothesized that both metal and oxide ions **diffused** through the metal oxide layer so as to build up the layer thickness **from both sides.** The following diagram is one representation of such a mechanism:

4.5.2.- Diffusion Mechanisms Prevalent in the Tarnishing Reaction

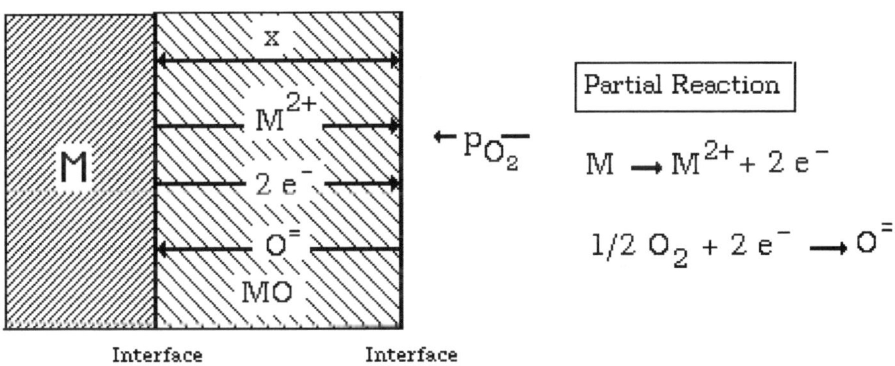


In this case, the partial pressure, p, of oxygen gas is a limiting factor in the overall reaction to form the oxide film, MO, whose thickness, x, is shown.

It should be apparent, from the above diagram, that **the more x increases,** the slower will be the change in the x-dimension (since it takes longer for the ions to diffuse), i.e.-

4.5.3.- $dx/dt \approx 1/x$ or: $dx/dt = K \cdot 1/x$

The integrated form is: $x^2 = Kt + C$ , or: $x = D\sqrt{t}$ , where D is the Diffusion Coefficient. Equation 4.5.3. is called the **Parabolic Law of Diffusion.** If the growth of a phase can be fitted to this equation, then it is likely that the primary reaction mechanism involves simple diffusion.

Diffusion mechanisms involve the following **defect reactions:**

4.5.4. - Interchange of Atoms
Vacancy Hopping
Interstitial Movement
Dissociative Exchange

We have already discussed most of these mechanisms. In the first, $M_X$ and $X_M$ interchange may be involved, whereas the last involves a mechanism similar to that given in 4.2.2. Note that the second type will only appear only if the individual ions involved are small enough to fit into the interstices of the lattice.

## 4.6. - FICK'S LAWS OF DIFFUSION

About 1942, Fick formulated laws which described diffusion processes in solids, similar to that already presented for the tarnishing reaction. To do this, Fick hypothesized that atoms (ions) would "hop" from site to site in the manner we have already described in 4.2.2. and 4.2.4.

Consider a lattice having a three-dimensional set of planes. If we have this three-dimensional lattice, it is easy to see that there are six (6) ways for an atom (ion) in the x-plane to move to the y-plane. This is shown in the following illustration, given as 4.6.1. on the next page.

By defining $n_A$ as the number of A atoms present per unit volume, $\tau$ as the time of stay at any given site, $1/\tau$ as the frequency of jumps and $J_{x \rightarrow y}$ as the number of jumps per unit time, Fick was able to derive two laws regarding diffusion in solids. To do this, he also defined:

$$I \equiv \text{Number of nuclei/Unit time/Unit volume.}$$

Using his definitions, we can derive a series of equations which results in Fick's First Law of Diffusion.

4.6.1.- Fick's Analysis of Diffusion Processes

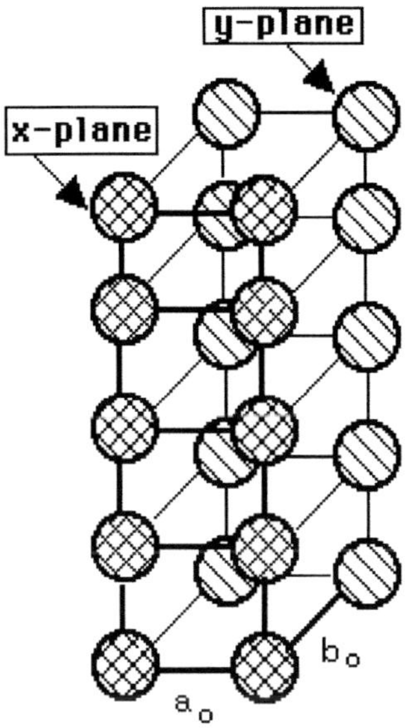


$\tau$ ≡ time of stay at given site

$1/\tau$ ≡ frequency of jumps

$n_A$ ≡ number of A atoms per unit volume

R ≡ cross-sectional area of a given plane

a ≡ lattice vector

$J_{x \to y}$ ≡ number of jumps per unit time from x-plane to y-plane

The diagram given above defines the parameters necessary to define the equations. How this is done is shown in Table 4-4, given on the next page, where the individual steps necessary are set forth.

The result is:

4.6.2.- FICK'S FIRST LAW: $\mathbf{J} = - D\, dc/dx$

where **J** is the as the total number of jumps that occur, and D is the diffusion coefficient for the process. If such exists, then dc/dx ≠ a constant, and we must then subtract two different jump rates, i.e.-

4.6.3.- $\mathbf{J_1} - \mathbf{J_2} = - \Delta x\, d\mathbf{J}/dx = - \Delta x\, (dc/dt)$

If the last term is taken **only** in respect to the x dimension, then we arrive at the second law, viz-

4.6.4.- FICK'S SECOND LAW: $[\partial c/\partial t]_x = \partial \{D\, \partial c/\partial x\}$ or $(\partial c/\partial t)_x = D(\partial^2 c/\partial x^2)$

Table 4-4

How Fick's First Law of Diffusion Is Derived

| Steps | | Equations leading to Fick's Laws |
|---|---|---|
| 1. | Here we have defined $\mathbf{J}_{x \to y}$ in terms of the number of possible "jumps", i.e.- 1 out of 6 ways, $\tau$ - the time that it takes to do so, the numbers of atoms present and the cross-sectional area of the lattice in question. | $\mathbf{J}_{x \to y} = 1/6 \cdot 1/\tau\ (n_A R \cdot a)$ |
| 2. | We then define the concentration of A-atoms in the y-plane to be: | $n_{A(y)} = n_A + a\, dn_A/dx$ |
| 4. | Combining these two equations gives us: | $\mathbf{J}_{x \to y} = (n_A + a\, dn_A/dx)(aR/6\tau)$ |
| 4. | If we assume that the opposite direction of hopping is also valid, i.e.- $\mathbf{J}_{y \to x}$, then we have: | $\mathbf{J} = \mathbf{J}_{x \to y} - \mathbf{J}_{y \to x}$ |
| 5. | Using **J** as the total number of jumps that occur, we come to: | $\mathbf{J} = aR/6\tau(n_A n_A + a dn_A/dx \cdot aR/6)$ <br> $= - a^2 R/ 6\tau \cdot dn_A/dx$ |
| 6. | We now define a diffusion coefficient "D" as: | $D \equiv a^2 / 6\tau$ |
| 7. | If we have the interstitial case (see 4.6.3.)., then we must include $\alpha$, which is the number of interstices in the lattice: | $D \equiv \alpha a^2 / \tau$ (interstitial) |
| 8. | This gives us: | $\mathbf{J} = - D R\, dn_A / dx$ |
| 9. | If we now define concentration in terms of $n_A$, i.e.- $c \equiv Rn_A$, so that: | $dc/dx = R\, dn_A/dx$ |
| 10. | **then we get FICK'S First Law:** | $\mathbf{J} = - D\, dc/dx$ |
| | | |

(Note that partial differentials are used in the 2nd Law (see 4.6.4. given above) since they relate only to the "x" dimension- similar equations may be derived for the "y" and "z" dimensions). Both of these Laws are used extensively in solid state chemistry.

We can also show that the diffusion coefficient, D, used in these Laws is temperature dependent, and can fit it to an Arrhenius equation:

4.6.5.- $D = D_o \exp -E/kT$

Sometimes, a more useful term is:

4.6.6.- $E_D = - k_1 \, d(\ln D)/d(1/T)$

where $E_D$ is the diffusion energy involved in the process. However, these equations are only used when the Fick Equations fail to fit the diffusion data obtained. It is possible to also show that the diffusion present in any compound is directly related to the fundamental (zero phonon line) of the lattice. The derivation is given in Appendix III. at the end of this chapter.

## **4.7.-** DIFFUSION MECHANISMS

To this point, we have examined diffusion growth in terms of nucleation and embryo formation. Let us now explore the actual species which diffuse in the lattice.

> WE WILL FIND THAT ALL DIFFUSION MOTION OCCURS BY DEFECT MOVEMENT IN THE LATTICE .

Consider the following diagram, given as 4.7.1. on the next page, showing two types of self-diffusion. Self diffusion can occur by at least two mechanisms, vacancy and interstitial. Both are "hopping" motions, as described above.

What we see is in both mechanisms is that one site is exchanged for another by the defect undergoing a "jump" in the lattice network.

in both directions. In our case, we will ignore charged species to simplify our explanation. Let us also suppose that the vacancies migrate much faster than the interstitials and that:

4.7.6.- $D_{V_A} >> D_{V_B}$ so that: $dc_{V_A}/dt >> dc_{V_B}/dt$

Given these restraints, we obtain a rather queer result. This is illustrated in the following diagram:

4.7.7.-

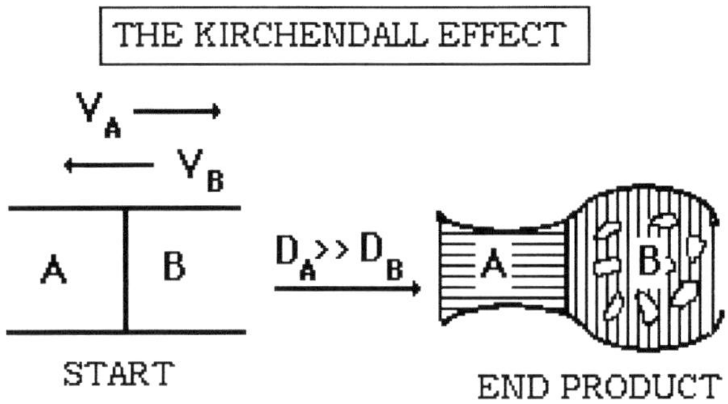


What we find is that the difference in diffusion mechanisms gives rise to the creation of **new sites** across the diffusion zone and actually causes a **deformation** in the solid because the $V_A$ defects pile up and finally become annihilated because of clustering. **Actual holes appear** in the solid due to vacancy diffusion. Note that in this case, we are not forming a newcompound through solid state reaction, but are forming a solid solution of A and B. The Kirchendall Effect has been observed many times, but occurs most often in reactions between metals which form alloys.

There are three (4) types of diffusion-controlled reactions possible for heterogeneous solid state reactions, viz-

4.7.8.- TYPES OF DIFFUSION REACTIONS

| | |
|---|---|
| Simple Diffusion: | $\beta = k_1 (t)^{1/2}$ |
| Phase-Boundary Controlled | $\beta = k_2 t$ |
| Material Transport | ------------- |

For simple diffusion-controlled reactions, we can show the following holds:

4.7.9.- $x^2 = 2 \, D \, c_o \, t \, V_M \; + (a)^{1/2}$

where $c_o$ = concentration of constituents at interface; V = volume of product AB per mole of reactants; a = surface layer thickness at interface when t = 0. It is also well to note that the final volume of the product, AB, may not be the same as that of the reactants vis.-

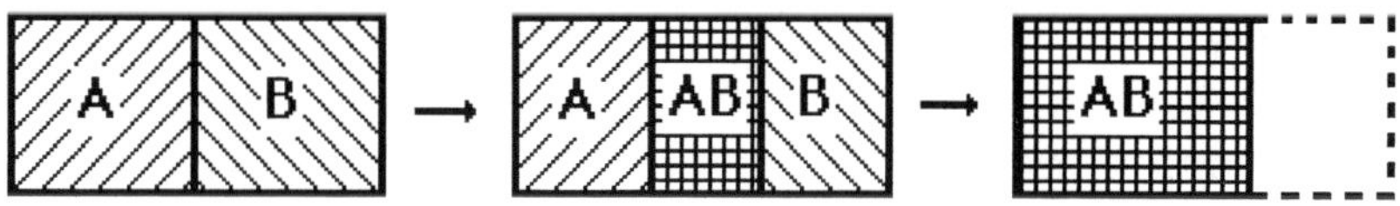


For phase-boundary controlled reactions, the situation differs somewhat. Diffusion of species is fast but the reaction is slow so that the diffusing species pile up. That is, the reaction to rearrange the structure is slow in relation to the arrival of the diffusing ions or atoms. Thus, a phase-boundary (difference in structure) focus exists which controls the overall rate of solid state reaction. This rate may be described by:

4.7.10.- $dx/dt = k_1 \, (s_t \, /V_o \,) \, - (1 - \, (1 - x)^n \,) \; = \; k_1 t \; / r_o$

where $s_t$ = instantaneous surface area; $V_o$ = original volume of particles reacting; $r_o$ = original radius of particles; and n = 1, 1/2, 1/3 for a 1-dimensional, 2-dimensional, or 3-dimensional reaction. Material transport frequently involves an external gaseous phase, and **a general formula has not evolved.**

The above equations, i.e.-4.7.9. and 4.7.10. summarize two of the three diffusion mechanisms, one of which usually predominates in any given case (see 4.7.8.). But these equations contain quantities that are hard to measure, or even estimate.

A much better way is to follow the method of **Hancock and Sharp** ( ~ 1948). If one can measure x, the amount formed in time, t, then the following equation applies:

4.7.11- Hancock & Sharp Equation: $\underline{-\log(\ln(1-x)) = m \ln T + \ln k}$

| Type of diffusion | range of m |
|---|---|
| simple diffusion | 0.57 - 0.62 |
| nuclei growth | 1.00 - 1.15 |
| phase boundary | 1.25 - 4.00 |

Here, we find it necessary to be able to measure the progress of a solid state reaction. If we can do so, then we can determine the type of diffusion involved. If $-\log(\ln(1-x))$ is plotted against ln t, one obtains a value for the slope, m, of the line which allows classification of the most likely diffusion process. Of course, one must be sure that the solid state reaction is primarily diffusion-limited. Otherwise, the analysis does not hold.

## 4.8. - ANALYSIS OF DIFFUSION REACTIONS

Let us now turn to diffusion in the general case, without worrying about the exact mechanism or the rates of diffusion of the various species. As an example to illustrate how we would analyze a diffusion-limited solid state reaction, we use the general equation describing formation of a compound with spinel (cubic) structure and stoichiometry:

4.8.1.- $AO + B_2O_3 \Rightarrow AB_2O_4$

where A and B are two different metallic ions and O is the oxygen atom (as the oxide). There are two (2) different cases that we can distinguish, both of which represent possible diffusion mechanisms in spinel. These involve either a gaseous or an ion diffusion mechanism.

Let us consider the ion-diffusion mechanism first. As shown in the following diagram, given as 4.8.2. on the next page, both A and B diffuse through $AB_2O_4$ at the same rate. We can show the probable diffusion conditions as specific solid state reactions, viz-

4.8.2.-

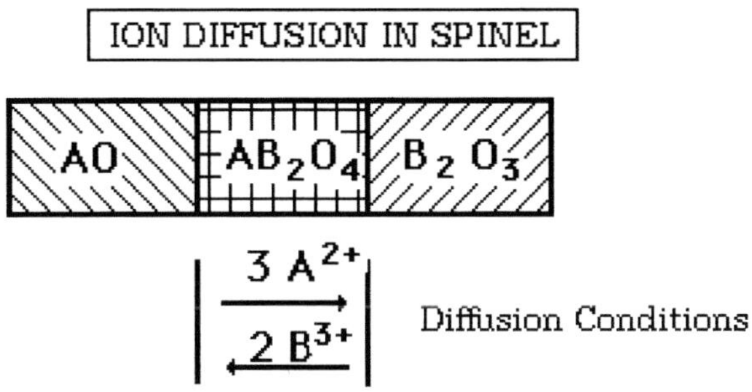


The actual diffusion reactions involved are:

4.8.3.-

$$3\ A^{2+} + 4\ B_2O_3 \Rightarrow 3\ AB_2O_4 + 2\ B^{3+}$$

$$2\ B^{3+} + 4\ AO \Rightarrow AB_2O_4 + 3\ A^{2+}$$

---

$$4\ AO + 4\ B_2O_3 \Rightarrow 4\ AB_2O_4$$

Notice that the partial reactions given in 4.8.3. are balanced both as to material and charge. These are the reactions which occur at the **interface** (or phase boundary) between the diffusing ions and the bulk of the reacting components, as we have already illustrated above. There are at least two other possible mechanisms, as shown in the following diagram:

4.8.4.-

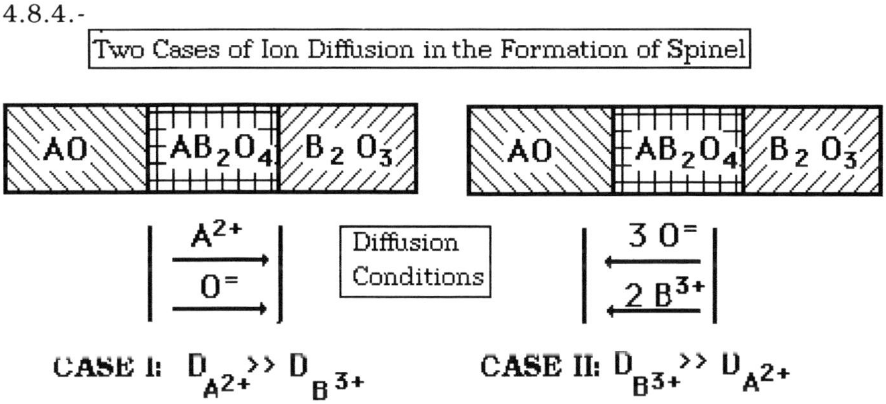

In one case, the diffusion of $A^{2+}$ is much faster than $B^{3+}$ and in the other the opposite is true. Note that charge-compensation of migrating species is maintained in all cases.

We can also illustrate a gaseous transport reaction mechanism in the same manner, as shown in the following diagram:

4.8.5.-

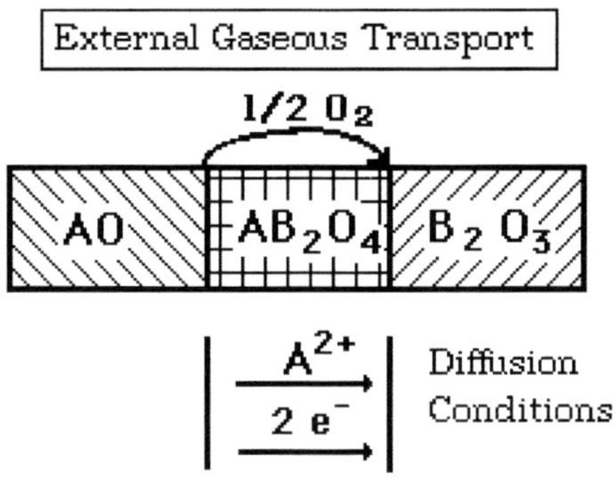


The reactions for this case (CASE III) are:

4.8.6.- $O^{=} \Rightarrow 1/2\ O_2 + 2\ e^-$

$$A^{2+} + 2\ e^- + 1/2\ O_2 + B_2O_3 \Rightarrow AB_2O_4$$

In this mechanism where $D_A{}^{2+} >> D_B{}^{3+}$, transport of external oxygen gas is involved in the overall solid state reaction, accompanied by electronic charge diffusion.

There are still two more mechanisms to consider, that where the diffusion rate of $B^{3+}$ is much greater than that of $A^{2+}$:

$$D_{B3+} >> D_{A2+}$$

$$D_A{}^{2+} >> D_B{}^{3+}$$

and the opposite case, as shown. In both of these cases, transport of $O_2$ is just opposite that shown in 4.8.6., and the reactions are:

4.8.7.- CASE IV: $3\ O^{=} \Rightarrow 3/2\ O_2 + 6\ e^-$

$AO + 3/2\ O_2 + 6\ e^- + 2\ B^{3+} \Rightarrow AB_2O_4$

or: $2\ O^{=} \Rightarrow O_2 + 4\ e^-$

$2\ A^{2+} + O_2 + 4\ e^- + 2\ B_2O_3 \Rightarrow 2\ AB_2O_4$

This mechanism is shown in the following:

4.8.8.-

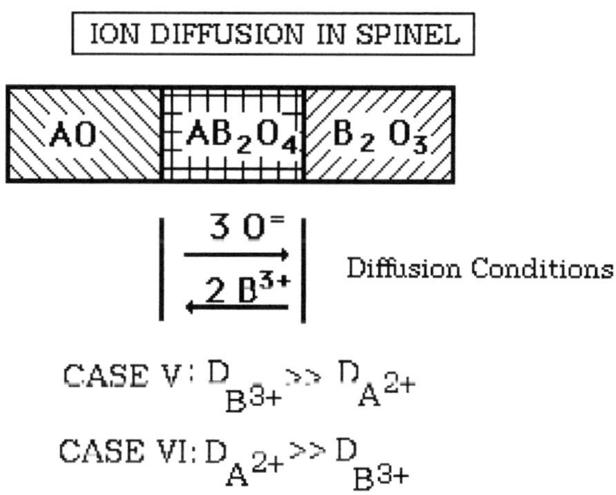


We have now presented **all of the possible** diffusion reactions in spinel synthesis. These are summarized as follows:

4.8.9.- <u>Possible Diffusion Mechanisms in Spinel</u>

<u>Ion Diffusion Reactions</u>:

$3\ A^{2+} + 3\ B_2O_3 \Rightarrow 3\ AB_2O_4 + 2\ B^{3+}$

$2\ B^{3+} + 4\ AO \Rightarrow AB_2O_4 + 3\ A^{2+}$

<u>Conceivable Cases</u>

1. $D_A{}^{2+} = D_B{}^{3+}$
2. $D_A{}^{2+} >> D_B{}^{3+}$ ($A^{2+} + O^{=}$ - for charge compensation)
3. $D_B{}^{4+} >> D_A{}^{2+}$ ($B^{3+} + O^{=}$ - for charge compensation)

4.8.9.- Possible Diffusion Mechanisms in Spinel (continued)

Gaseous Transport Reactions:

$O^{=} \Rightarrow 1/2\ O_2 + 2\ e^-$

$A^{2+} + 2\ e^- + 1/2\ O_2 + B_2O_3 \Rightarrow AB_2O_4$

Possible Cases

1. $D_B^{2+} >> D_A^{3+}$ (transport of charge only)
2. $D_A^{3+} >> D_B^{2+}$ (transport of charge only)

It should be clear, by examining 4.8.9. carefully, that a number of possibilities exist for the diffusion of reacting species through spinel during the solid state reaction used to form it. Which of these is more likely will depend upon the exact nature of "A" as well as that of "B".

It would be too much to expect that the same diffusing mechanism would exist for all types of spinels (yes, there are many types of spinels, since "spinel" is a particular composition, i.e.- $AB_2O_4$ , having a cubic structure. What we need to do is to determine which of these cases is the most relevant. To do this, we need to generate some experimental data.

Thus, we measure formation rate in air, pure oxygen gas and then in an inert gas. If the rates do not differ significantly, then we can rule out gaseous transport mechanisms. There are other tests we can apply, including electrical conductivity, transference numbers and thermal expansion. Although these subjects have been investigated in detail, we shall not present them here.

It should be clear that a number of mechanisms exist. Which of these dominates will, as we stated before, will depend upon the nature of both "A" and "B". However, it has been observed throughout many investigations that one mechanism seems to dominate. We will not delve through these works.

> It suffices to say that: "**For the most part, in oxygen-dominated hosts, diffusion by small cations prevails and that diffusion of charge-compensated pairs predominates**".

It is for this reason that we write the solid state diffusion reaction for the magnesium aluminate spinel as follows:

4.8.10.- $3\ Mg^{2+} + 4\ Al_2O_3 \Rightarrow 3\ MgAl_2O_4 + 2\ Al^{3+}$

$$2\ Al^{3+} + 4\ MgO \Rightarrow MgAl_2O_4 + 3\ Mg^{2+}$$

$$4MgO + 4\ Al_2O_3 \Rightarrow 4\ MgAl_2O_4$$

Note that the ions given in 4.8.10. cancel out in the equations, and that only the overall solid state reaction remains. However, it has been determined that diffusion is limited by:

4.8.11.- Diffusion Mechanism:

**Frenkel Pair:** $Al_{Al} + \infty V_i \leftrightharpoons V_{Al} + Al_i^{3+}$

In this case, the Frenkel pair diffusion, i.e.- $V_{Al} + Al_i^{3+}$, predominates and **is faster** than any other possible mechanism.

**4.9. -** DIFFUSION IN SILICATES

We have presented spinel synthesis because these systems were studied first and can be understood in a simple manner. Various silicate systems have also been studied and it has been determined that they are more representative of the general case involving solid state synthesis reactions than spinel. Let us examine the following simple silicate reaction:

4.9.1.- $2\ CaO + SiO_2 \Rightarrow Ca_2SiO_4$

Calcium oxide, in proper proportion, reacts with silica to form calcium orthosilicate. In terms of the spinel case, we would expect to see the following diffusing species:

4.9.2.- $Ca^{2+}\ ,\ Si^{4+}\ ,\ O^{=}\ ,\ e^{-}\ ,\ p^{+}$

Using our model, we would illustrate the diffusion reactions with the following diffusion conditions:

4.9.3.- Mechanisms of Diffusion and Reaction Between CaO and $SiO_2$

| Diffusion Mechanism | Diffusion Reaction |
|---|---|

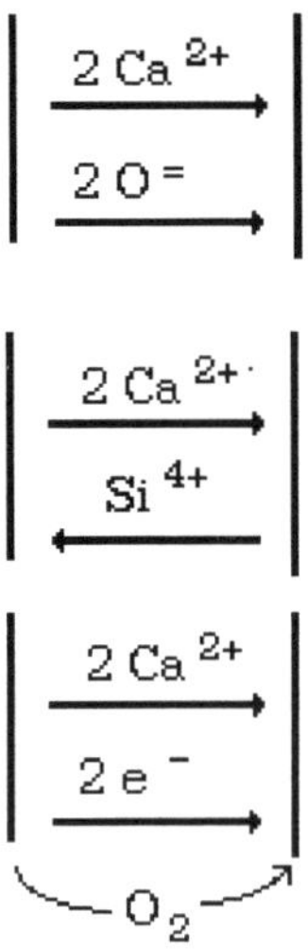


$$2\,Ca^{2+} + 2\,O^{=} \rightarrow Ca_2SiO_4$$

$$2\,Ca^{2+} + 2\,SiO_2 \rightarrow Ca_2SiO_4 + Si^{4+}$$

$$Si^{4+} + 4\,CaO \rightarrow Ca_2SiO_4 + 2\,Ca^{2+}$$

$$2\,O^{=} \rightarrow O_2\uparrow + 4\,e^-$$

$$2\,CaO + O_2 + 4\,e^- \rightarrow Ca_2SiO_4$$

In this case, we have two concomitant materials, CaO and $SiO_2$, reacting together to form the compound, calcium orthosilicate, which exists as a phase boundary between the five diffusing species. We can hypothesize at least three cases involving diffusion. In the first case, both $Ca^{2+}$ and $O^{=}$ diffuse together in the same direction.

In the second case, $Ca^{2+}$ and $Si^{4+}$ diffuse in opposite directions while the third case involves $O_2$ transport along with the diffusion of $Ca^{2+}$ which then reacts with the $SiO_2$.

These three mechanisms might seem to be valid but further investigation reveals that actual diffusion is controlled by the nature of the lattice structure which involves the silicate tetrahedron. Additionally, we find that the actual diffusion reaction which occurs is quite different than that of the spinel case. This is shown in the following diagram:

4.9.5.-

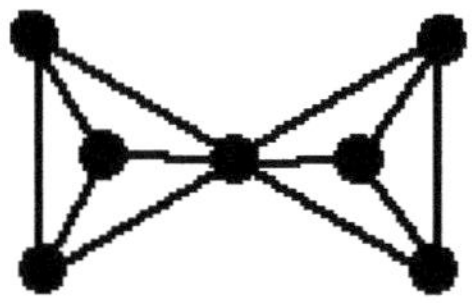

Thus, the rate of diffusion has been ascertained to be:

RATES OF DIFFUSION FOR SILICATE-BASED COMPOUNDS

Rate of species diffusion: $Ca^{2+} \cong O^{=} >> Si^{4+}$

This is caused by the fact that the $Si^{4+}$ is tied up in the form of silica tetrahedra where some of the oxygen atoms are shared within the structure and are not free to move. A representation of this is given as follows:

4.9.6.-

| Diffusion Mechanism | Diffusion Reaction |
|---|---|
| 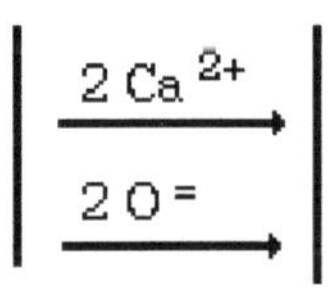 | $2\,Ca^{2+} + 2\,O^{=} + SiO_2 \rightarrow Ca_2SiO_4$ |

These tetrahedra are tied together at the corners so that a silicate "backbone" forms the structure. The metal cations form "bridges" between backbone-layers and are much more free to move. However, it is well to note that a small amount of silicate **does move**, but the exact nature of the diffusing specie cannot be quantitatively defined (It may depend upon the nature of the compounds being formed. Most probably, the diffusing specie is actually $SiO_n$ but the charge of each actual specie may vary). In order to avoid confusion, we will continue to use $Si^{4+}$ to indicate silicate diffusion.

**Thus, we find that the predominating diffusion mechanism for ion diffusion in $Ca_2SiO_4$ synthesis is that of $Ca^{2+}$ and the oxide ion, $O^{=}$.**

We now come to the most important point of this Section. Up to now, we have assumed that the phase within the phase boundary, as given in the diagrams, is **inviolable** and not subject to change. That is, once the $Ca_2SiO_4$ has formed, no further reaction can occur. While this may be true for some solids, it certainly is not true for $Ca_2SiO_4$, and indeed for most of the other systems that we might study.

The way we determine if further reaction is possible is to consult the phase diagram of the system. If a compound is stable, then it has a finite probability of forming during the solid state reaction. Let us now suppose that further reaction **does take place** at the phase boundary of CaO and $Ca_2SiO_4$, and also at $Ca_2SiO_4$ and $SiO_2$. We will call these **new phases** "X" and "Y". This gives us the situation shown in the following:

4.9.7.-

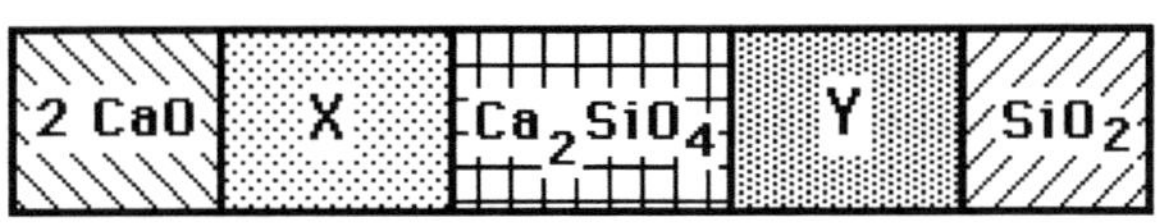


$Ca^{2+}$ → , $O^{=}$ →

$$Ca^{2+} + O^{=} + Ca_2SiO_4 \rightarrow Ca_3SiO_5$$

$2\ Si^{4+}$ ←, $2\ O^{=}$ ←

$$Si^{4+} + 2\ O^{=} + 3\ Ca_2SiO_4 \rightarrow 2\ Ca_3Si_2O_7$$

Here, we have reacted our diffusing species, $Ca^{2+}$ and $O^{=}$, with the initial product of our reaction, $Ca_2SiO_4$ in one direction, and $Si^{4+}$ and $O^{=}$ in the other to derive two entirely new reaction species. It would thus appear that "X" should be $Ca_2SiO_5$ and "Y" should appear as $Ca_3Si_2O_7$. A further look at the phase diagram shows that the former stoichiometry does not exist, but the latter does. Because the diffusing species, $Ca^{2+}$ and $O^{=}$, cannot react according the above-given reaction, **they continue to diffuse until they reach the vicinity of the pyrosilicate**, $Ca_3Si_2O_7$. There, the reaction at this phase-boundary is:

4.9.8.- $Ca^{2+} + O^{=} + Ca_3Si_2O_7 \Rightarrow 2\ Ca_2SiO_4$

But there is nothing to stop further diffusion of these species and they continue to diffuse to the vicinity of the $SiO_2$ phase boundary, where the reaction is:

4.9.9.- $Ca^{2+} + O^{=} + SiO_2 \Rightarrow CaSiO_3$

Thus, the **final** possible reaction is that which forms the *metasilicate* stoichiometry. In the opposite direction, we also have the diffusing species, $Si^{4+}$ and $O^{=}$. These react with the nearest phase to give $CaSiO_4$, as shown in the following diagram:

4.9.10.-

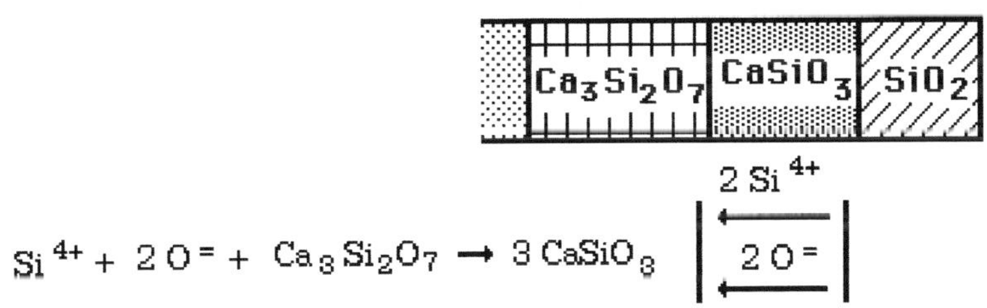


Note that we have **two diffusion reactions** which form the metasilicate, one proceeding from the cationic species and the other from anionic species. This has an amazing effect on the overall reaction, as shown in the following diagram, given as 4.9.11. on the next page.

Note carefully the sequence of intermediate reactions that have occurred, as depicted in this diagram. We see that the first solid state reaction involves formation of the orthosilicate. But this is very quickly transformed into meta- and pyro-silicates. However, it is the metasilicate which predominates as the solid state reaction reaches its maturity. Finally, it predominates over all other forms of silicate present.

4.9.11.-

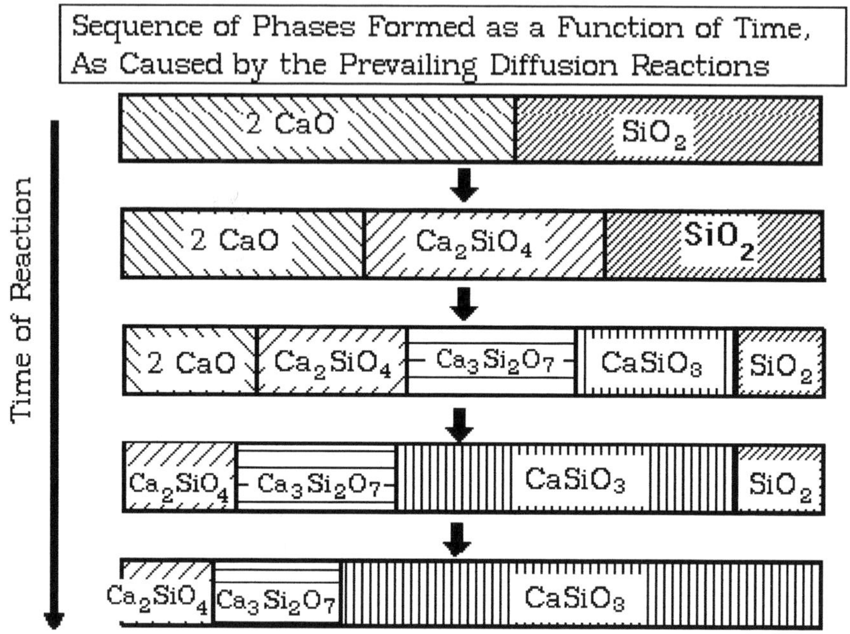


But, we still have small amounts of the others present. Let us now recall the prevailing diffusion conditions, as given in 4.9.4., i.e.-

4.9.12.- $Ca^{2+} \cong O^{=} >> Si^{4+}$

Because of this diffusion mechanism, we observe the above **sequence** of diffusion reactions which occur with time, as given in 4.9.12. It is actually possible to observe these reactions by the use of a polarizing microscope hot stage wherein the reactions are caused to occur by heating while observing the final structures formed via the x-ray diffraction patterns. The above diagram is somewhat cumbersome and it is much easier to illustrate these reactions by a graph. This is shown in the following diagram, given as 4.9.13. on the next page.

In the sequence of diffusion reactions, we note that $Ca_2SiO_4$ is formed immediately, followed by $Ca_4Si_2O_7$. Both **begin to disappear** when $CaSiO_3$

4.9.13.-

Reactions Which Occur in the Calcium Silicate System Between Calcium Oxide and Silica

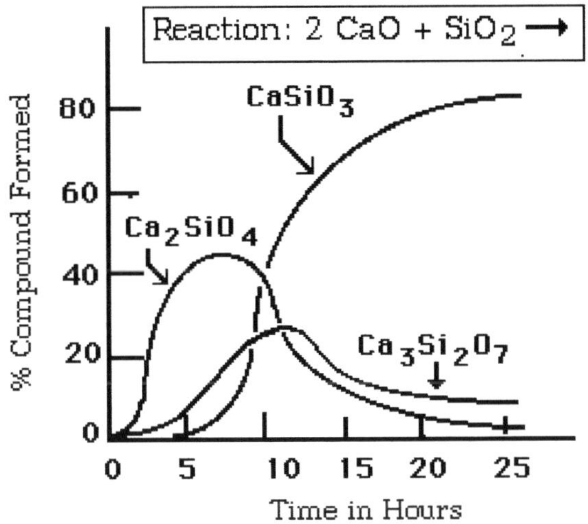


begins to form. Near the end, $CaSiO_3$ becomes the major phase present, but we never reach the point where just ONE COMPOUND remains. We **always** obtain a mixture. This point cannot be overstressed and is one of the most important points to be made in this chapter.

The overall reaction mechanisms are diffusion controlled, and the total solid state reaction can be summarized as follows:

4.9.14.-

$$\begin{array}{lcl} 8\ CaO + 4\ SiO_2 & \Rightarrow & 4\ Ca_2SiO_4 \\ Ca_2SiO_4 + SiO_2 & \Rightarrow & 2\ CaSiO_3 \\ 3\ Ca_2SiO_4 + SiO_2 & \Rightarrow & 2\ Ca_3Si_2O_7 \\ \underline{2\ Ca_3Si_2O_7 + 2\ SiO_2} & \underline{\Rightarrow} & \underline{6\ CaSiO_3} \\ 8\ CaO + 8\ SiO_2 & \Rightarrow & 8\ CaSiO_3 \end{array}$$

Although these reactions are written to show the formation of the metasilicate, we already know that none of these reactions ever comes to completion. There are competing **side-reactions.** We start with a 2:1

stoichiometry of CaO and $SiO_2$, but end up with a stoichiometry which is mostly 1:1, that of the metasilicate.

Another way to look at this phenomenon is that the diffusion conditions favor the formation of the **metasilicate**, and that the "excess Ca" is taken up in the formation of compounds having a "calcium-rich" stoichiometry (in comparison to that of the metasilicate). However, if we start with a stoichiometry of 1:1 CaO to $SiO_2$, we discover that the same compounds form as before, and a mixture of compounds is **still obtained.** The only difference is that there are smaller amounts of $Ca_2SiO_4$ and $Ca_3Si_2O_7$ present!

The only conclusion that we can draw is that diffusion-controlled solid state reactions tend to produce mixtures of compounds, the relative ratio of which is related to their thermodynamic stability **at the reaction temperature.** Obviously then, if we change the temperature of reaction, we would expect to see somewhat different mixtures of compounds produced.

Let us now look briefly at another similar system where we will start with a stoichiometry of :

1.00 BaO to 1.00 $SiO_2$ (in moles).

This mixture would be expected to react as:

4.9.15.- $BaO + SiO_2 \Rightarrow BaSiO_3$

with the diffusion condition: $Ba^{2+}$, $O^{=}$ >> $Si^{4+}$ . We can show the solid state reaction behavior again by a chart, given as 4.9.16. on the next page.

One might think that since $Ba^{2+}$ is a much larger ion than $Ca^{2+}$, it would diffuse slower. Such is not the case as can be seen by comparing the x-axis of 4.9.14. with that of 4.9.16. in terms of reaction time.

4.9.16.-

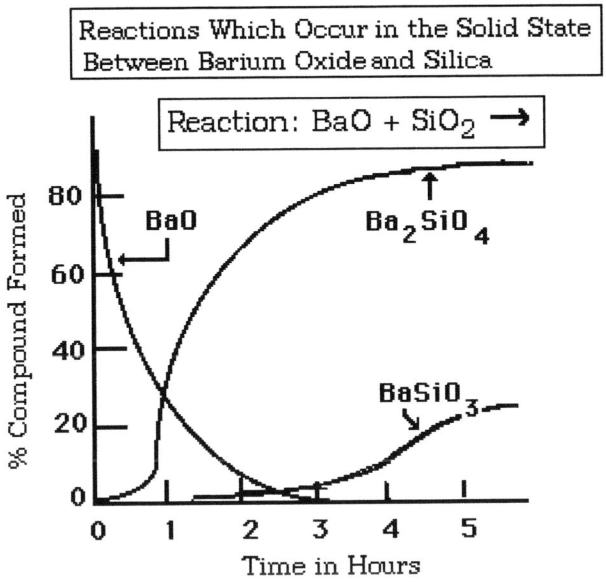


It is remarkable that in this case , we started with a 1:1 stoichiometry and ended up with a compound that has a 2:1 stoichiometry !

It should be clear by comparing the examples for calcium silicate and barium silicate that one cannot predict how the diffusion-controlled solid state reactions will proceed since they are predicated upon the relative thermodynamic stability of the compounds formed in each separate phase.

The series of diffusion-controlled reactions are for the case of the solid state reaction between $BaO + SiO_2$ as given in 4.9.16. above. These solid state reactions can be summarized as:

4.9.17.- (a) $2\ Ba^{2+} + 2\ O^{=} + SiO_2 \Rightarrow Ba_2SiO_4$

(b) $Ba_2SiO_4 + Si^{4+} + 2\ O^{=} \Rightarrow BaSiO_3$

Because the rate of Reaction (a) is so much faster than that of (b), we end up with $Ba_2SiO_4$ **as the major product.** But, we always end up with a **mixture** of barium silicate stoichiometries.

THIS BRINGS US TO A MAJOR AXIOM FOR CHAPTER-4 REGARDING SOLID STATE REACTIONS, VIS-

> **"DIFFUSION-CONTROLLED REACTIONS BETWEEN REFRACTORY OXIDES ALWAYS RESULTS IN MIXTURES OF COMPOUNDS "**

WHAT THIS MEANS IS THAT WE CANNOT PREPARE PURE COMPOUNDS BY REACTING OXIDES TOGETHER!

By now, you are probably wondering how to avoid such a mess. There ought to be a way to evade this situation, and indeed there is. We can find at least four (4) methods to do so. They include:

4.9.18.-
1. Using a gaseous reactant
2. Using a flux
3. Promoting a super-reactive component
4. Using a precipitated product to act as the reaction base

Note that all of these methods attempt to bypass the dependence of the solid state reaction upon diffusion. But, using a gaseous reactant may not be practical in all cases. And, sometimes it is hard to find a flux which does not interfere with the reaction. A flux is defined as follows:

4.9.19.- DEFINITION OF A FLUX

> "A material which melts lower than the solid state reaction temperature, dissolves one or more of the components and allows material transport to the reaction zone, **without** entering into the solid state reaction. Preferably, the end-product should be insoluble in the flux".

We can also employ a super-reactive product so as to obtain 100% of the desired product. Examples are given in 4.9.20. on the next page as follows. In this case, we are varying the ratios of reacting molar proportions in a deliberate manner to produce separate compounds.

4.9.20.-

$$2\ BaCO_3 + SiO_2 \Rightarrow Ba_2SiO_4\ (100\%\ \text{yield})$$

$$BaCO_3 + SiO_2 \Rightarrow BaSiO_3\ (100\%\ \text{yield})$$

$$3\ BaCO_3 + SiO_2 \Rightarrow Ba_3SiO_5\ (100\%\ \text{yield})$$

The mechanism involved is as we have already described in 4.2.8., i.e.-

4.9.21.-

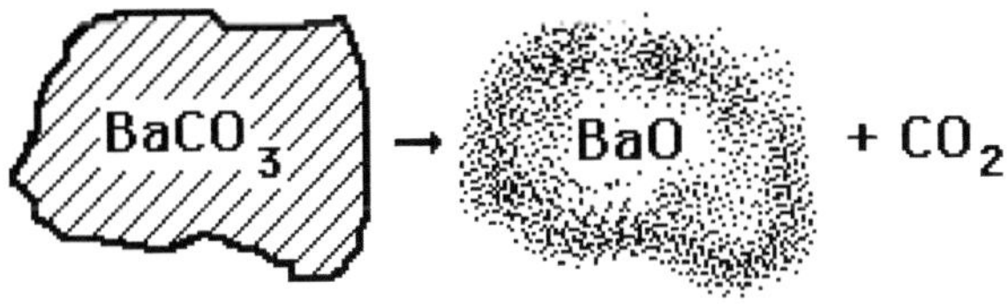


The very fine particles of nearly atomic proportions react almost immediately. Because the BaO product has an extremely large surface area, the solid state reaction is **not** diffusion limited. In the last method (see 4.9.18.), we might use a precipitated product to use as the basis to form the desired compound. We might precipitate:

4.9.22. $$Ba^{2+} + HPO_4^{=} \rightarrow BaHPO_4 \Downarrow$$

which is then reacted to form the orthophosphate as the desired product, e.g. -

4.9.23.- $$2\ BaHPO_4 + BaO \Rightarrow Ba_3(PO_4)_2 + H_2O \Uparrow$$

If barium orthophosphate is formed in this manner, we can guarantee that the final solid state reaction product will consist of 100% of the desired product.

**4.10.-** DIFFUSION MECHANISMS WHEN A CATION CHANGES VALENCE

Before we leave the subject of diffusion, let us examine the case where a change of valence state of the cation occurs. This mechanism is quite common for those solid state reactions involving the transition metals of the Periodic Table. In studies of solid state reactions, this situation continues to be the subject of much investigation because of the unusual situations which are encountered. Consider the reaction:

4.10.1.- $NiO + Al_2O_3 \Rightarrow NiAl_2O_4$

This would give us the following diffusion diagram:

4.10.2.-

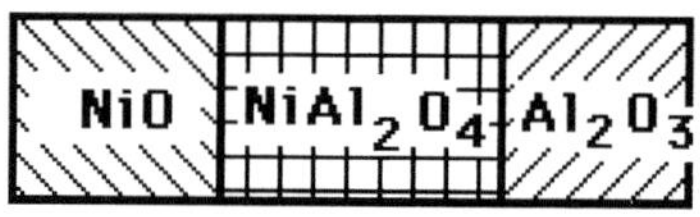


| Diffusion Mechanism | Diffusion Reaction |
|---|---|
| $3\ Ni^{2+}$ → <br> $2\ Al^{3+}$ ← | $3\ Ni^{2+} + 4\ Al_2O_3 \rightarrow 3\ NiAl_2O_4 + 2\ Al^{3+}$ <br> $2\ Al^{3+} + 4\ NiO \rightarrow NiAl_2O_4 + 3\ Ni^{2+}$ |
| $Ni^{2+}$ → <br> $O^{=}$ → | $3\ Ni^{2+} + O^{=} + Al_2O_3 \rightarrow NiAl_2O_4$ |
| $2\ Al^{3+}$ ← <br> $3\ O^{=}$ ← | $2\ Al^{3+} + NiO + 3\ O^{=} \rightarrow NiAl_2O_4$ |

In this case, we have 4 possible diffusion mechanisms all producing $NiAl_2O_4$. We cannot say which of the diffusion processes is faster at this point, or which mechanism prevails. From what we have stated previously, it is likely to be that of $Al^{3+}$ and $Ni^{2+}$ diffusing in opposite directions.

But, suppose some of the $Ni^{2+}$ oxidizes to $Ni^{3+}$ , vis-

4.10.3.- $Ni^{2+} = Ni^{3+} + e^{-}$

This is a case where a trivalent cation is present in a divalent lattice (see

Chapter 3).The following defect reactions would then be operative as shown:

4.10.4.-

$$2\ Ni^{2+}_{Ni} = Ni^{3+}_{Ni} + V_{Ni} + p^{+}$$
$$V_{Ni} + e^{-} = V_{Ni}^{-}$$
$$2\ Ni^{2+}_{Ni} + e^{-} = Ni^{3+}_{Ni} + V_{Ni}^{-} + p^{+}$$

This would give us the following diffusion diagram:

4.10.5.-

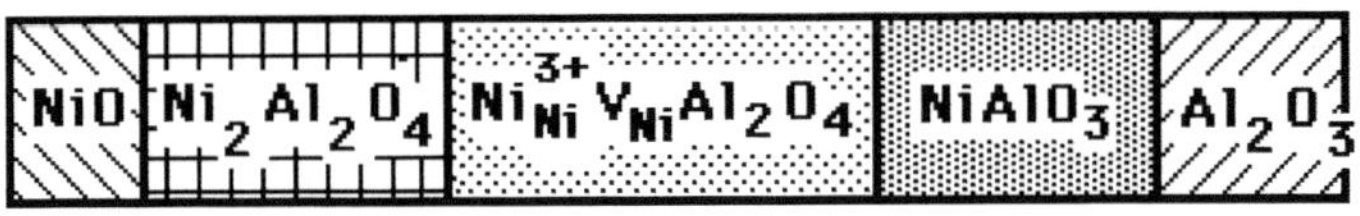


| Diffusion Mechanism | Diffusion Reaction |
|---|---|
| $Ni^{3+}$ →, $O^{=}$ →, $V^{-}_{Ni}$ → | $Ni^{3+} + O^{=} + Al_2O_3 + V^{-}_{Ni} \rightarrow NiV^{-}_{Ni}Al_2O_4$ |
| $Ni^{3+}$ →, $Al^{3+}$ ← | $2\,Ni^{3+} + 2\,Al_2O_3 \rightarrow 2\,NiAlO_3 + 2\,Al^{3+}$<br>$2\,Al^{3+} + 4\,NiO \rightarrow Ni_2Al_2O_4 + 2\,Ni^{2+}$ |

Note that $Ni^{2+}$ does not migrate but gives rise to a trivalent species plus a vacancy. This situation gives rise to a complicated set of diffusion conditions:

a. The charged vacancy is one of the migrating species.

b. The charged vacancy combines with $Ni^{3+}$ to form a defect compound, $NiV^{-}_{Ni}Al_2O_4$.

c. The divalent ion, $Ni^{2+}$, is also produced in the diffusion reactions.

d. Since $D_{Ni^{3+}} >> D_{V_{Ni}}$ , $Ni^{3+}$ diffuses oppositely to $Al^{3+}$ , as shown in the above set of actual diffusion reactions, another new compound arises, i,e,- $Ni_2Al_2O_4$.

Note that the charged **vacancy** diffuses as one of the **reacting** species to form the defect compound. This situation is quite common in the solid state chemistry of compounds containing multivalent cations. The trivalent $Ni^{3+}$ also gives rise to a new compound, $NiAlO_4$ . Yet the same reactions given in 4.10.2. are also OPERATIVE. **Thus, the overall reaction actually taking place** is a combination of 4.10.2. and 4.10.5.

LET US NOW SUMMARIZE THE AREAS WE HAVE COVERED SO FAR:

a. Types of Solid State Reactions
b. Formation of Phase Boundaries between Reacting Solids
c. Nucleation Rate Processes
d. The Critical Radius and Transformation from Embryo to Nucleus.
e. Diffusion Laws and Mechanisms
f. Analysis of Diffusion-Limited Solid State Reactions

These subjects include the majority of the cases we will encounter in our study of solid state reactions and their complexity. Let us now summarize the information we have covered:

1) We can define the exact nature of a solid state reaction.

2) Before any solid state process can occur, nuclei must form so as to allow the reaction to go to completion.

3) In some cases, an embryo must form before the nucleus formation is complete.

4) Nucleation can occur within any given solid (homogeneous nucleation) or be induced by any outside factor (heterogeneous), including the presence of a foreign solid.

5) In a diffusion-limited reaction, movement of species occurs by a "hopping" motion.

6) If a solid state reaction is diffusion-limited, if is unlikely that we can obtain 100% of any product, and will always obtain a mixture of compounds whose relative ratio will depend upon their thermodynamic stability at the firing temperature.

7) Several different types of species, including various solid state defects, diffuse and form a phase boundary of reaction, which may further react to form specific compositions.

Although we have covered mechanisms relating to solid state reactions, the formation and growth of nuclei and the rate of their growth in both heterogeneous and homogeneous solids, and the diffusion processes thereby associated, there exist still other processes after the particles have formed. These include sequences in particle growth, once the particles have formed. Such sequences include:

Impingement of nuclei
Ostwald Ripening (coarsening)
Sintering
Formation of Grain Boundaries

We will reserve a discussion of these subjects to the appendices of this chapter. Any student who has interest in these subjects amy brouse in the descriptions given.

## PROBLEMS FOR CHAPTER 4

1. Write equations for all of the phases changes given in 4.1.1. Use $\Delta$ for heat added or subtracted, as needed.

2. Write equations for the four (4) primary types of heterogenous solid state reactions usually encountered. Do not use those already given in Table 4-1.

We can distinguish between the interface structure of the nucleus, **N**, and the crystal being transformed, **A**. This can be seen in the following diagram:

Ap.1.12.-

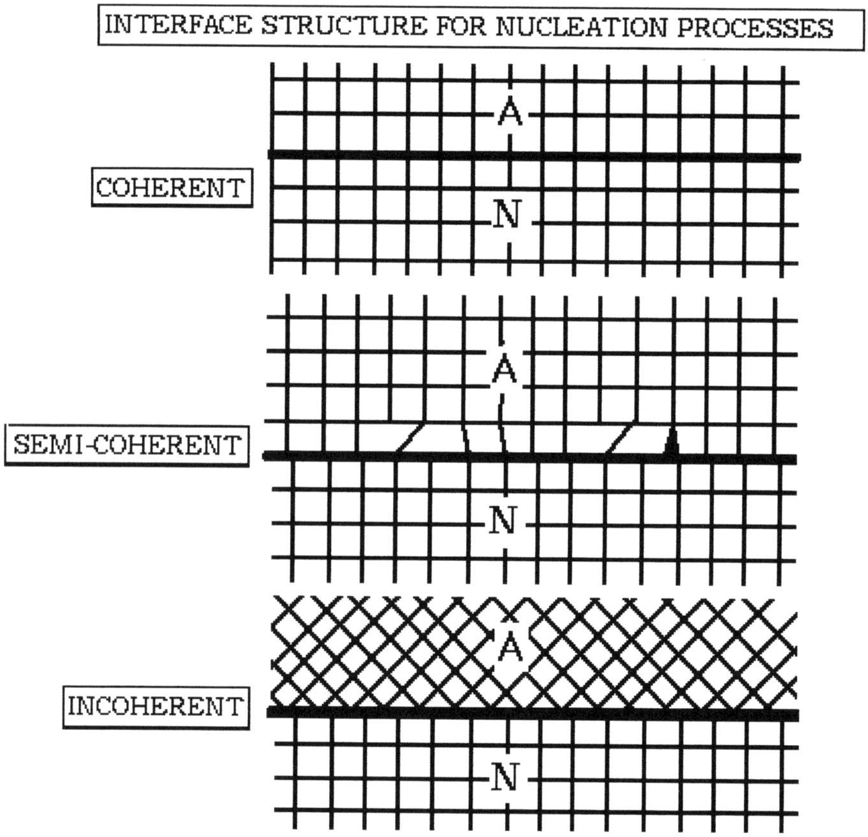


Here, we show three differences in the interface between the nucleus, **N**, and the original crystal, **A**. We find that in the first case, the lattices match fairly closely and are coherent. In the second case, there is some correspondence between the lattices. But the incoherent case shows little matching of the two lattices.

It is therefore not surprising that the interface structure has a large effect on the interfacial energy, but also upon the thermodynamics of change which influences the rate of formation of embryos, **E**, which then form nuclei.

The internal energy ($\Delta G_{max}$) of the embryo is found in the same way:

Ap.1.13.- $W \equiv \Delta G_{max} = 1/3\ (4\pi\ r_{crit}\ \gamma) = 16/3\ \{\ \pi\ \gamma^3\ /\ \Delta G_V\}$

where $G_V$ is the difference of free energy between the nucleus within the crystal; N is the number of nuclei; $\lambda$ is the nuclei volume; $\beta$ is a shape factor; $\sigma$ is the interfacial tension or surface energy; and $\gamma$ is the interface **exchange** energy. Using thermodynamics, we can also define:

Ap.1.14.- $\Delta G_V \equiv RT \ln P/P_{equil} \cdot 1/V_{crit}$ , then:

$$\Delta G_V = H_V \left\{ (T - T_{equil}) / T_{equil} \right\}$$

where equil. refers to the equilibrium state, $V_{crit}$ is the critical volume of the embryo, and both $G_V$ and $H_V$ still refer to the difference between the **states** of B and C. Accordingly, it is easy to arrive at:

Ap.1.15. $I = Z \exp.(- E_o\ /\ RT)\ \exp.(- W/RT)$

where $E_o$ is the embryo motion energy. If $N_o$ is the number of original embryo nuclei present, then:

Ap.1.16.- $I = dN_n/dt = k_1\ (N_o - N)^{\delta}$

where $k_1$ is a rate constant and $\delta$ is equal to the number of steps involved in the interface change of structure. Thus, we have:

Ap.1.17.- $N_n = N_o\ [\ 1 - \exp.\ (-k_1\ t)]$ or: $I = k_1 N_o\ \exp.\ (-k_1\ t)$

These equations allow us to calculate both numbers of embryos being formed and the energy involved in their formation. Thus, the interface structure affects the **rate** of nuclei formed and the rate of transformation of **A** to **N** (which then grows to form the "B"-phase). We can now correlate Ap.1.12. and $\delta$ of Ap.1.16. with the cases given in Ap.1.12. Note that we have assumed that $N^{2/3} / N \cong 1$ in the calculation.

## Appendix II

Mathematics Associated with Incipient Nuclei Growth

Consider the transformation to form a nucleus in terms of the rate instead of a volume. We define:

Ap.2.1.- Rate of nucleation(I) $\equiv$ the number of nuclei formed per unit time per unit volume.

If the nucleation is random, we will have the **homogeneous** case; if it is specific (with foreign or phase boundary walls), we have the **heterogeneous** case, vis-

Ap.2.2.

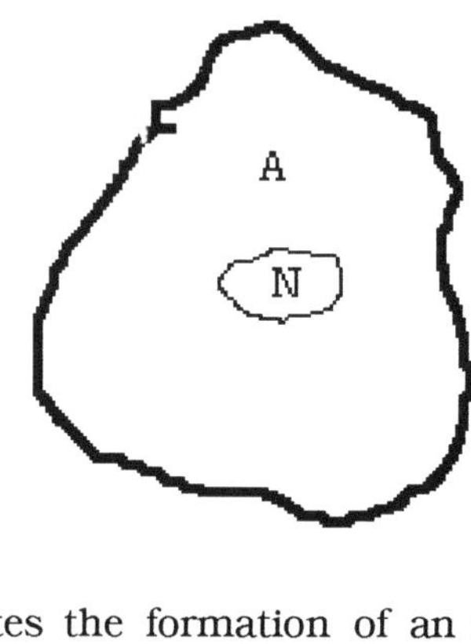


The above drawing illustrates the formation of an nucleus **N** within the crystallite, **A**. (not drawn to scale in the above diagram). The change in free energy in going from A to N will be a function of both volume and surface area (shape). We must discriminate between the interfacial surface energy between A and N, and the internal energy of N. Thus, the change in free energy will be:

Ap.2.3.- $\Delta G = N_n (G_V + \sigma) \lambda + \beta N^{2/3} \gamma$

where $G_V$ is the difference of free energy of N within A; $N_n$ is the number of nuclei; $\lambda$ is the nuclei volume; $\beta$ is a shape factor; $\sigma$ is the interfacial tension or surface energy; and $\gamma$ is the interface **exchange** energy. The equation Ap.1.3., given above is used here, i.e.-

Ap.2.4.- $$V(t) = \int \{ \theta [\int xdt ]^2 (dN/dt)_{t=y} \} dy$$

This is the general form in case we do not have spherical nuclei. If we do:

Ap.2.5.- $$V(t) = \int \{ \theta \int \{ 4\pi/3 [ \int xdt ]^3 (dN/dt)_{t=y} \} dy$$

We will assume a spherical shape factor and use the latter of the two equations. Since the first part of Ap.2.4. is the volume factor while the last part is the surface factor, we can do this with some justification. For a spherical nucleus, i.e.- Ap.2.5., we get:

Ap.2.6.- $$\Delta G = N_i \, 4/3 \, \pi r^3 (G_V + \sigma) + 4\pi r^2 N^{2/3} \gamma$$

We now differentiate Ap. 2.6. so as to obtain zero surface energy, i.e.-

Ap.2.7.- $$d(\Delta G)/dr = 0 = 4\pi N_i r^2 (G_V + \sigma) - 8 \pi r N^{2/3} \gamma$$

If we have **zero surface energy**, then $d(\Delta G)/dr = 0$ and $\sigma = 0$. Thus we have:

Ap.2.8.- $$4\pi r^2 n_i G_v + 8 \pi r N^{2/3} \gamma = 0$$

or: $$r_{crit} = - 2 \gamma / \Delta G_V$$

which is the Gibbs-Thompson equation defining the critical radius required for an embryo to form. Note that we have assumed that $N^{2/3}/N \cong 1$ in the calculation. The internal energy ( $\Delta G_{max} = W$) is found in the same way:

Ap.2.9.- $$W = 1/3 (4\pi r_{crit} \gamma) = 16/3 \{ \pi \gamma^3 / \Delta G_V \}$$

where we have used similar criteria to simplify the resulting equation. Since from thermodynamics:

Ap.2.10.- $\Delta G_V = RT \ln P/P_{equil} \cdot 1/V_{crit}$ , then:

$$\Delta G_V = H_V \left\{ (T - T_{equil} / T_{equil} \right\}$$

where equil. refers to the equilibrium state, $V_{crit}$ is the critical volume, and both $G_V$ and $H_V$ still refer to the difference between the **states** of A and B. Using equations already developed, it is easy to arrive at:

Ap.2.11.- $I = Z \exp(- E_o / RT) \exp( - W/RT)$

where $E_o$ is the embryo motion energy. If $N_o$ is the number of original embryo nuclei present, then:

Ap.2.12.- $I = dN/dt = k_1 (N_o - N)^{\delta}$

where $k_1$ is a rate constant and $\delta$ is equal to the number of steps involved in the interface change of structure. Thus:

Ap. 2.13 - $N = N_o [ 1 - \exp(-k_1 t)]$ or: $I = k_1 N_o \exp(-k_1 t)$

We note that the growth of nuclei is associated with the motion of a phase boundary.

This completes our analysis of the growth of embryos.

## Appendix III
### HOMOGENEOUS NUCLEATION PROCESSES

In a manner similar to that used for heterogeneous reactions, we can define diffusion and nuclei growth for a homogeneous solid. We have already stated that homogeneous nucleation can be contrasted to heterogeneous nucleation in that the former is random within **a single** compound while the latter involves more than one phase or compound.

Homogeneous nucleation might apply to precipitation processes where homogeneous nucleation must occur **spontaneously** before any precipitation process can occur. Since precipitation of compounds from solution is a common industrial process, we include it here.

To begin, we know that a change in free energy must occur simultaneously as a change in phase (nucleation) occurs. This will be related to the total volume, V, and the total surface area of both nuclei and particles, Σ, vis-

Ap.3.1.- $\Delta G = V_{\Delta g} + \Sigma\sigma$

where $V_{\Delta g}$ is the free energy per unit volume and σ is the surface energy of each nuclei. If we have spherical nuclei, then:

Ap.3.2.- $\Delta G = 4/3\, \pi\, r^3\, \Delta g + 4\pi\, r^2\, \sigma$

If the total free energy, ΔG, does not change with nuclei radius, i.e.- $d(\Delta G)/dr$ is defined as zero, then:

Ap.3.3.- $r^*_{crit} = -2\,\sigma / \Delta g$

where $r^*_{crit}$ is the critical radius for nucleus formation (Note the similarity of this equation to the Gibbs-Thompson of Ap.1.10. which describes that of the critical radius of nuclei growth between heterogeneous solids).

In the following table, given on the next page, we summarize the methods used to obtain the necessary equations describing nuclei growth and the kinetics of growth.

In the last equation, I is still defined as:

I ≡ Number of nuclei per Unit time per Unit volume.

We find that the quantity, $NS^* f_e$, is approximately equal to $10^{46}$ per cc. per second.

Table 4-3

Derivation of Formulas Describing Homogeneous Nuclei Formation

| | |
|---|---|
| 1. Using the equations in Ap.3.1. & Ap.3.2., i.e.- $\Delta G = V_{\Delta g} + \Sigma\sigma$ and $\Delta G = 4/3\ \pi r^3\ \Delta g + 4\pi r^2 \sigma$ we combine them to obtain: | $\Delta G^* = 16\pi\ \sigma^3 / 3(\Delta g)^2$ (where $\Delta G^*$ = change in free energy at the critical radius, $r^*$, for nucleus formation). |
| 2, We then define: $\Delta g \equiv \Delta h(T-T_o)/T_o = \Delta h(\Delta T/T_o)$ to obtain: | $\Delta G^* = 16\ \pi\ \sigma^3\ T_o^2 / 3\ (\Delta h)^2 \Delta T^2$ ($\Delta h$ is the enthalpy of the individual nuclei) |
| 3. The number of nuclei can be described by a Boltzmann distribution, i.e.- | $N^* = N \exp. \Delta G^* / kT$ |
| 4. By defining dN/dt as a frequency (t is time), we get: | $dN/dt = S^* f_e \exp. \Delta G^* / kT$ ($G^*$= free energy of the nucleus, $S^*$ = critical surface interface $f_e$ = **interfacial energy** |
| 5. Combining these two equations finally gives us: | $I \equiv dN^*/dt = NS^* f_e \exp.(-\ \Delta G^*/kT + \Delta G_D / kT)$ |
| | |

We can solve this graphically, as follows:

Ap.3.4.-

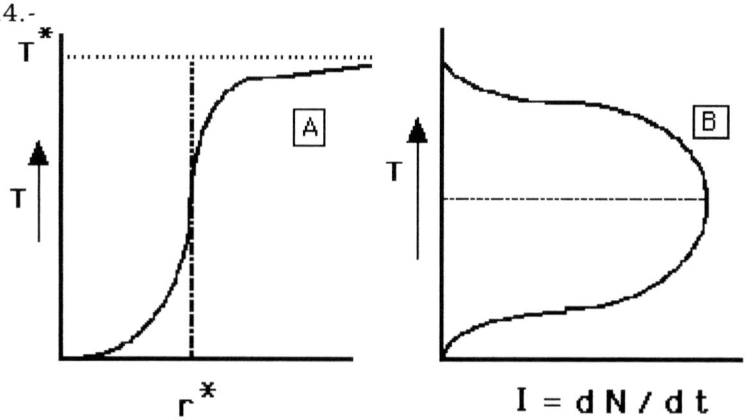

In this case, we plot T vs: r* so as to determine the value of T* and T vs: I = dN/dt, i.e.- the change in numbers of nuclei with time so as to determine the temperature of maximum formation of nuclei.

It can be seen in **A** that the value of r* is remarkably constant over a wide range of temperatures, but that it starts to approach infinity at some critical temperature, T*. In contrast, the number of nuclei produced in **B** is maximum at some particular temperature. This shows us that although the critical radius does not change significantly over a wide range of temperature, the numbers of homogeneous nuclei produced is very dependent upon temperature.

Let us now examine transformation kinetics for the general case, that is, the homogeneous case for **nuclei growth** after they have formed. We assume:

a. The nucleation is random.

b. The nucleation and growth rate are independent of each other.

c. The spheres grow until impingement.

The general equations we will use are:

Ap.3.5.- $dx/dt = k_1 (1-x)^n$ and: $x = 1 - \exp. (-k_1 t)^t$

where all terms are the same as already given in Section **3.3.** We solve these equations by a Taylor series, i.e.-

Ap.3.6.- $-\ln(1-x) = 1 + kt/2! + kt^2/3! + kt^3/4!$

If we now define $\tau$ as an instantaneous **moment of time**, we can get:

Ap.3.7.- $(dx/dt)_{t=\tau} = K n e^{-k_1 t}$

This can be arranged to:

Ap.3.8.- $(dx/dt)_{t=\tau} = nk_1 t^{n-1} (1-x)$

If x is small, then:

Ap.3.9.- $dx/dt = nk_1 t^{n-1}$

By defining spheres which are nucleated at time, $\tau$, the time of growth becomes: $(t - \tau)$. If we now define $x_{ext}$ as the **extrinsic amount** per nucleus before growth starts (note that it is not a radius; in fact, it is the same x given in Ap.3.5.), then we have:

Ap.3.10.- $x_{ext} = \int 4/3 \pi U^3 (t - \tau) I \, d\tau$

where U ≡ volume of the nuclei. If we then further define $x_f$ as the **final amount of volume per sphere** at impingement where growth stops, i.e.- $x_f = 1 - x_{ext}$, then we get the Johnson-Mehl equation for **slow nucleation**:

Ap.3.11.- $x_f = 1 - \exp. \{( - 1/3 \pi I U^3 ) t^4\}$

If the nucleation is **fast,** the Avrami equation is operative:

Ap.3.12.- $x_f = 1 - \exp. \{( - 1/3 \pi N_v U^3 ) t^3\}$

where $N_v$ is defined as the number of nuclei per unit volume. It is these two equations which have been used extensively to describe homogeneous nucleation, particularly in precipitation processes.

## Appendix IV

### Diffusion Equations Related to Fundamental Vibrations of the Lattice

Referring to the Fick Equations of diffusion, let us now reëxamine $1/\tau$, the jump frequency, so as to relate diffusion processes to lattice vibration processes. The **reciprocal** of the time of stay is the jump frequency and is related to the diffusion coefficient by:

AP.4.1.- $D = a^2 / 6\tau = \alpha a^2 / \tau = \alpha \varpi a^2$

where $\varpi = 1/\tau$ , and $\tau$ is the jump frequency (The other factors remain the same as already defined). $\varpi$ can be related to the free energy of the lattice via a Boltzmann distribution, i.e.-

AP.4.2.- $\varpi = \varpi_o \, \alpha \exp(-\Delta G_o / rt) = \varpi_o \, \alpha \exp(-\Delta S/R - \Delta H/RT)$

by a series of mathematical manipulations. It turns out that $\varpi_o$ **is the atomic vibrational frequency** or the **zero phonon mode** of the lattice, and that:

AP.4.3.- $\varpi_o = 1/2\pi \sqrt{(K/m)}$

where K is the elastic constant of the lattice and m is the **reduced** mass of the atoms composing the lattice. Note that we have just related the diffusion coefficient, D, to the zero-phonon mode of the lattice. This is reasonable since the jump frequency ought to be a function of the phonon vibrational constant as well.

# Chapter 5

## Particles and Particle Technology

Solids appear in one of two forms, either as crystals or powders. The difference is one of size, since many of the powders we use are in reality very fine crystals. This, of course, depends upon the manner in which the solid is prepared. Nevertheless, most solids that we encounter in the real world are in the form of powders. That is, they are in the form of discrete small particles of varying size. Each particle has its own unique diameter and size. Additionally, their physical proportions can vary in shape from spheres to needles. For a given powder, all grains will be the same shape, but the particle shape and size can be altered by the method used to create them in the first place. Methods of particle formation include:

Precipitation
Solid state reaction
Condensation

In this Chapter, we intend to investigate mechanisms of particle growth. We will, in the next chapter, address the methods used to form (grow) single crystals and how they have been utilized. Certain properties of particles, including measurement of size, will be discussed later in this chapter. But first, we need to define exactly what we mean by a particle.

In the previous chapter, we covered mechanisms relating to solid state reactions, formation of nuclei and the rate of their growth in both heterogeneous and homogeneous solids. Still other processes exist **after** particles have formed, including sequences in particle growth. We have already mentioned **precipitation** as one method for obtaining discrete particles. As indicated in the previous chapter, nucleation in precipitation processes can be homogeneous (no outside influences) or heterogeneous (by specific outside coercion). In precipitation processes to form a distribution of particle sizes, it is probably a combination of both mechanisms.

## 5.1. - SEQUENCES IN PARTICLE GROWTH

We have already shown that embryo formation leads to nucleation, and that this nucleation **preceeds** any solid state reaction or change of state. In a like manner, nuclei must form in order for any precipitation process to proceed. Once formed, these nuclei then grow until impingement of the growing particles occurs. Impingement implies that all of the nutrient supplying the particle growth has been used up. This mechanism applies to both solid state reaction and precipitation processes to form product particles. Then a process known as **Ostwald Ripening** takes over. This process occurs even at room temperature. The sequences in particle growth are:

5.1.1.- Embryo formation
Nucleation
Nuclei Growth
Impingement
Ostwald Ripening (coarsening)
Sintering
Formation of Grain Boundaries

We have already covered the first three in some detail. **Impingement** involves the point where the growing particles actually touch each other and have used up all of the nutrient which had originally caused them to start growing.

**Ostwald ripening** usually occurs between particles, following impingement, wherein larger particles grow at the expense of the smaller ones. One example of this would be if one had a precipitate already formed in solution. In many cases, the smaller particles redissolve and reprecipitate on the larger ones, causing them to grow larger. Interfacial tension between the particles is the driving force, and it is the surface area that becomes minimized. Thus, larger particles, having a total lower surface area, **increase** at the expense of numerous smaller particles which have a relatively high surface area.

Ostwald ripening differs from nuclei growth in that the relative size and numbers of particles change, whereas in nuclei growth, the numbers of particles growing from nuclei do not change. Sintering, on the other hand, is an entirely different process, and usually occurs when external heat is applied to the particles.

## 5.2. - SINTERING, SINTERING PROCESSES AND GRAIN GROWTH

Sintering of particles occurs when one heats a system of particles to an elevated temperature. It is caused by an interaction of particle surfaces whereby the surfaces fuse together and form a solid mass. It is related to a solid state reaction in that sintering is governed by diffusion processes, but no **solid state reaction,** or change of composition or state, takes place. The best way to illustrate this phenomenon is to use pore growth as an example.

When a system of particles, i.e.- a powder, is heated to high temperature, the particles do not undergo solid state reaction, unless there is more than one composition present. Instead, the particles that are touching each other will sinter, or fuse together, to form **one** larger particle. Such a mechanism is illustrated in the following diagram, given as 5.2.1. on the next page.

As can be seen, voids arise when the particles fuse together. The void space will depend upon the original shape of the particles. In 5.2.1., we have shown spherical particles which produce only a few voids. Additionally, an overall change in the total volume of the particles, and that of the fusion product, occurs. Mostly, the change is negative, but in a few cases, it is positive. Experimentally, this change is a very difficult problem to measure. What has been done is to form a long thin bar or rod by putting the powder into a long thin mold and pressing it in a hydraulic press at many tons per square inch. One can then measure the change of volume induced by sintering, by measuring a change in length as related to the overall length of the rod.

5.2.1.-

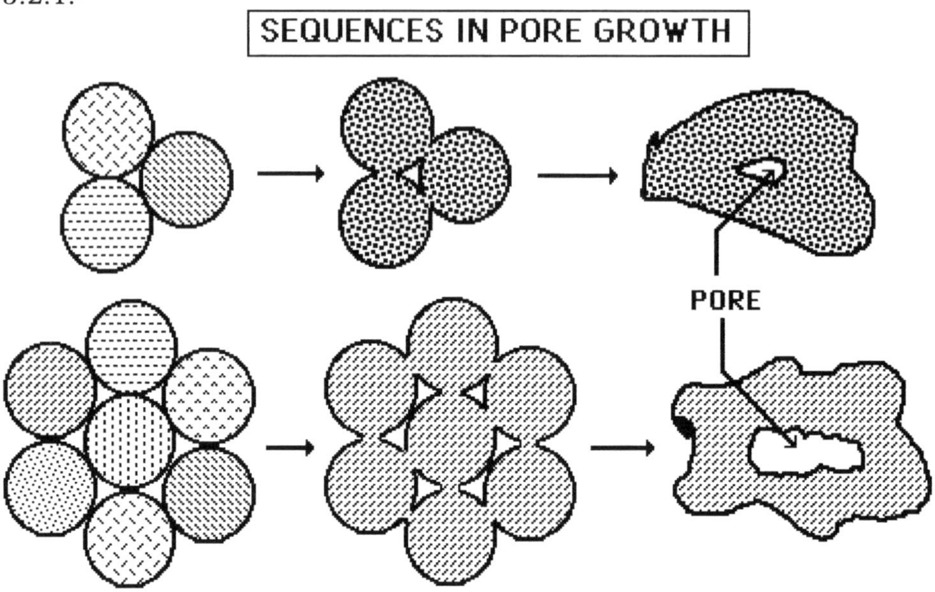


It has been found that the sintering of many materials can be related to a power law, and that the **shrinkage** , $\Delta L$, can be described by:

5.2.2.- $$\Delta L / L_o = k\, t^m$$

where k is a constant, $L_o$ is the original length, t is the time of sintering, and the exponent, m , is dependent upon the material being investigated. It has also been further demonstrated that a considerable difference exists between the sintering of **fine** and **coarse** particles, as shown in the following diagram, given as 5.2.3. on the next page.

In this case, the "fine" particles and the "coarse" particles were separated so that the difference in size between individual particles was minimized. That is, most of the individual particles in each fraction were almost the same size. Both the fine and coarse particles have a sintering slope of 1/2 but it is the coarse particles which sinter to form a solid having a density closest to

5.2.3.-

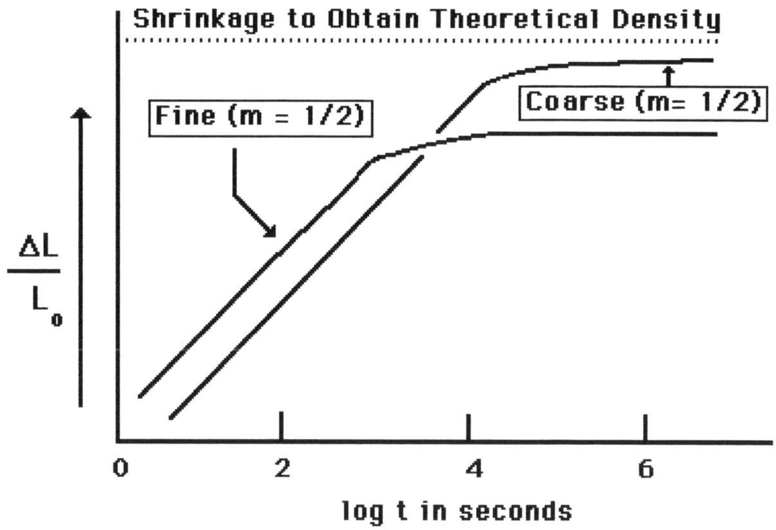


theoretical density. This is an excellent example of the effect of pore volume, or void formation, and its effect upon the final density of a solid formed by powder compaction and sintering techniques. Quite obviously, the fine particles give rise to many more voids than the coarser particles so that the attained density of the final sintered solid is much less than for the solid prepared using coarser particles. It is also clear that if one wishes to obtain a sintered product with a density close to the theoretical density, one needs to start with a particle size distribution having particles of varied diameters so that void volume is minimized.

The next subject we will discuss is that of grain growth. The simplest way to illustrate this factor is through the sintering behavior of aggregates. An aggregate is defined a large particle, composed of many small particles, as shown in the following diagram, given as 5.2.4. on the next page.

In this diagram, two steps are implicit. It is the aggregates, composed of very fine specks, which sinter to form larger grains (particles).

5.2.4.-

FORMATION OF GRAIN BOUNDARIES BY SINTERING

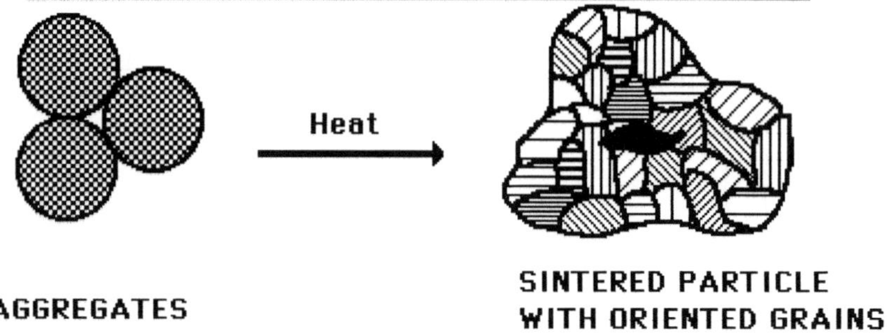


But, since many particles are growing at the same time, growth occurs until impingement. The formation of boundaries between the particles (grains) growing thus results. It is these grains which form the final sintered whole. Note that the crystallographic orientation of each grain differs from that of its neighbors, as illustrated in the above diagram.

When a system of **very fine** particles is formed, the interfacial tension is high due to the very high surface area present. **Agglomerates** (weak surface energy interchange) will form, or **aggregates** (strong surface energy interchange) can result, especially if the Ostwald ripening mechanism is slow, or is inhibited. Immediate removal of a precipitate from its "mother liquor" is one example where the likelihood of aggregate formation is enhanced. Sintering then produces both pore growth and grain growth. This mechanism also applies to powder compaction processes where aggregates may be present in the powder. Sintering then leads to grain growth as well.

It should be clear, then, that the precipitation process needs to be controlled carefully in order to produce a material composed of particles of a desired configuration. Both Ostwald riping and sintering can be utilized to obtain a particle of desired size, dimensions and particle habit. Industrial technologists have taken advantage of these particle forming and altering mechanisms. One example of this type of particle growth is described as follows.

In fluorescent lamps, a layer of phosphors is applied to the inside of a glass tube by means of a suspension of particles, i.e.- the halophosphate phosphor having a composition of: $Ca_5F,Cl(PO_4)_3$: $Sb^{3+}$ : $Mn^{2+}$ (plus minor amounts of other phosphors to achieve certain "colors"). It has been determined that the lamp brightness and duration of light output (maintenance) is highly dependant upon how well the the internal surface of the glass is covered by the particles. By maintaining precipitation conditions so that small thin squares of $CaHPO_4$ result from the precipitating solution, a maximum coverage of the glass is achieved. Precipitation occurs by adding a solution of $(NH_4)_2HPO_4$ to a solution of $CaCl_2$. The resulting precipitate is consists of very fine particles and is not very crystalline. As a matter of fact, the particles were usually ill-defined and bordered upon amorphous. The solution temperature is then raised so that Ostwald ripening can occur. As the larger crystallites begin to grow, the smaller ill-defined crystallites, having a much larger surface area, dissolve and reprecipitate upon the larger ones. The resulting single-crystal squares, i.e.- □, have an average size of about 25μm. The solid state reaction to form the halophosphate phosphor is:

5.2.5.- $6\ CaHPO_4 + 3\ CaCO_3 + CaF_2 \Rightarrow 2\ Ca_5\,F\,(PO_4)_3$

where we have not shown the $Sb_2O_3$ and $MnCO_3$ added as "activators". The halophosphate thus produced follows the crystal habit of the major ingredient, in this case that of $CaHPO_4$ habit. When applied, the thin squares lie flat and overlap on the glass surface. During sintering at about 1200 °C, the $CaHPO_4$ does not disintegrate but undergoes an internal rearrangement to form $Ca_2P_2O_7$ while maintaining the same crystal habit. The $CaCO_3$ disintegrates into small particles (like $BaCO_3$) while $CaF_2$ exhibits a sublimation pressure at the firing temperature. The resulting solid state reaction to form the halophosphate product thus depends upon the crystal habit of the major ingredient, $CaHPO_4$.

The same type of mechanisms apply to other materials. Even where a metal is melted and then cast, nucleation leads to formation of many fine particles in the sub-solidus (partially solidified) state. This leads to grain growth in the solid metal, thereby lowering its strength. Sometimes, special additives are added to the melt to slow nucleation during cooling, thereby increasing the strength of the metal product.

Another example is our old friend, $BaCO_3$. If we fire this solid compound in air at a very high temperature for a long time, we get several changes. First, it decomposes to very fine particles of BaO. These fine particles have a large surface area, and with continued firing sinters to form larger particles. Eventually, the particles get big enough, and the porosity decreases to the point where grain boundaries begin to form between particles. The grains sinter together to form a large particle with many grain boundaries. It should be again be emphasized that each grain in the large particle is essentially a small single crystal with its lattice oriented in a slightly different direction from that of it neighbors (see 5.2.4.).

Let us now consider the thermodynamics of sintering. There are two types of sintering which are distinguished by the change in volume which occurs. These are:

5.2.6.- NO SHRINKAGE : $dV/dt = 0$

WITH SHRINKAGE : $dV/dt = f(V)$

As we have already said, the change in volume, from initial state to final state, can be positive or negative, but is usually negative. The driving force is a **decrease** in Gibbs free energy, $\Delta G$. It is related to both the interfacial tension (surface energy), $\gamma$ , and the surface energy of the particles (which is related to their size), viz-

5.2.7.- $dG = \gamma \, dA$

Consider a more familiar example, that of a droplet sitting upon the surface of a **liquid.** The droplet has a radius, r, and there are n-moles of liquid within it with a molal volume, V. To form the droplet requires an amount, nV, of the liquid, where V is the fractional **molar** volume of the droplet. This gives us:

5.2.8.- $nV = 4/3\ \pi\ r^3$

and the change in free energy to form the droplet is:

5.2.9.- $dG = \Delta G\ dn = \gamma\ dA$

If we now differentiate these equations, i.e.-

5.2.10.- Volume of Droplet: $n = 4\pi\ r^3 / 3\ V$ so that: $dn = 4\pi\ / V\ r^2\ dr$

and:

Surface Area of Droplet: $A = 4\pi\ r^2$ so that: $dA = 8\pi\ r dr$

we can put all of the equations together so as to yield the Kelvin equation for change in free energy as a function of the radius of the **spherical** particle:

5.2.11.- $\Delta G = 2\ \gamma\ V / r$

One can do the same for a cubic particle, in fact for any shape factor. Since the chemical potential, $\mu$ , is related to $\Delta G$, we use the following equations:

5.2.12.- $\mu - \mu_o = \Delta G = RT \ln p/p_o$

$$2\ \gamma\ V / r = RT \ln p/p_o$$

or: $p = p_o \exp\ (2\ \gamma\ V / r\ RT)$

If we now define $\Delta p = p - p_o$ , then:

5.2.13.- $$p/p_o = 1 + \Delta p/p_o$$

Mathematically, $\ln(1 + \Delta p/p_o) \cong \Delta p/p_o$ (within 5% if $\Delta p/p_o \geq 0.1$). Thus we get the approximate equation:

5.2.14.- $$\Delta p/p_o = 2\gamma V / r \cdot 1/RT$$

**It is this equation which has been used more than any other to evaluate sintering.** To evaluate its use, consider the following example.

Alumina , $Al_2O_3$ , is a very refractory compound. It melts above 1950 °C. and is not very reactive when heated. If we attempt to sinter it at 1730°C., we find the following values to apply:

5.2.15.-
$$\Delta p/p_o \cong 0.1$$
$$\gamma \cong 2000 \text{ dyne/cm.}$$
$$V = M/d = 25.4 \text{ cc./mol}$$
$$R = 8.3 \times 10^{-7} \text{ erg/mol}$$
$$r = 6 \times 10^{-6} \text{ cm} = 0.06 \text{ micron}$$

What this means is that if the alumina particles are smaller than about 0.06 micron, they **will not sinter.** Even though the above is an approximation, the specific case for alumina has been confirmed experimentally.

THUS, WE HAVE SHOWN THAT IF THE PARTICLES ARE TOO SMALL, THEY WILL NOT SINTER. NORMAL NUCLEI GROWTH THROUGH DIFFUSION PROCESSES REMAINS THE NORM UNTIL THE PARTICLES GET LARGE ENOUGH TO SINTER.

Let us now examine why this mechanism might be true. In the case of sintering of spheres, we can define two cases as before, that of no shrinkage and that of shrinkage, both as a function of volume. If we have two spheres in direct contact, we can define certain parameters, as shown in the following diagram:

5.2.16.-

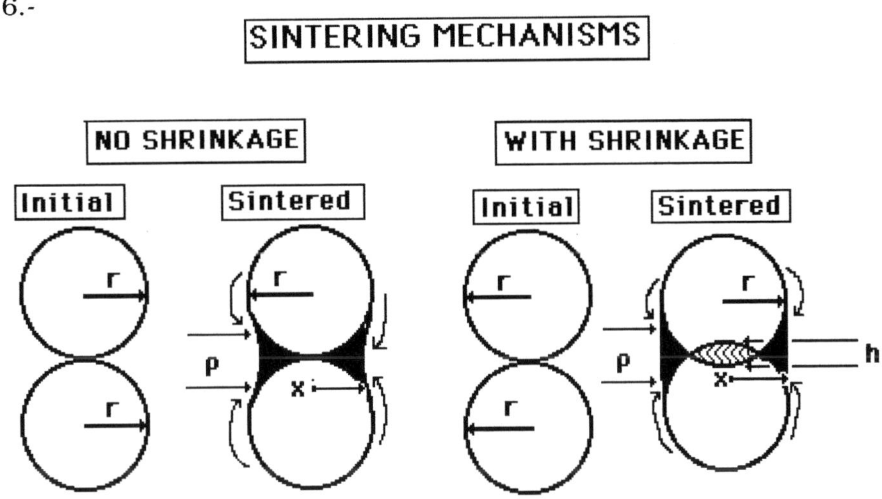


We start with spheres of radius, r, in direct contact. The two cases shown are:

1) **no shrinkage**
2) **with shrinkage.**

Actual sintering occurs by flow of mass from each sphere to the mutual point of contact, which gradually thickens. We can estimate the volume of mass, V, at the contact area, A, in terms of the following parameters: r, the radius of the spheres; ρ , the thickness of the layer buildup; and x, the radius of contact of the built-up layer. If we have shrinkage, then we must also evaluate h, the amount of shrinkage, shown above as the height of interlinking layer.

This model brings us to an important point, vis: If the spheres are too small, there can be little mass flow to the area of joining (sintering) of the spheres. Actually, it is the rate of flow of mass to the joining area that is important and the area of touching of the spheres will determine this.

***It is therefore logical that a size limitation should apply to the case of sintering and its mechanisms.*** If the spheres are too small, then there is not enough touching area and volume for the sintering mechanism to occur.

The actual values of the sintering parameters have been found to be:

5.2.17.-

VALUES OF SINTERING PARAMETERS

| | V | h | A | ρ |
|---|---|---|---|---|
| No Shrinkage | $\pi x^2 / 2r$ | 0 | $\pi^2 x^2 / r$ | $x^2 / 2r$ |
| | | | | |
| With Shrinkage | $\pi x^2 / 2r$ | $x^2 / 2r$ | $\pi^2 x^3 / 2r$ | $x^2 / 4r$ |

**Most** systems exhibit shrinkage in sintering and it has been found that the following equation applies:

5.2.18.- $\Delta L / L_o = h / r$ and $x = c\ t^m$

where t is the time of sintering, h is a characteristic sintering constant for **shrinkage**, and m is an exponent dependent upon the mechanism of sintering.

We have indicated that flow of mass is important in sintering. There are several operative mechanisms which are determined by the type of material involved. Actually, the study of sintering deserves a separate chapter, but we will only summarize the major mechanisms that have been observed for the constants of 5.2.18. These are given in 5.2.19. on the next page.

Note that the values of h depends upon the material and the mechanism of sintering. Also, these values depend upon r and x, the radii of the particles and the radius of joining. These equations apply only to spheroids and a shape factor must be considered as well.

5.2.19.- SINTERING MECHANISMS

| Mechanism | m | h | Material |
|---|---|---|---|
| Viscous flow | 1/2 | 1 | Glass |
| Evaporation-Condensation | 1/3 | $\Delta L/L_o = 0$ | NaCl |
| Grain boundary diffusion by Vacancy Formation | 1/4 | 1/2 | ----- |
| Volume diffusion | 1/5 | 2/5 | Cu, Ag |
| Grain boundary diffusion by Intersitial Formation | 1/6 | 1/3 | ------- |
| Surface diffusion | 1/7 | $\Delta L/L_o = 0$ | Ice |

The derivation becomes much more difficult and is beyond the scope of this chapter. The reader is referred to specific references on sintering and sintering mechanisms.

## 5.3. - PARTICLE SIZE

Most solids that we actually deal with are powders. The powder is composed of discrete particles, and each particle may be single crystal or contain grain boundaries. The size of the particles are of interest to us because many of their chemical and physical properties are dependent upon particle size. For example, the "hiding power" of pigments is dependent upon the size of the pigment particles, and in particular the distributions of sizes of the particles. There is an optimum particle size distribution to obtain maximum hiding power, depending upon the specific application. Pharmaceuticals are another product where particle size (PS) and particle size distribution (PSD) are important. The effectiveness of dosage and ingestion depends upon the proper PSD. Particles are usually defined by their diameter since most are spheroid. We use the micron as a base, i.e.-

5.3.1.- 1 μ (micron) = $10^{-6}$ meter = $10^{-4}$ centimeter.

The following defines particle ranges that we usually encounter in solid state chemistry:

5.3.2.-

PARTICLE RANGES

| Range | Centimeters | μ | Description |
|---|---|---|---|
| Macro | 1.0 - 0.05 | $10^4$ - 500 | Gravel |
| Micro | 0.01- 0.0001 | 100 - 1.0 | "Normal" |
| Sub-micro | 0.0001- $10^{-7}$ | 1.0 - 0.001 | Colloidal |

It is easier to describe PS in terms of microns rather than meters or centimeters.

In the real world, we encounter particles in a variety of forms, although we may not recognize them as such. In the following diagram, given as 5.3.3. on the next page, are summarized many of the particles of interest. At the top of the diagram, the particle diameters in microns is shown. Immediately below are standard screen sizes, including both U. S. and Tyler standard mesh. Screens are made by taking a metal wire of specific diameter and cross-weaving it to form a screen with specific hole sizes in it. Thus, a 400 mesh screen will pass 37μ particles or smaller, but hold up all those which are larger. A 60 mesh screen will pass up to 250μ particles, etc. (U. S. Screen Mesh).

This diagram places and defines the size of most of the particles that we are likely to encounter in the real world. On the left is a comparison of angstroms (Å) and microns up to 1 μ = 10, 000 Å . In the middle of the section, "Equivalent Sizes" , is a comparison of sizes and "theoretical mesh". Why the latter term is sometimes used will be discussed later. In "Technical Definitions", size ranges are given for various types of solids in gases, liquids in gases, and those for soil. In addition, we have a separate classification for water in air (fog). Finally, there is a classification for various commercial products and by-products we might encounter, including viruses and bacteria. Note the varied sizes of particles that we normally encounter. This listing is not meant to be all- inclusive but outlines many areas of technical interest where the sizes of particles are important.

5.3.3.-

## Size Range of Typical Particles and Dispersoids

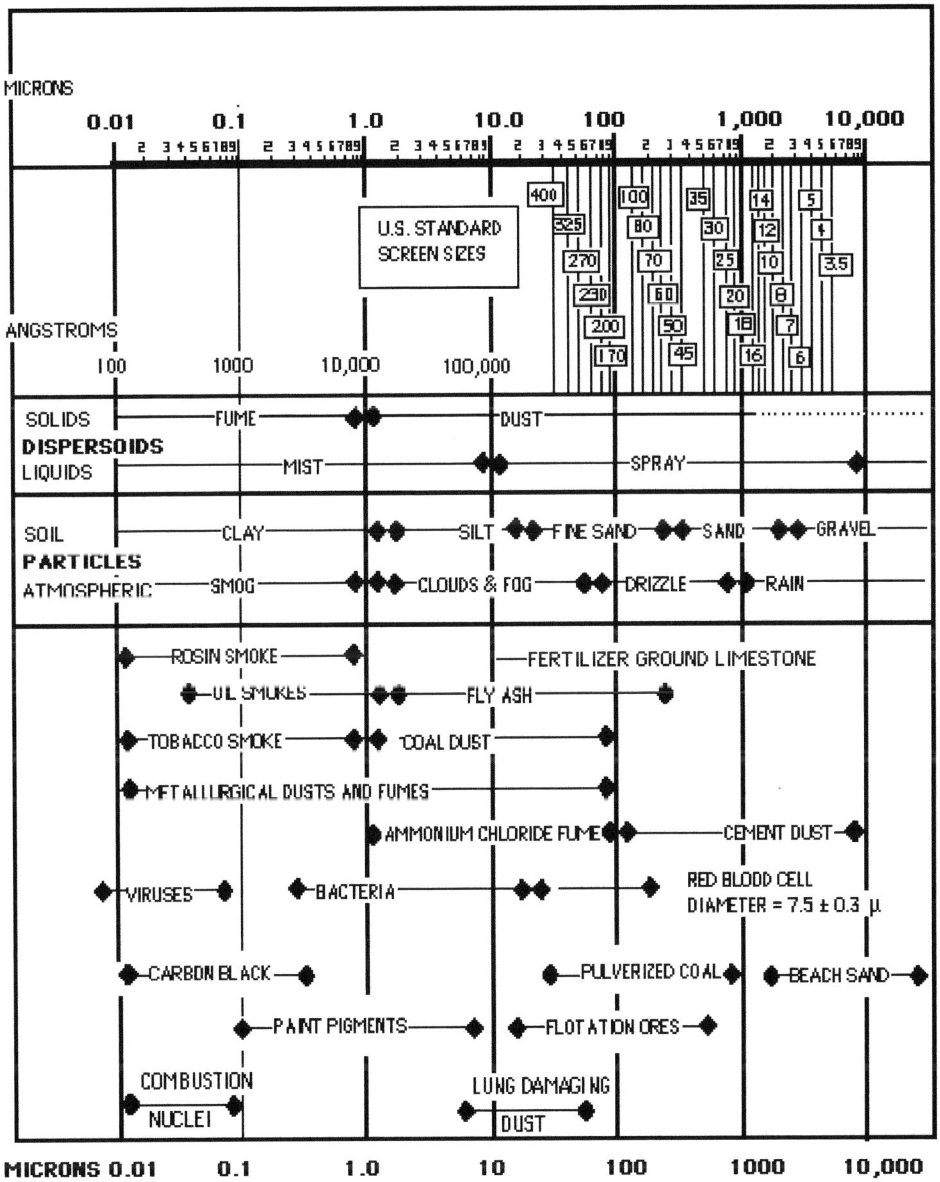

Now let us begin to more clearly define the aggregate properties of particles. Let us take a 1.00 cm. cube and cut it into **fourths**, viz-

5.3.4.-

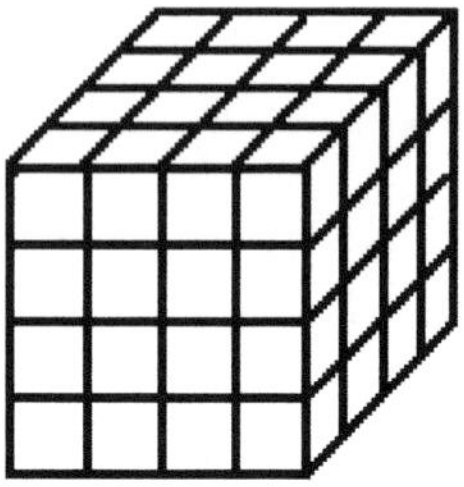

NUMBERS = $4^3$ = 64

SURFACE = $(64)(6)(1/4)^2$ = 0.0096 sq. m.

Now we take the same cube and divide it into 1.0 μ cubes, with the following result:

5.3.5.- 1μ cubes = $10^4$ cuts in 3-dimensions

Diameter = $1.0 \times 10^{-6}$ $m^3$ for each particle

Number = $(10^4)^3$ = $1.0 \times 10^{12}$ particles

Surface = $(6)\ (1.0 \times 10^{12})\ (1.0 \times 10^{-6})^2$ = 6.0 sq. meters

We have discovered two facts about particles:

a) There are large numbers present in a relatively small amount of material (we started with a 1-cm. cube).
And:
b) The total surface area can be quite large and depends upon the size of the particles.

Since the rate of solid state reaction depends upon surface area, we can see that very small particles ought to react much faster than large particles.

Another observation is that solid state reactions involving very fine particles ought to be very fast in the beginning, but then will slow down as the product particles become larger. There are other properties of particles which we can think of, as follows:

5.3.6.- PROPERTIES OF PARTICLES

| Size | Surface area | Porosity | Aggregates | Agglomerates |
|---|---|---|---|---|
| Shape | Density | Pore size | Numbers | Size distribution |

At this point, we are most interested in the size of particles and how the other factors relate to the question of size. The next most important factor is shape. Most of the particles that we will encounter are spheroidal or oblong in shape, but if we discover that we have needle-like (acicular) particles, how do we define their average diameter? Is it an average of the sum of length plus cross-section, or what?

**5.4.** - PARTICLE DISTRIBUTIONS

Fundamental to particle technology is the factor,*size distribution.* By this, we mean **the number of each size** in a given collection of particles. From a close examination of 5.3.3., it is evident that we must deal with large numbers of particles, even when we have a small starting sample weight. This problem becomes clearer if we correlate numbers of particles with size, starting with a cube, one (1) cm in size, as in 5.3.4. , and make the cuts indicated in 5.4.1, given on the next page.

Having done so, we mix the seven (7) separate products. We could then plot the particle distribution as shown in 5.4.2., also given on the next page, where we have plotted the log of the size of particles in microns vs: the log of the number of particles created. Obviously, there is a linear relationship between the two variables. The other factor to note is that this distribution consists of specific (discrete) sizes of particles.

5.4.1.- Various Sized Cuts of a One Centimeter Cube

| Size | Number/ $cm^3$ |
|---|---|
| 1 μ | $1.0 \times 10^{12}$ |
| 10 μ | $1.0 \times 10^{9}$ |
| 50 μ | $8.3 \times 10^{6}$ |
| 100 μ | $1.0 \times 10^{6}$ |
| 250 μ | $6.4 \times 10^{4}$ |
| 500 μ | $8.3 \times 10^{3}$ |
| 1000 μ | $1.0 \times 10^{3}$ |

5.4.2.- A DISCRETE POPULATION OF PARTICLES

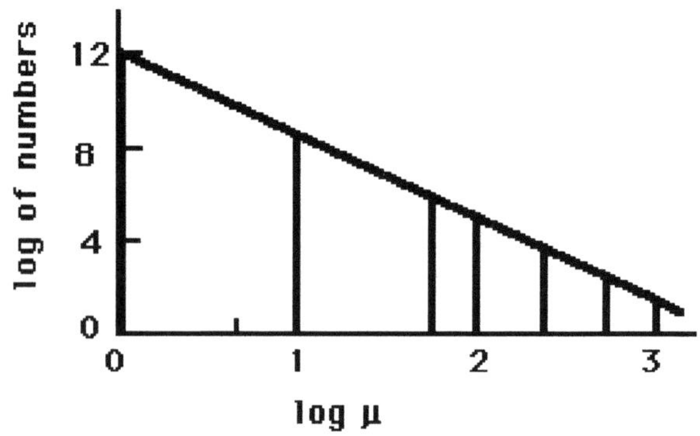


In Nature, however, we always have a **continuous** distribution of particles. This means that we have **all** sizes, even those of fractional parentage, i.e.- 18.56μ, 18.57μ, 18.58 μ, etc. (supposing that we can measure 0.01 μ differences). The reason for this is that the mechanisms for particle formation, i.e.- precipitation, embryo and nucleation growth, Ostwald ripening, and sintering, are ***random*** processes. Thus, while we may speak of the "statistical variation of diameters", and while we use **whole** numbers for the particle diameters, the actuality is that the diameters **are** fractional in nature. Very few "particle-size" specialists seem to recognize this fact. Since the processes are random in nature, we can use statistics to describe the properties of a population of particles, and the particle size distribution. Statistical calculation is well suited for this purpose since it was originally designed to handle large numbers in a population. It should be clear that we do have a population of particles in any given particle distribution.

## 5.5.- PARTICLE DISTRIBUTIONS AND THE BINOMIAL THEOREM

To describe particle distributions, we will use our own nomenclature, but will soon find that it is related to the science of statistics as well. We begin by the use of the Probability Law.

If we are given n-things where we choose x variables, taken r at a time, the individual **probability** of choosing (1 + x) will be:

5.5.1.- $$p_i = (1 + x)^n$$

This is the BINOMIAL THEOREM. Using a Taylor Expansion, we can find the total probability for items taken r at a time as:

5.5.2.- $$P(r) = \frac{\{n\}}{\{r\}} \; p^r \, (1 - p)^{n-r}$$

where:

5.5.3. $$\frac{\{n\}}{\{r\}} \equiv {}_nC_r = n! \,/\, r!\,(n\ r)!$$

If we now let n approach infinity, then we get:

5.5.4.- $$P(r) = \Sigma\, p_i = 1 \,/\, [\,2\pi\, np\,(1-p)]^{1/2} \cdot \exp\,(-\,x^2 \,/\, 2np\,(1-p)$$

One will immediately recognize this as a form of the BOLTZMANN equation, or the GAUSSIAN LAW. We can modify this equation and put it into a form more suitable for our use by making the following definitions.

First, we define a mean (average) size of particles in the distribution as $\bar{\mathbf{d}}$ , and then define what we call a "standard deviation" as σ, for the distribution of particles. From statistics, we know that this means that 68% of the particles are being counted (34% on either side of the mean, $\bar{\mathbf{d}}$), i.e. -

5.5.5.- $\Sigma\,(\bar{\mathbf{d}} \pm d_\sigma) = 0.68$

where $d_\sigma$ is the diameter measured to give ± 34% of the total number of particles. In a like manner:

5.5.6.-
$$2\,\sigma \equiv 0.954 = \Sigma\,(\bar{\mathbf{d}} \pm d_{2\sigma})$$
$$3\,\sigma \equiv 0.999 = \Sigma\,(\bar{\mathbf{d}} \pm d_{3\sigma})$$

Note that these standard deviations **only** apply **if** we have a Gaussian distribution. In this way, we can specify what fraction we have of the total distribution, or locate points in the distribution. By further defining:

5.5.7.-
$$n\,p_i \equiv \bar{\mathbf{d}}$$
$$\sigma \equiv (\,np\,(1\text{-}p\,)^{1/2} = \text{std. dev.}$$
$$r \equiv np + x = \bar{\mathbf{d}} + x = (d - \bar{\mathbf{d}})$$

we can obtain expressions to describe a particle distribution where we have used x as a deviation from the mean.

This then brings us to the expression for the GAUSSIAN PARTICLE SIZE DISTRIBUTION, as shown in the following:

5.5.8.-
$$P_r = \frac{1}{(2\pi)^{1/2}\,\sigma}\ \exp\left(\frac{-\,[\,d - \bar{\mathbf{d}}\,]^2}{2\,\sigma^2}\right)$$

$$\textbf{if:}\ \{\,-\infty < d < +\infty\,\}$$

Let us now examine a Gaussian distribution. A plot is shown as follows:

5.5.9.-

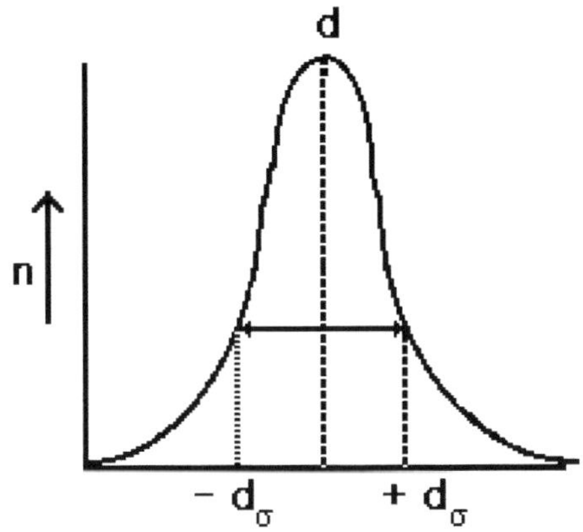


What we have is the familiar "Bell-Shaped" curve. This distribution has been variously called:

5.5.10.-

Gaussian
Log Normal
Boltzmann
Maxwell - Boltzmann

We shall use the term "Log Normal" for reasons which will clear later. It should be now apparent that we use terms borrowed from statistics and a statistical approach to describe a distribution of particles. The two disciplines are well suited to each other since statistics is easily capable of handling large assemblages, and the solid state process with which we deal are random growth processes which produce large numbers of particles.

Let us now give some examples of the log-normal distribution. A representation of several types of these distributions is given on the next page as 5.5.11.

In this diagram, the parameters of the various log normal distributions are

5.5.11.- Properties of Some Log Normal Particle Distributions

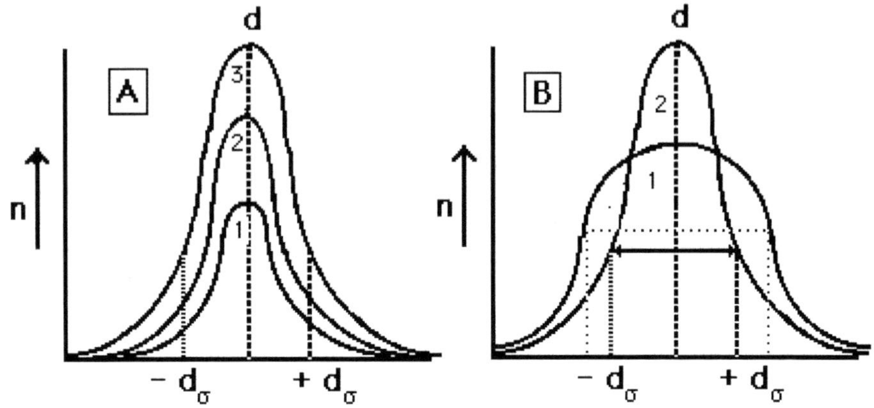


| Case A | | Case B |
|---|---|---|
| $d_1 = d_2 = d_2$ | | $d_1 = d_2$ |
| $\sigma_1 = \sigma_2 = \sigma_3$ | | $\sigma_1 \neq \sigma_2$ |
| $A_1 \neq A_2 \neq A_3$ | $(A \equiv \Sigma nd)$ | $A_1 = A_2$ |

compared. Here, d is the average particle size, is the spread of the particle size distribution and A is the area, i.e.- the total number of particles present in the particle distribution. Note that each distribution can look quite different from the others, depending upon the values of the parameters involved. Thus, in Case A, the three distributions differ in numbers only, the average diameters of the distribution being equal as well as the spread of the distribution, i.e.- σ. But in Case B, the two distributions have the same numbers of particles but not the same spread of particles. Yet, in all of these cases, each particle distribution remains a log-normal distribution and conforms to the definition of a Gaussian distribution. Thus, it should be clear that one cannot determine if a particle distribution is truly log-normal just by the **appearance** of plotted data. That is only decided by plotting the particle data on a log-normal plot and determining if the distribution conforms.

## 5.6. - MEASURING PARTICLE DISTRIBUTIONS

Our next task is to determine how to measure a PSD. We know that we can specify a mean diameter, but how do we obtain it? The following is a simple example illustrating one method of doing so.

Suppose we obtain a sample of beach sand. From 5.3.3., we can see that the particles are liable to range from about 8μ to 2400 μ. In order to generate a PSD, we must separate the particles. One way we can separate such particles is by seiving them. But we find that seives are only available in certain sizes, as shown in the following:

5.6.1.- COMPLETE LISTING OF ALL SEIVE SIZES AVAILABLE - (U.S. STD.)

| Seive No. | Microns | Permissible Variation(%) | Seive No. | Microns | Permissible Variation(%) |
|---|---|---|---|---|---|
| 3.5 | 5660 | ± 3 | 40 | 420 | ± 5 |
| 4 | 4760 | ± 3 | 45 | 350 | ± 5 |
| 5 | 4000 | ± 3 | 50 | 297 | ± 5 |
| 6 | 3360 | ± 3 | 60 | 250 | ± 5 |
| 7 | 2830 | ± 3 | 70 | 210 | ± 5 |
| 8 | 2380 | ± 3 | 80 | 177 | ± 6 |
| 10 | 2000 | ± 3 | 100 | 149 | ± 6 |
| 12 | 1680 | ± 3 | 170 | 88 | ± 6 |
| 14 | 1410 | ± 3 | 200 | 74 | ± 7 |
| 16 | 1190 | ± 3 | 230 | 62 | ± 7 |
| 18 | 1000 | ± 5 | 270 | 53 | ± 7 |
| 20 | 840 | ± 5 | 325 | 44 | ± 7 |
| 25 | 710 | ± 5 | 400 | 37 | ± 7 |
| 30 | 590 | ± 5 | | | |
| 35 | 600 | ± 5 | | | |

There are two types of standard screens, the U.S. and the Tyler standard screens. We have given the U.S. screen values. Those of the Tyler are very similar, e.g. - #5 Tyler = 3962 μ , # 20 = 833 μ, and #400 = 38 μ.

To screen our sand sample, we choose to employ the following seives:

#10, # 18, #20, #30, #40, #50, #80, #100,
#170, #200, #270, #325 & #400.

Now if we screen the sand with the #10 screen, all particles larger than 2000 ± 60 μ will be **retained** upon the screen. The next step is to rescreen the part that **passed through** the #10 screen, using the #18 screen, to obtain a **fraction** which is > 1000 μ ± 50μ. We then repeat this procedure to obtain a series of fractions. The results are given in the following:

5.6.2.-

| Fractions Obtained | Mean diameter $\bar{d}$ | |
|---|---|---|
| $f_o$ > 2000 μ | ? | $d_o$ |
| 1000μ < $f_1$ < 2000 μ | 1500 μ | $d_1$ |
| 840μ < $f_2$ < 1000 μ | 920 | $d_2$ |
| 590μ < $f_3$ < 840 μ | 715 | $d_3$ |
| 420μ < $f_4$ < 590 μ | 505 | $d_4$ |
| 297μ < $f_5$ < 420 μ | 358 | $d_5$ |
| 177μ < $f_6$ < 297 μ | 237 | $d_6$ |
| 149μ < $f_7$ < 177 μ | 163 | $d_7$ |
| 88 μ < $f_8$ < 149 μ | 118 | $d_8$ |
| 77 μ < $f_9$ < 88 μ | 83 | $d_9$ |
| 53 μ < $f_{10}$ < 77 μ | 68 | $d_{10}$ |
| 44 μ < $f_{11}$ < 53 μ | 49 | $d_{11}$ |
| 37 μ < $f_{12}$ < 44 μ | 41 | $d_{12}$ |
| $f_{13}$ < 37 μ | ? | $d_o$ |
| | | |

We now have 12 fractions that we can use. The mean diameter is apparent from the screens used to separate the fractions. In the last case that we can use, $\bar{d}_{12}$ = 41 ± 2.8 μ . We can now weigh each fraction and calculate the % weight in each fraction, providing we use the total weight of all 14 fractions. Knowing the density of the sand, and assuming the particles to be spheres, we can then obtain the number of particles in each fraction. (Note that we have assumed that all particles small enough to pass through a given screen have done so. In many cases, this is not true, and we have to be cognizant of this error, i.e.- this factor is the main source of error in the SIEVE METHOD of particle size analysis). We now find that there are two general methods for data-reporting, namely:

5.6.3.- METHOD I = % of total weight

METHOD II = Cumulative weight-%

Each has its advantages. In the first method, we calculate the total weight and assign each fraction a %-value. Mean diameters , $\bar{d}$, are easy to obtain in the first method because we know the screen diameters used to separate the fractions. In the second, we add $f_2$ to $f_1$ , then $f_3$ to $f_2 + f_1$ , then $f_4$ to $f_3 + f_2 + f_1$, etc. To obtain $\bar{d}$, we take the average of the added fractions, shown as follows:

5.6.4.-

| | $\bar{d}$ |
|---|---|
| $f_1$ | 1500 μ |
| $f_2 + f_1$ | 1210 |
| $f_3 + f_2 + f_1$ | 963 |
| $f_4 + f_3 + f_2 + f_1$ | 734 |

We continue with all the fractions we have. If we now plot the data, we get the diagram, given as 5.6.5. on the next page.

5.6.5.-

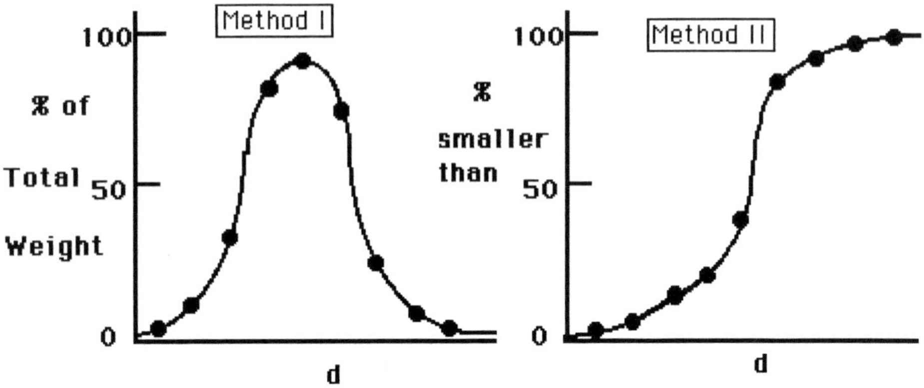


We can see the spread of particles in Method I, but cannot determine the mean accurately, since $\bar{d}$ is a guess. Method II allows us to obtain the mean from the 50% point, that is, the point where 50% of the particles are smaller than, and 50% are larger than.

Actually, the data of Method I are better plotted as shown in the following diagram:

5.6.6.-

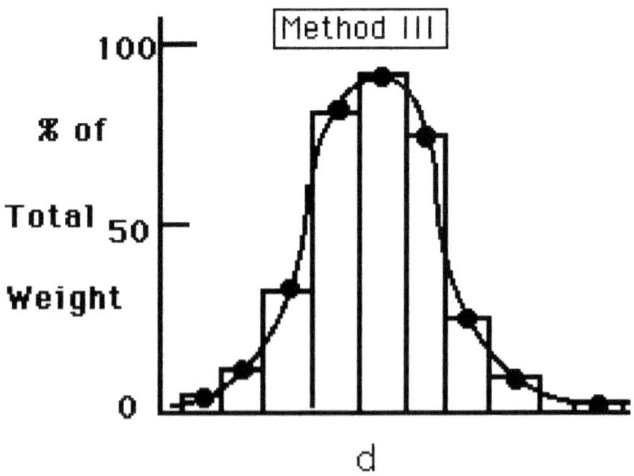

Method III is called a "Histogram Plot" while Methods I & II are called "Frequency Plot" and "Cumulative Frequency Plot", respectively. There is one important point which needs to be emphasized. That is:

"ALL METHODS FOR PRESENTING DATA FROM THE MEASUREMENT OF PARTICLE SIZE DISTRIBUTIONS, WHETHER INSTRUMENTAL, SEIVING, SEDIMENTATION, OR PHOTOMETRIC METHODS, MEASURE **FRACTIONS** OF THE TOTAL PARTICLE DISTRIBUTION. IF THE METHOD IS SENSITIVE, THE FRACTION-SEGMENTS CAN BE SMALL, AND THE MEASURED PARTICLE DISTRIBUTION WILL BE CLOSE TO THE ACTUAL ONE. IF THE MEASUREMENT IS LESS SENSITIVE, THERE MAY BE SIGNIFICANT DEVIATIONS FROM THE CORRECT PSD.

The data for the fraction-steps can be in terms of numbers of particles, weight, %-weight or even a packed volume. We shall next investigate parameters of distributions as a function of method of reporting data.

## 5.7. - ANALYSIS OF PARTICLE DISTRIBUTION PARAMETERS

Each of the methods of data presentation has its own special problems in terms of the amount of extractable information one can obtain. We will address each of these in turn, starting with the least complicated one, which is the "Histogram".

### A. THE HISTOGRAM

A typical histogram is shown in the following diagram, given as 5.7.1. on the nexty page, along with an analysis of the parameters which can be calculated.

The first step is to calculate the individual areas of the steps, $A_i$. These are summed and then divided by the number of individual areas, and then multiplied by the sum of $(d_2 - d_1) / n_i$ to find the mean, $\overline{\mathbf{d}}$. Admittedly, the method is cumbersome, but sometimes this is all the data we have.

5.7.1.-

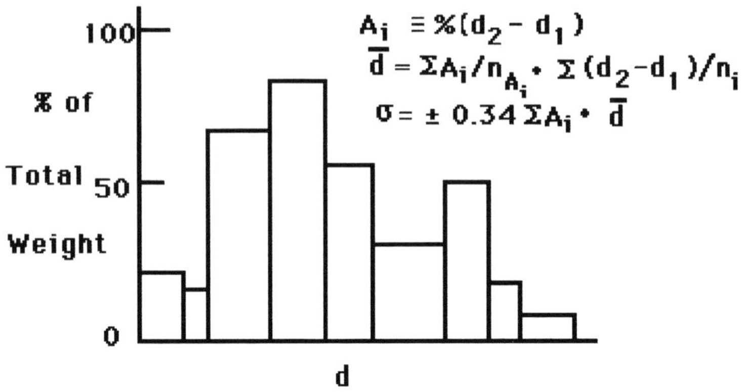


B. FREQUENCY PLOTS

Many particle-measuring methods use STOKE'S LAW to determine particle distributions. By suitable manipulation(see below), we obtain an equation relating the Stokes diameter, M, with the particle density, $\rho_1$, and the liquid density, $\rho_2$, namely:

5.7.2.- $$M = 18\ \eta h\ /\ \alpha t\ (\rho_1 - \rho_2)g$$

where $\eta$ is the liquid viscosity, h the distance settles, $\alpha$ is a shape factor ($\alpha$ = 1 for a spherical particle), and g is the gravitational constant. The photometric method (which we shall discuss in more detail below) gives % at a specific particle diameter and hence can be directly plotted as frequency, as given in the following diagram which is labelled 5.7.3. on the next page.

As shown in the diagram, $d_\sigma$ can be calculated from the mean diameter if the distribution is symmetrical, **but they rarely are.** Note also that $\sigma$, which is defined as the diameter limits equal to 68% of the total particle population, cannot be obtained, only $d_\sigma$ .

5.7.3.-

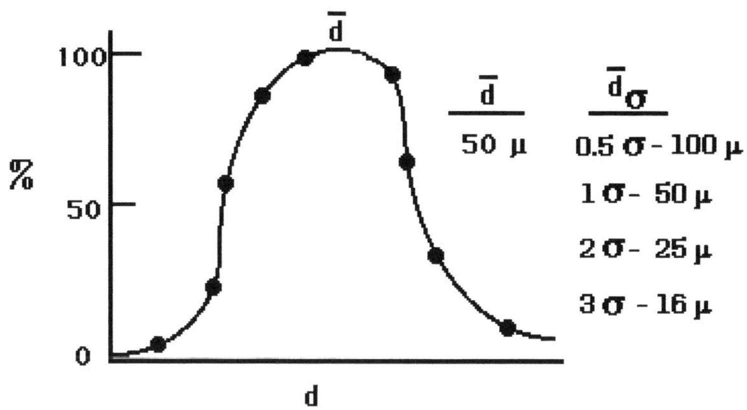


C. CUMULATIVE FREQUENCY

The most popular method of data presentation is that of cumulative frequency. An example of this is shown as follows:

5.7.4.-

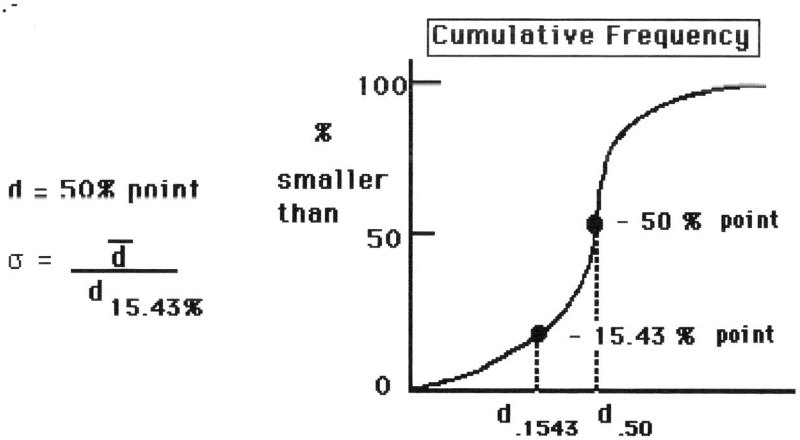


Note that both $\bar{d}$ and σ are easily obtained. But no other information can be derived. What we need is a more versatile method. This can be accomplished

particles grow at the expense of small ones, is one mechanism that can give rise to growth limits. Growth limits are denoted by:

5.7.7.-

$\overrightarrow{\mathfrak{L}}$ - Upper Growth Limit

$\underrightarrow{\mathfrak{L}}$ - Lower Growth Limit

$\mathfrak{X}_{\infty}$ - Upper Discontinuous Limit

$\mathfrak{X}_{0}$ - Lower Discontinuous Limit

The discontinuous limit is that in which all particles beyond a specific size have been removed, or do not exist. The diameter of the particles is found on the y-axis and plotted at the proper point on the x- axis as "% less than" (see Method II of 5.6.3. and 5.7.4., given above).

## **5.8** - TYPES OF LOG NORMAL PARTICLE DISTRIBUTIONS

We shall present some examples of particle distributions and how to interpret them in order to show the versatility of the method. As you will see, the method of plotting particle distributions via a log-normal method allows one to interpret particle size in a manner not feasible by other methods. Yet you will find that most particle size specialists do not take advantage of the method.

### A. UNLIMITED PARTICLE DISTRIBUTIONS

The following diagram, given as 5.8.1. on the next page, shows a typical PSD where the distribution does not have limits.

5.8.1.-

**LOG PROBABILITY - UNLIMITED PARTICLE DISTRIBUTION**

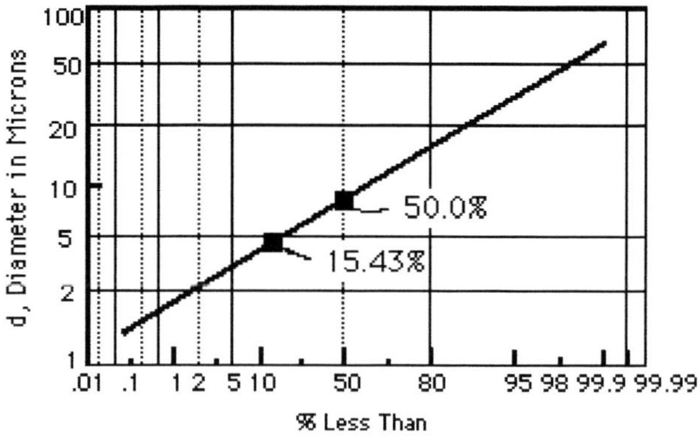


$\bar{d}$ = 50% prob.

$\sigma = \bar{d} / d_{15.43}$

$\vec{\mathfrak{L}} = +\infty$

$\underleftarrow{\mathfrak{L}} = -\infty$

Following this example are distributions that **do** have limits to the "normal" distribution. What this means is that the distributions conform to the limits defined in 5.7.6. Note that in 5.8.1., a straight line is evident. This is the type of distribution usually found as a result of most precipitation processes. But as we shall see, this is not true for the other types of log-normal distributions.

B. LIMITED PARTICLE DISTRIBUTIONS

If a precipitate is allowed to undergo Ostwald ripening, or is sintered, or is caused to enter into a solid state reaction of some kind, it will often develop into a distribution which has a size limit to its growth. That is, there is a maximum, or minimum limit (and sometimes both) which the particle distribution approaches. The distribution remains continuous as it approaches that limit. The log-probability plot then has the form shown in 5.8.2. on the next page.

In this case, both upper and lower limits are shown. However, one may have **one** or the other, or both, in the *general* case. Ostwald ripening tends to use

5.8.2.-

LOG PROBABILITY - PARTICLE DISTRIBUTION WITH LIMITS

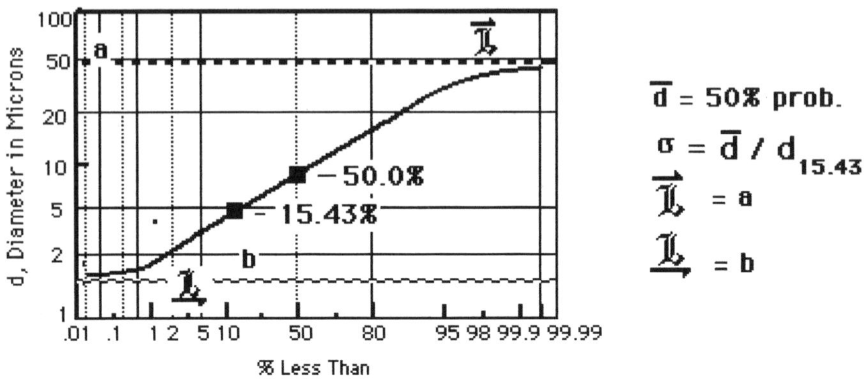


up **all** of the small particles, without limiting the upper size. Then we would have the lower limit but not the upper.

C. PARTICLE DISTRIBUTIONS WITH DISCONTINUOUS LIMITS

Suppose we use a 200-mesh screen to remove particles larger than 74μ and a 400-mesh screen to remove those smaller than 37μ. The PSD is then:

5.8.3.-

LOG PROB.-PARTICLE DISTRIBUTION WITH DISCONTINUOUS LIMITS

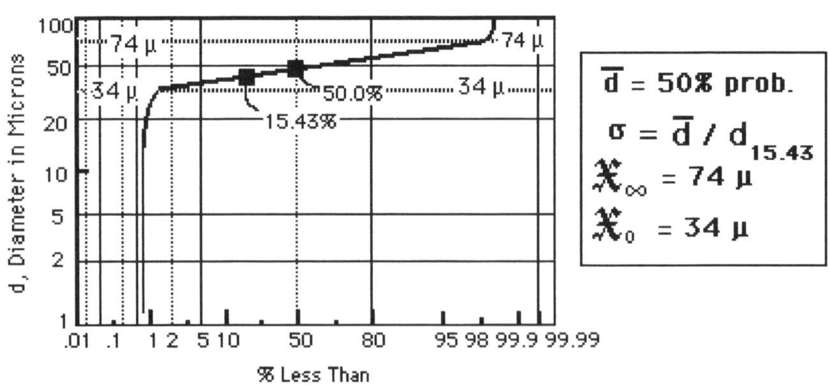

Note that the distribution is linear between 37 and 74 μ but that it abruptly shifts to ± ∞ at these points. This is a particle distribution for which it is impossible to obtain an accurate picture by any other means.

A **frequency plot** of the same data would look like then look like this:

5.8.4.-

FREQUENCY PLOT- DISCONTINUOUS LIMITS

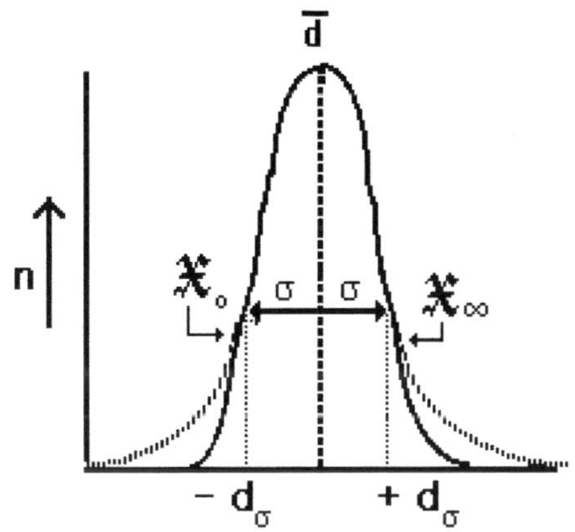


If we were using the frequency plot method to present this data, we would think that the curve was just assymmetrical and that the distribution was not log-normal. Yet, it is obvious from 5.8.3. that it is **log-normal.**

D. MULTIPLE PARTICLE DISTRIBUTIONS

Log probability plots are particularly useful when the distribution is **bimodal**, that is, when two separate distributions are present. Suppose we have a distribution of very small particles, say in suspension in its mother liquor. By an Ostwald ripening mechanism, the small particles redissolve and reprecipitate to form a distribution of larger particles. This would give us the distribution shown in 5.8.5. on the next page.

5.8.5.-

**LOG PROBABILITY – BIMODAL PARTICLE DISTRIBUTION**

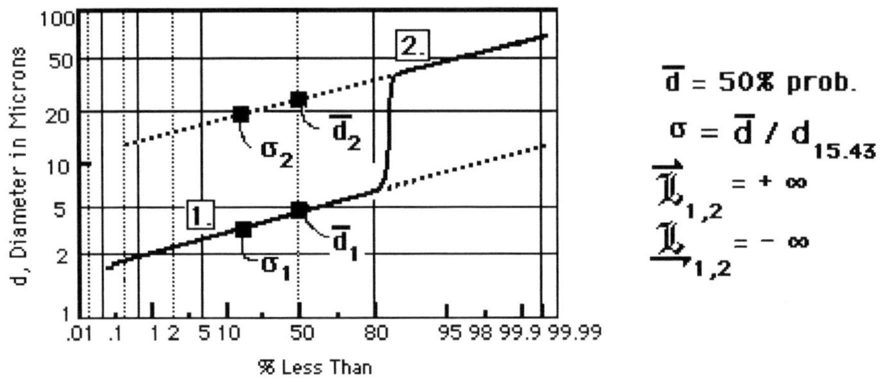


In this illustration, the size distribution parameters of both distributions are readily apparent. A similar determination is almost impossible with other types of data presentation methods. Although a large **discontinuous** gap between the two distributions is shown in 5.8.5., it is rarely the case.

As examples of the usefulness of particle distributions, consider the following. Two actual cases encountered in Industry are described in the following examples:

CASE I - FLUORESCENT LAMP PHOSPHOR PARTICLES

Fluorescent lamps are manufactured by squirting a suspension of phosphor particles in an ethyl cellulose lacquer upon the inner surface of a vertical glass tube. Once the lacquer drains off, a film of particles is formed. The lacquer is then burned off, leaving a layer of phosphor particles. Electrodes are sealed on; the tube is evacuated; Hg and inert gas is added; and the lamp ends are added to finish the lamp. Lamp brightness and lifetime are dependent upon the particle size distribution of the phosphor particles. The number of small particles is critical since they are low in brightness output but high in light scattering and absorption. A lamp brightness improvement of 10% is easily achieved by removal of these particles. In 5.8.6. is shown the PSD that resulted when a certain mechanical method was used to remove the "fines", viz-

5.8.6.-

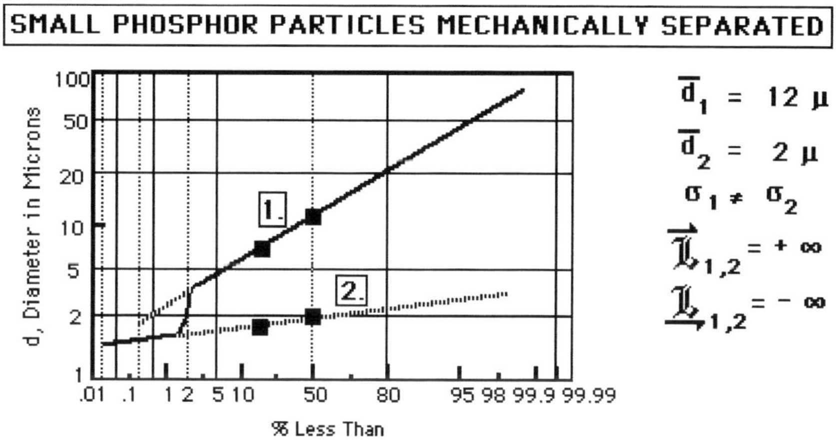


In this application of the log-normal plot, note that the "mechanical" separation of "fines" has created a new particle distribution with $\bar{d}_2 = 2\ \mu$. Even the value of $\sigma_2$ differs from that of the major particle distribution. In the "fines" fraction, it appears that the largest particle does not exceed about 5 μ. Needless to say, lamps prepared from this phosphor were inferior in brightness. Armed with this information, one could then recommend that the method of "fines" removal be changed.

Another case is presented on the next page.

CASE II. - TUNGSTEN METAL POWDER

The lifetime of a tungsten filament in an incandescent light bulb depends a great deal on the grain size of the wire used to make the filament. W-metal powder is pressed into a bar at pressures which ensure that the density is as close as possible to the theoretical density. The bar is sintered in an inert atmosphere and then "swaged" , i.e.- "hammered" into a long rod. This rod is then drawn into a fine wire (~ 2-5 mil) which is wound into the filament form. Metal powder particle size has a major effect upon wire quality since fine particles form **small grains** in the wire while large particles will form **large grains.**

When the filament is operated at 3250 °K to produce light, the wire is hot enough that vacancy migration will occur in the wire. Simultaneously, grain growth plus an increase in grain size also occurs. The grain boundaries between large grains will "pin" vacancies, but grain boundaries between small grains do not appear to do so. If they do, the effect is at least an order of magnitude smaller. As the vacancy concentration builds up at the large grain boundaries (see the Kirchendall effect given in Chapter 4), local resistance to electron flow increases. Eventually, a "hot spot" develops and the tungsten wire melts at that point, with consequent failure of the filament and the lamp. If only small metal particles are used to make the wire, then the small grains produced must grow into large grains before "hot spots" cause its eventual failure . Therefore, incandescent lamp manufacturers exercise very close control of the tungsten-metal powder PSD and the sintering processes used to produce the wire to eliminate, as much as possible, pores and control actual density. The diagram shown as 5.8.7. on the next page illustrates a typical PSD for a tungsten metal powder where the large particles have been removed. In this case, the PSD is suitable and particles larger than about 5 μ appear to have successfully removed.

This completes the types of particle distributions that we might encounter. It is now time to show how particle size counting-data are used. To do this, we must select an instrument that produces counts of size of particles correlated with numbers of particles in each size range. There are several types of such instruments whose nature will be delineated below. But, first, we must show how this is done. Let us now examine a method of calculating a particel size distribution.

5.8.7.

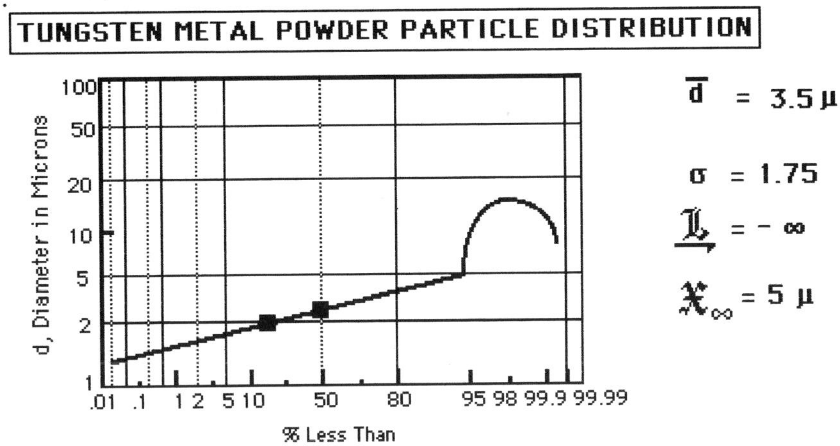


## 5.9 - : A TYPICAL PSD CALCULATION

Let us suppose that we have a particle counting instrument which sorts and counts the number of particles at a given particle size. The experimental data that we collect are:

5.9.1.- EXPERIMENTAL COUNTS AT PARTICLE DIAMETER

| d | n Counts | d | n Counts |
|---|---|---|---|
| 2.2 µ | 2 | 10 µ | 902 |
| 3.0 | 4 | 15 | 1402 |
| 5.0 | 28 | 25 | 1902 |
| 7.0 | 232 | 30 | 1982 |
| 9.0 | 602 | 40 | 1998 |
| | | 50 | 2002 |

Having obtained the data as given in 5.9.1., our next step is to calculate the **average** size in each particle **interval.** For particles less than 2.2 µ, we have 2 particles. At 3.0 μ, we have 4 particles. Therefore, **in the range** 2.2 - 3.0 μ, we have **two (2)** particles whose average size is 2.6 μ.

In the range 3.0 - 5.0 μ, we have **four (4)** particles whose average size is 4.0 μ. We therefore continue to calculate Δn, as shown in **Table 5 - 1**. for each size range that we have measured:

TABLE 5 - 1

A TYPICAL PARTICLE SIZE DISTRIBUTION CALCULATION

| Diameter | n-Counts | Δn | $\bar{d}$ | % | Σ% |
|---|---|---|---|---|---|
| 2.2 μ | 0 | | | | |
| | | 2 | 2.6 μ | 0.1 | 0.1 |
| 4 | 2 | | | | |
| | | 24 | 4.0 | 1.2 | 1.3 |
| 5.0 | 28 | | | | |
| | | 204 | 6.0 | 10.2 | 11.5 |
| 7.0 | 232 | | | | |
| | | 370 | 8.0 | 18.5 | 30.0 |
| 9.0 | 602 | | | | |
| | | 300 | 9.5 | 15.0 | 45.0 |
| 10 | 902 | | | | |
| | | 500 | 12.5 | 25.0 | 70.0 |
| 15 | 1402 | | | | |
| | | 500 | 20.0 | 25.0 | 95.0 |
| 25 | 1902 | | | | |
| | | 80 | 27.5 | 4.0 | 99.0 |
| 30 | 1982 | | | | |
| | | 18 | 35.0 | 0.9 | 99.9 |
| 40 | 2000 | | | | |

We have a total of 2000 particles that we have measured and we can calculate the percent (%) of total counts for each size range. This gives the following log-normal PSD plot, viz-

5.9.2.-

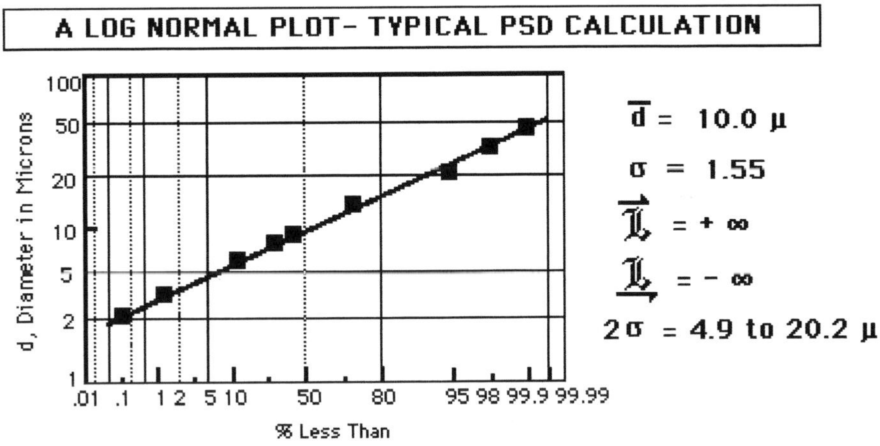


Here, the parameters of the distribution are given. This material appears to have been precipitated, or it may have been obtained by a grinding process, since the parameters indicate a Gaussian unlimited distribution. Particles grown by solid state reaction, including Ostwald ripening, generally have PSD characteristics of the original PSD from which they were formed. That is, there will be limits according to the original distribution from which it has arisen.

If the original PSD had limits, then the progeny PSD will also have such limits. The basis for this behavior lies in the fact that if one starts with a certain size range of particles as the basis of particle reaction and growth, one will end with the same size range of particles in the PSD of the particles produced by the solid state reaction. Such a case is shown in the following diagram, given as 5.9.3. on the next page. Here, the oxalate was prepared by addition of oxalic acid to $Gd(NO_3)_3$ solution to form a precipitate.

5.9.3.-

**COMPARISON OF PRECIPITATED OXALATE AND OXIDE PRODUCED**

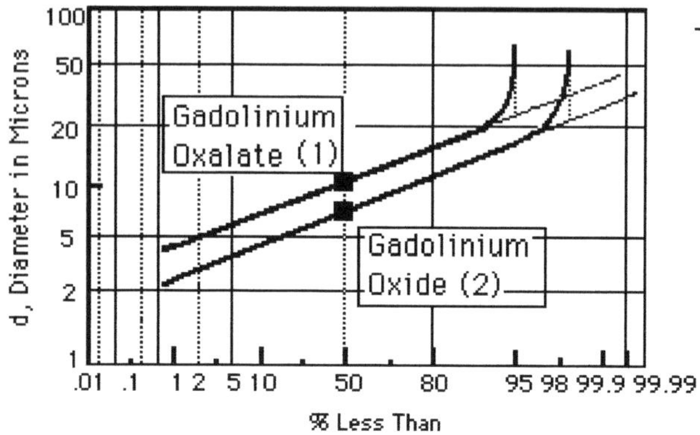


**Oxalate Used**

$\bar{d} = 11.3\ \mu$

$\sigma = 1.59$

**Oxide Produced**

$\bar{d} = 8.0\ \mu$

$\sigma = 1.51$

$\mathfrak{X}_{Oxal} = 23\ \mu$

$\mathfrak{X}_{Oxide} = 20\ \mu$

$\mathfrak{L} = -\infty$

There is an upper limit of about 23 microns in size. This may be due to the fact that the precipitation was accomplished at 90 °C., or from the fact that rare earth oxalates tend to form very small particles during precipitation which then grow via Ostwald ripening and agglomeration to form larger ones. Nevertheless, it is clearly evident that when the oxalate is heated at elevated temperature (~ 900 °C), the oxide produced **retains** the same PSD characteristics of the original precipitate.

## **5.10.-** METHODS OF MEASURING PARTICLE DISTRIBUTIONS

Having defined particles and partical distributions, we now examine methods by which we can measure particles and properties of powders. There are four (4) primary methods used to obtain data concerning particle size distributions. These include:

| | |
|---|---|
| OPTICAL | ELECTRICAL |
| SEDIMENTATION | ABSORPTION |
| PERMEATION | |

A. The Microscope - Visual Counting of Particles

At the proper magnification and with a special eye piece, one can directly measure particle diameters. The eye-piece must have internally-marked concentric circles so that a given particle will fall within one or more of the circles. The diameter then can then be read and/or estimated directly as shown in the following:

5.10.1.-

MICROSCOPIC COUNTING - PARTICLE DIAMETER EYEPIECE

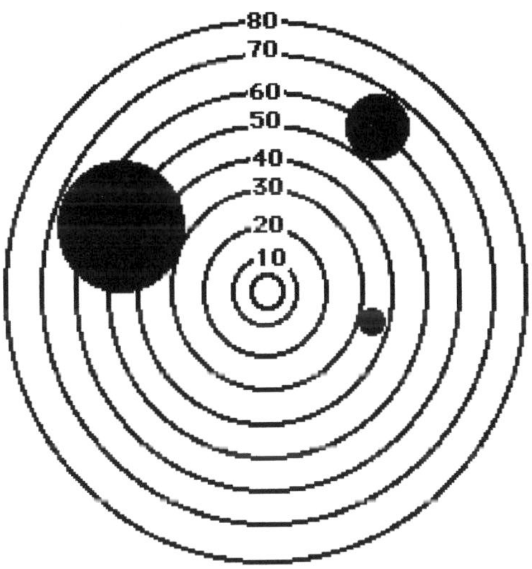


In this case, three particles are shown, a 40 μ, 20 μ and a 10 μ particle. The most important step is sample preparation on the microscope slide. since only a pinch of material is used, one must be sure that the sample is uniform and **representative** of the material. Also, since most materials tend to agglomerate due to accumulated surface charge in a dry state, one adds a few drops of alcohol and works it with a spatula, spreading it out into a thin layer which dries. Too much working breaks down the original particles, particularly the larger ones, and too little working leaves agglomerates. With a little practice, the preparation of a proper slide of particles becomes easy.

One then proceeds to count particles. It has been found that if one wishes to determine an accurate value of the mean diameter, one must count at least 450 particles. as the number of counts increases, the mean comes closer and closer to the **true** value, as shown in the following:

5.10.2.-

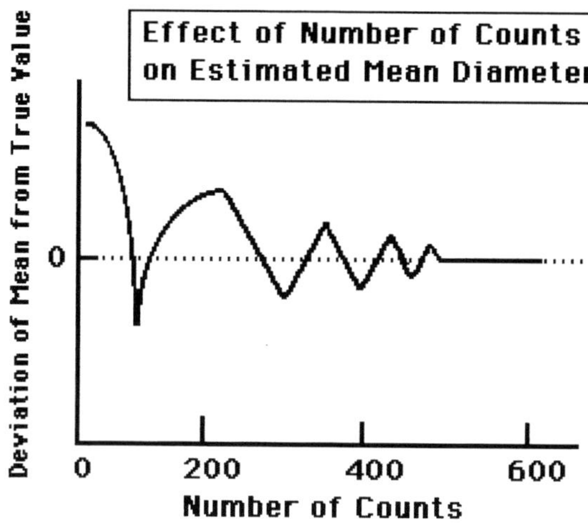


To calculate the mean diameter, we count the number of particles at $d_1$, then those at $d_2$, etc., sum these and average the value, i.e.-

5.10.3.- $$\bar{d} = \Sigma n_i d_i \;/\; \Sigma n_i$$

Another method is to count particles between a given range and then sum the counts as in 5.9.1. Alternately, one can purchase an automatic particle counting instrument for about $ 50-60,000. The instrument consists of a microscope, a scanning device (usually a flying-spot scanner), a television display and a pre-programmed microprocessor. All of the particles within a given frame can be counted and grouped automatically. Thus, several frames per minute can be counted by an operator, or the instrument can be set to operate automatically according to a preset program. Since the instrument scans **images** of the particles, a problem of orientation arises. This is shown in the following diagram:

5.10.4.-

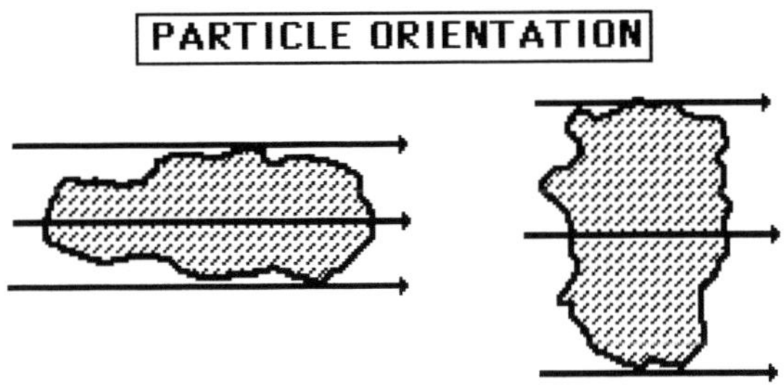


The microprocessor is programmed to recognize length and width, but the question of the correct diameter of irregular particles has remained controversial. For example, Martin's diameter is defined as the shortest line which divides the area of the image in half. A depiction of this is shown as follows:

5.10.5.-

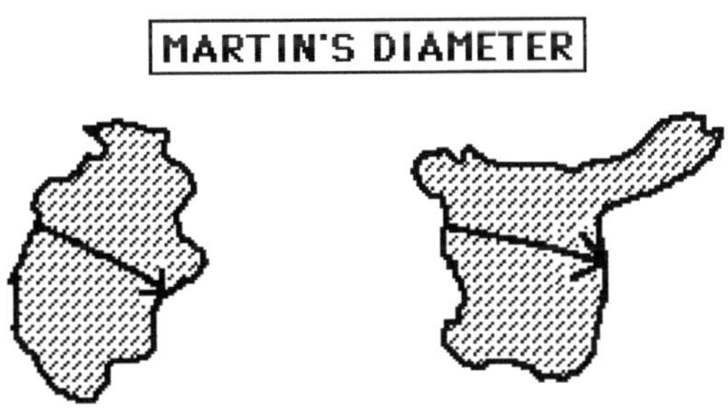


Yet, it is obvious that this is not the correct diameter for the these particles. Another school of thought would use the **average** of the length and width, but for the above particles, the actual value of the length that ought to be used remains questionable. This problem has remained the most serious barrier to obtaining PSD by optical microscopy until recently.

With the advent of computers which can process data at the rate of 700 billion bits per second, this problem has been solved satisfactorily. This PSD instrument (Malvern Instruments, 10 Southville Road, Southborough, Ma. 01772 - Sysmex FPIA-2100) is a fully automated particle **size and shape** analyzer. It uses CCD, i.e.- "charge-coupled device", technology (the optical basis of a digital video camera) to capture a series of images of particles suspended in a liquid medium. Particles are sampled from a dilute suspension which is forced through a "sheath-flow" cell. This insures that the largest area of the particle is oriented toward the video camera and that all of the particles are in focus. The cell is illuminated via a stroboscopic light source and images are captured at 30 Hz. per second. A computer program then processes the images in real time by the following steps: digitization, edge highlighting, binarization, edge extraction, edge tracing and finally storage. The following diagram shows how this is accomplished:

5.10.6.-

Automated Analysis of Particle Outline and Diameter

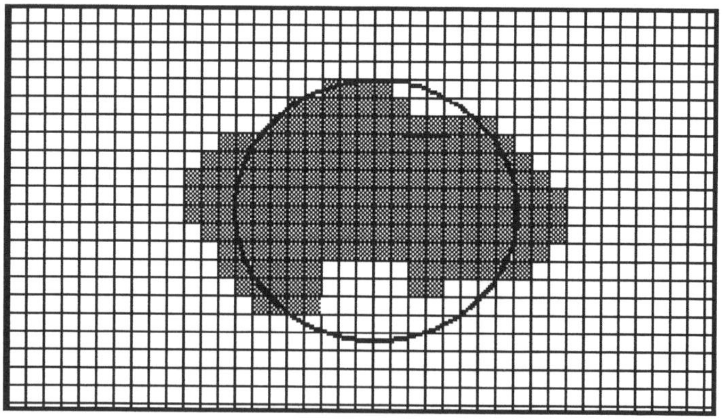

All this requires a high speed computer with large memory storage capacity and RAM (something not possible until recently). Image analysis software then calculates the area and perimeter of each of the captured images and then calculates the particle diameter and circularity. The particle shape is easily recognized in this diagram. The area of the circle is equal to that of the digitized particle so that the particle circularity (shape) can be classified. Circularity, C, is determined by:

5.10.7.- C = perimeter of circle / perimeter of particle

Once the measurement is complete, the particle size and circularity can be displayed in both graphical and table form. A typical report includes: particle size distribution, circularity distribution and a scattergram of particle size vs: circularity. All these can be displayed and printed in hard copy. Individual particle images can be displayed and then classified into categories including uni-particle, bi-particle, and agglomerate. This rather new instrument has solved the problem of particle shape and corresponding size through optical digitization methods.

B. SEDIMENTATION METHODS

There are several particle sizing methods, all based upon sedimentation and Stokes Law. If a particle is suspended in a fluid (which may be gas, or any liquid), the force of resistance to movement by the particle will be proportional to the particle's velocity, v, **and** its radius, r, vis-

5.10.8- $f = 6\pi\, r\, \eta\, v$

where $\eta$ is the viscosity of the fluid, providing the particle is spherical. If the particle settles under the influence of gravity, then we can write:

5.10.9.- $f = 4/3/\, \pi\, r^3\, (\rho_s - \rho_l)\, g = 6\pi\, r\, \eta\, v$

where $\rho_s$ and $\rho_l$ are the densities of the solid particle and the liquid, respectively. Since the distance settled is: h = v t, then:

5.10.10- $$d = \{ 18 \eta \ h \ / \ \alpha \ t \ (\rho_s - \rho_l) \ g\}^{1/2}$$

This is Stokes Law for sedimentation where we have added $\alpha$, a shape factor, just in case we do not have spherical particles. For spheres, $\alpha = 1$. It is fractional otherwise.

One way to measure particles is to weigh them as they settle out from suspension. such an apparatus is called a "sedimentation balance" and is designed as shown in the following diagram:

5.10.11.-

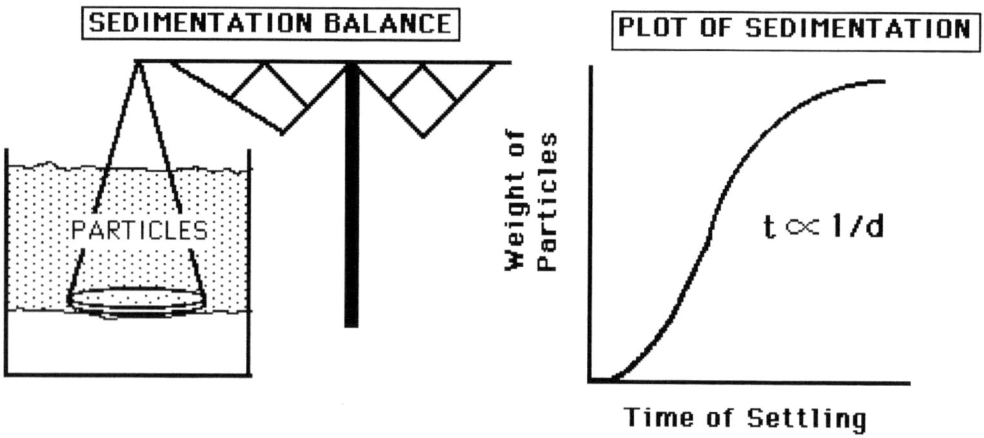


The weight of the particles builds up with time and is proportional to 1/d. If we assume spherical particles, then we can convert the above curve to particle diameter from Stokes Law. Although we have added the particle suspension to a "water cushion" as shown above, it might not seem that the settling of the particles would strictly adhere to Stokes Law, which assumes the terminal velocity to be constant.

But, as shown in the following Table, the approximate distance a particle needs to travel to reach terminal velocity in a given liquid **is very short.**

**Table 5 - 1**

Rate of Fall in Water at 25 °C. for Particles Having a Specific Gravity of 2.0

| Size (microns) | Approximate Distance in cm. Traveled prior to Reaching Terminal Velocity | Approximate Terminal Velocity (cm/sec) |
|---|---|---|
| 2000 | $1.2 \times 10^{2}$ | 240 |
| 200 | $1.2 \times 10^{-2}$ | 2.4 |
| 20 | $1.2 \times 10^{-6}$ | $2.4 \times 10^{-2}$ |
| 2 | $1.2 \times 10^{-10}$ | $2.4 \times 10^{-4}$ |
| 0.2 | $1.2 \times 10^{-14}$ | $2.4 \times 10^{-6}$ |
| 0.02 | $1.2 \times 10^{-16}$ | $2.4 \times 10^{-8}$ |

These data show that the particles separate very fast due primarily to their differing terminal velocities. Thus, in the graph shown in 5.10.9., the initial weight is due to the large particles, while the upper part of the curve is due to the smaller sizes. For particles less than about 2.0 microns, settling times under gravity becomes extremely long and this size remains a lower limit for sedimentation methods of determining particle size. 2000 μ is probably the upper limit since one cannot adjust the balance to operate quickly enough before these large particles are lost. Major problems encountered with the sedimentation method include:

- Proper Design of the Apparatus
- Using the Correct Liquid with the Specific Powder
- Proper dispersion of the Particles
- Operator Technique

The latest commercial apparatus incorporates a microprocessor to automatically plot the size distribution as a frequency or cumulative plot.

Another sedimentation method used involves the ANDREASEN PIPETTE. A typical design is shown in the following diagram:

5.10.12.-

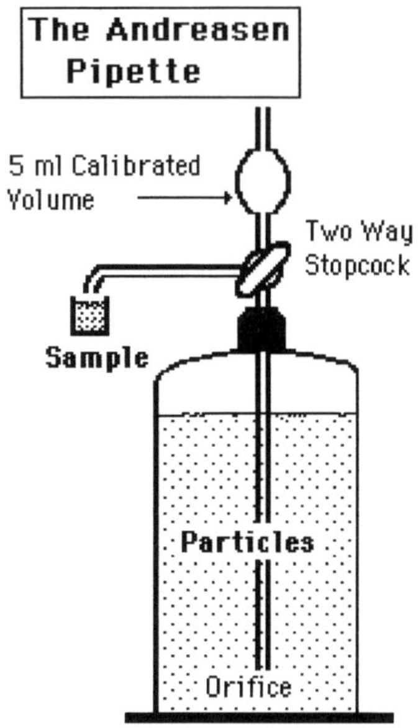


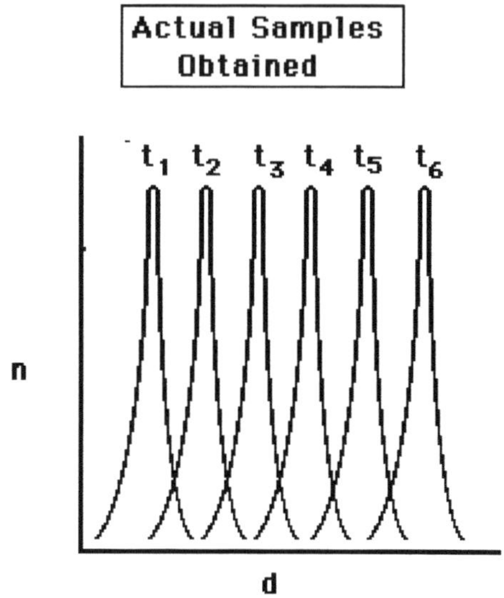


This glass apparatus is inexpensive and consists of a bottle having an internal sampling tube and calibrated sampling volume (5 ml). One draws a sample and then expels it into an external 5 ml beaker.

Using the times calculated for successive diameters to settle past the orifice, one withdraws samples corresponding to that diameter. A series of samples are obtained at $t_1$ , $t_2$ , etc., which are dried and weighed. The actual sample obtained is a narrow particle distribution rather than one **specific size**. Thus, the diameter measured (by calculation) does not exactly correspond to the **true** Stoke's diameter, but to the peak of a narrow distribution.Yet, the Andreasen Pipette method continues to be the most widely used sedimentation method since it involves inexpensive equipment, and is reasonably accurate if one will take the time to develop a correct experimental technique. The reproducibility can be excellent, the accuracy less so.

Another sedimentation method used is the so-called MSA-analyzer. If the value of "g" in 5.10.8. is increased (such as the use of a centrifuge) one can analyze the very small particles in any given distribution in a short time. The problem of course lies in accurate determination of the weight accumulated at a given time under a specific centripetal force. This problem has been neatly solved by careful design of the sedimentation tube, as shown in the following diagram:

5.10.13.-

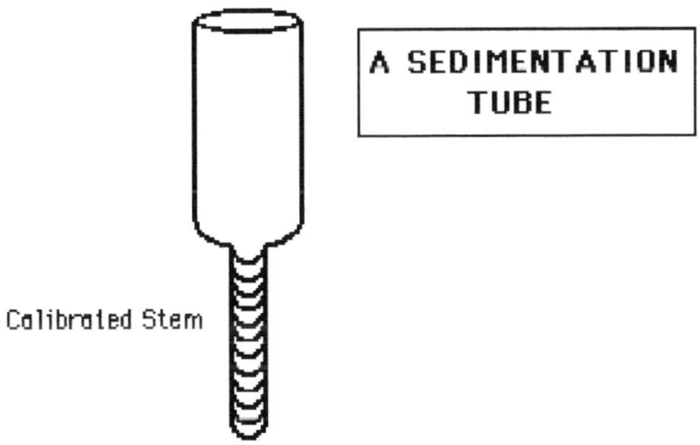


The particles build up **by layers** because it has been found that all monosized particles can be removed from suspension by rotating at a specific speed. Thus, one runs the instrument at a series of rotational speeds, measuring the weight of the build-up layers in between each run. The overall analysis is run at specified rpm's which correspond to selected particle diameters, resulting in data sufficient to characterize the particle distribution.

C. ELECTRICAL RESISTIVITY - THE COULTER COUNTER

Perhaps the most useful method for determining particle distributions is that of electrical conductivity. the most widely used instrument is the Coulter Counter (named after the Inventors), although there are now other similar instruments on the market. Originally, this instrument was designed to measure blood corpuscles which are 2-8 μ in size. It has proven to be very suited to measure micron and submicron particles easily, accurately, and reproducibly.

Consider a conductive solution consisting of water with a soluble salt, i.e.- 1% NaCl, and a dispersing agent used to prevent agglomeration of particles in suspension. Two electrodes are placed in solution with a non-conducting orifice between them, as shown in 5.10.14, given on the next page:

5.10.14.-

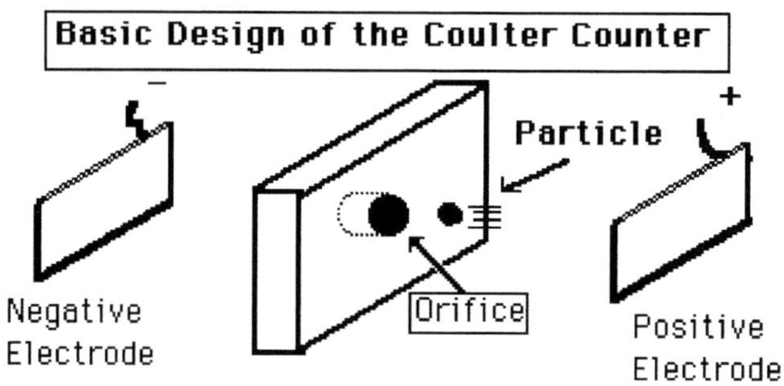


A ceramic block such as alumina is generally used because of its chemical inertness and the orifice is bored or drilled to precise dimensions. The block is placed so that two (2) chambers result. Then if a DC voltage is applied across the electrodes, a current will flow **only through** the orifice, and an effective resistance arises which depends upon the voltage applied and the size (volume of the conducting solution) of the orifice. Particles are added to one side of the two chambers created by the ceramic block, and the suspension is pumped through to the other side. There will be an electrical pulse as each particle passes through the orifice.

A depiction of the pulses obtained is shown in the following diagram, given as 5.10.15. on the next page. Each pulse will be proportional to the size of the particle because the particle volume displaces part of the conducting solution during its passage through the orifice. If we have properly established steady state conditions,

5.10.15.-

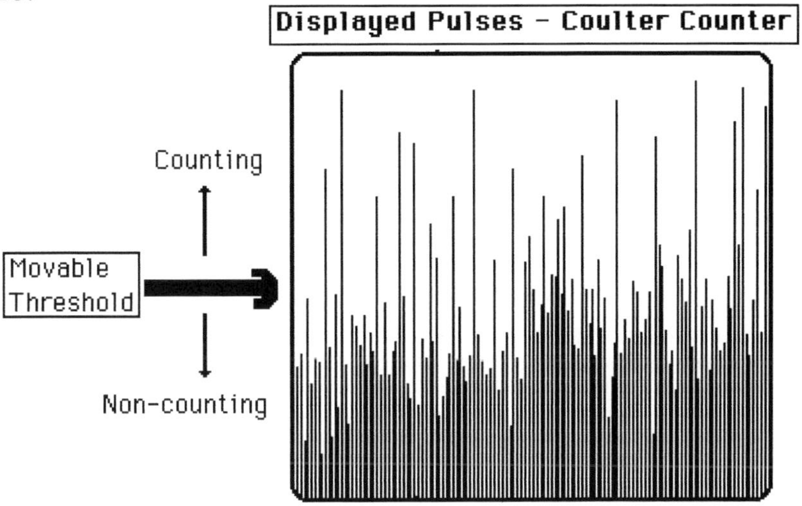


we will obtain a drop in current as each particle passes through. Electronically, this can be converted to a positive pulse whose intensity is proportional to each particle volume. **We assume sphericity of the particles.** If V is constant, then:

5.10.16.- $I = V / \Delta R$ and: $\Delta R = r_o V_p / A^2 \cdot 1/ \{1 / 1 - (r_o / r_e)\} - a/A$

where $r_o$ is the actual particle radius, $r_e$ is the **effective** radius (incase the particle is not spheroid), $V_p$ is the volume of the particle, A is the cross-sectional area of the orifice, and a is a constant which depends upon the solvent and solute used to make the conducting solution.

The size of the orifice must be fitted to the **range** of particles present. If the particles are too large, they will not pass through the orifice. Too small and the pulse is not easily detected, Since large numbers of particles can be present, the powder content must be controlled. Too high a particle concentration-density will result in "coincidence counting". That is, two small particles can be counted as one larger one. This phenomena has been addressed and tables are provided to correct for incidence counting. The electronic part of the Coulter Counter has several hundred "channels" to accept pulses. The "threshold" is electrical, **and movable,** so that one counts the large particles first (large pulses) and then moves the threshold lower to count the smaller particles. The threshold is actually a pulse height analyzer which converts heights to numbers. The data obtained are easily converted to a frequency plot or a cumulative plot. Since ΔR response is linear, one can readily calibrate the instrument at one point, using monosized particles such as pollen or specially prepared plastic particles. The only requirement is that the calibrating particles ought to be in the range of the particles to be measured.

The Coulter Counter is particularly useful and unique in that steps as small as 1.0 micron may be used if so desired. Results by this author have demonstrated that some distributions which appear to Gaussian may actually consist of more than one subdivision where the peaks are separated by only a few microns. an example is given as follows:

5.10.17.- A PARTICLE DISTRIBUTION WITH SUB-DISTRIBUTIONS

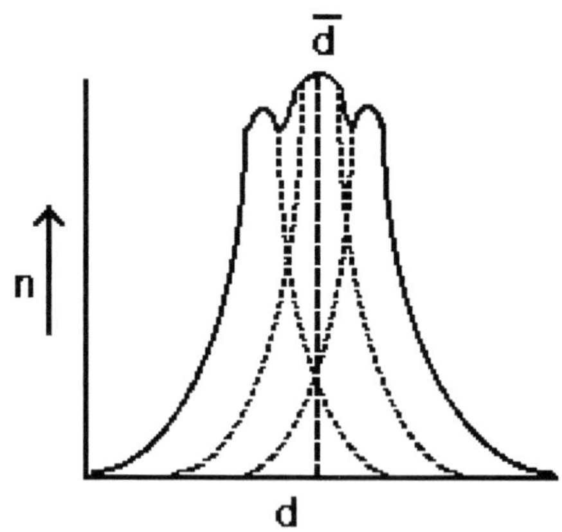


This result was achieved with a selected particle distribution which originally appeared to be a single Gaussian distribution. When the particle measuring steps were changed so that a range of about 1-2 microns was covered, the above result was obtained. Even so, the measurement must be done carefully to reveal the substructure. To the author's knowledge, no other method has the capability of such precision in particle diameter measurement. The latest Coulter Counter instruments incorporate a microprocessor and the operator can specify the form of the output data to be obtained. The instrument converts count-data to practically any format desired, with the exception of log-probability plots.

### D. OTHER METHODS OF MEASURING PARTICLE SIZE

**Permeability** is another method for obtaining information about particle diameters. If one packs a tube with a weight of powder exactly equal to its density, and applies a calibrated gas pressure **through** the tube, the pressure drop can be equated to an average particle size. The instrument based on this principle is called the "Fisher Sub-Sieve Sizer™". Only one value can be obtained but the method is fast and reproducible. The instrument itself is not expensive and the method can be applied to quality control problems of powders. Permeametry is useful in the particle range of 0.5 to 50 μ.

**Gas adsorption** is one other method sometimes used for determining average particle size. In this case, one is usually interested in the surface area of a powder and calculates the average size of the particles secondarily. The method is called the BET-method after its developers, Brunauer, Emmett and Teller. The procedure is time-consuming and is accomplished as follows. A gas analysis train is used in which gas volumes can be recycled and measured very accurately. A weighed sample is placed in the sample tube and allowed to come to equilibrium. Since all solid materials have a monolayer of water on the surface of the particles (we live in a wet world), the sample is heated to 300 °C. to expel the water, in a nitrogen gas flow. Next, it is cooled to liquid nitrogen temperature. At this point, the sample will adsorb a **monolayer** of $N_2$-gas molecules on the surface of each particle. By allowing the temperature of the sample to return to room temperature, while measuring the volume of nitrogen gas released, one obtains a value which can be converted to a surface area of the particles being measured. In practice, one recycles the system several times, measuring adsorption and desorption volumes successively so as to obtain an average value. It is found that the **unoccupied** fraction of the surface (1- y) approaches a constant value where it is assumed that y = 0. Nitrogen gas is essential since it is the only gas which forms monolayers easily without the tendency to form more than one layer of gas molecules on the surface. If the gas pressure is kept between strict limits, the BET equation is:

5.10.18.- $p_g \,/\, (V_g [\, 1 - p_g \,] = 1 / (V_{mono} \cdot C) + (C - 1)\, p_g / (V_{mono} \cdot C)$

where: $C = \exp \{ (F_{Ads.} - F_{Cond.}) \,/\, RT\}$ , and $p_g$ is the gas pressure. $V_g$ is the gas volume measured; $V_{mono}$ is the gas volume of the monolayer calculated from the known dimensions of the nitrogen-gas molecules; $F_{Ads.}$ is the Gibbs free energy of adsorption, and $F_{Cond.}$ is the free energy of condensation. $p_g$ is measured as the **ratio** of the equilibrium pressure at specific temperatures, i.e.- room and liquid nitrogen temperatures.

It has been found that a plot of $p_g \,/\, \{V_g\, [1 - p_g]$ vs: $p_g$ is linear for the pressure range of 0.05 to 0.4, with a slope of $(C - 1) \,/\, (V_{mono} \cdot C)$ and intercept of $1 / (V_{mono} \cdot C)$. Let us now do a simple calculation using BET data obtained. Suppose we have a 20 gm. sample having a density of 2.0. We measure the surface area as 6 $m^2$ . From the area of a sphere, $A = \pi D^2$ , and the volume of a sphere, $V = 4/3\, \pi\, D^3$ ·, we find the total volume of **n** spheres to be 10 cc, i.e.- $n\{4/3\, \pi\, D^3\} = 10$. The surface area of $n\{\pi D^2\}$ spheres is 6 $m^3$. The total number of spheres present, n, is the same in both formulas. Therefore, by substitution, we find D= 10 μ. If we obtain a particle diameter by some other method and find that it is much smaller than that of the BET method, we infer that the particles are porous. We thus speak of the porosity and need to correct for the pore surface area if we are to make a reasonable estimate of the true diameter by the BET method.

**Particle Size by Laser Refractometry** is based upon Mie scattering of particles in a liquid medium. Up until about 1985, the power of computers supplied with laser diffraction instruments was not sufficient to utilize the rigorous solution for homogeneous spherical particles formulated by Gustave Mie in 1908. Laser particle instrument manufacturers therefore used approximations conceived by Fraunhofer.

The hypotheses made in Mie Theory include:

1. The particle is an optically homogeneous smooth sphere whose real and imaginary refractive indices are both known.
2. The sphereical particle is illuminated by a plane wave of infinite extent and of defined wavelength.
3. The real and imaginary refractive indices of the medium surrounding the particle are both known.

The Fraunhofer approximation includes:

1. The incident light-wave is plane, of infinite extent and of known wavelength.
2. The scattering is from a circular aperature in a thin opaque screen.
3. The extinction coefficient for all sizes is 2.0

Fraunhofer rules do not include the influence of refraction, reflection, polarization and other optical effects. Early laser particle analyzers used Fraunhofer approximations because the computers of that time could not handle the storage and memory requirements of the Mie method. For example, it has been found that the Fraunhofer-based instrumentation cannot be used to measure the particle size of a suspension of lactose (R.I. = 1.533) in iso-octane (R.I. = 1.391) because the relative refractive index is 1.10, i.e.- 1.533/1.391. This is due to the fact that diffraction of light passing through the particles is nearly the same as that passing around the particles, creating a combined interference pattern which is not indicative of the true

4. Plot your screen-data via log-probability; are there any limits to the distribution?

5. You have obtained the following particle data:

| Particle Count | Measured Diameter |
|---|---|
| **4** | **2.8** μ |
| **12** | **3.2** |
| **84** | **3.7** |
| **804** | **7.7** |
| **2404** | **9.3** |
| **4824** | **15.7** |
| **6428** | **20.3** |
| **7610** | **76.7** |
| **7930** | **29.3** |
| **8002** | **36.0** |

A. Calculate the parameters of the distribution.

B. Classify the distribution as to its probable origin.

C. Plot the data as:

I. Histogram;

II. Frequency;

III. Cumulative Frequency.

Make sure that you specify all of the distribution parameters available for esch method. Comment on the efficiacy of the methods as applied to the study of particle size analysis.

# Chapter 6

## Growth of Crystals

Having examined particle size and particle growth, i.e.- powders, in some depth in the last chapter, let us now turn to an investigation of the growth of single crystals. Many of the devices that we use on a daily basis employ a single crystal element as the heart of the mechanism. A good example is the "quartz-crystal" watch that you wear on your wrist. Here, a quartz crystal is made to vibrate under an applied voltage. Its vibrations are coupled to a sensing circuit and translated into seconds, minutes and hours by using and counting the known resonant frequency of vibration of the crystal. The resonant vibrational frequency is determined by the angle of "cut" (or angle in relation to the crystallographic axes) and the size of the crystal. There are many other sensing devices based upon a crystal component such as a device which controls heating, translates force into electrical voltage, modulates light, or regulates motion. We use them as electrical heaters, strain gauges, laser controllers or piezoelectric devices. We will explore some of these later in this chapter, particularly single crystal silicon used to fabricate integrated circuits used in computers.

As we stated previously, particles usually grow from a nucleus to which atoms are added in a regular manner to form a three-dimensional structure. Such crystals cease growing when the "nutrient" (the material which serves to form the particle) becomes depleted. Such particles are known as "crystallites". Each will consist of several grains, having a differing orientation of the crystal lattice, within each individual particle, namely:

A Particle with Grain Boundaries

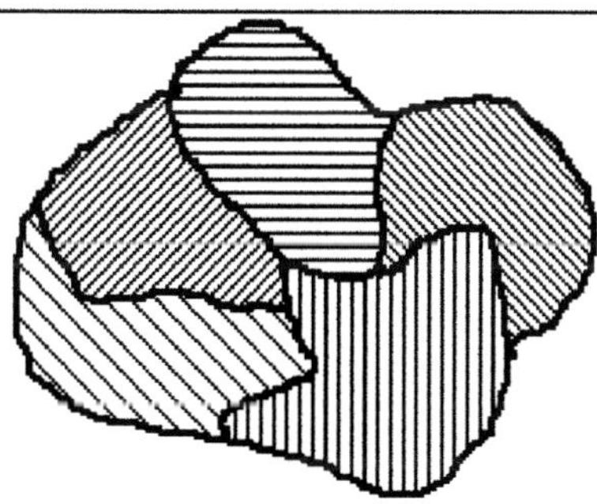

Grain boundaries form junctions between grains within the particle, due to vacancy and line-defect formation. This situation arises because of the 2nd Law of Thermodynamics (Entropy). Thus, if crystallites are formed by precipitation from solution, the product will be a powder consisting of many small particles. Their actual size will depend upon the methods used to form them. Note that each crystallite can be a single-crystal but, of necessity, will be limited in size.

However, if one desires a large **single-crystal**, then the methods needed to grow **it** differ considerably. This is the subject of this Chapter. We will examine crystal-growing methods and the equipment needed to do so. We will then examine the different type of crystals used both as sensors and as the basis of specific devices used in the electronics industry.

## 6.1- METHODS OF GROWTH OF CRYSTALS

Obviously, the major difference in the single-crystal and polycrystalline (crystallite) state is a matter of size. For the single-crystal, the size is large (≥ 10 cm), whereas in the polycrystalline state, the size of the crystals is small (10 μm = 0.001 cm.) The methods for obtaining one or the other differ considerably. They include formation from:

| 6.1.1- | Single Crystal | Crystallites (Powders) |
|---|---|---|
| | Liquid solvent | Vapor |
| | Vapor | Melt |
| | Melt | Flux |
| | Molten salt | Precipitation from solution |

The method of choice for growing a single crystal depends upon many factors. Most relate to the physical properties of the compound whose single crystal we desire. Some of the more important properties of such a compound are given in 6.1.2. on the next page.

For the most part, single crystals are grown from a melt of the compound, provided that melting the compound does not cause it to decompose, i.e.-

6.1.2.- Melting point
Partial vapor pressure of melt
Thermal stability of solid phase
Reactivity of liquid or gaseous phase
Thermal conductivity of both liquid and solid phases.

melt incongruently. Before we consider crystal growth in detail, let us examine the **hardware needed to obtain a melt.** The reason for this is that many of the furnace components available have temperature limitations, and many of the crystals we might wish to grow have high melting points, i.e.- >1600 °C. The first apparatus we may need is a furnace, which is simply a closed space, heated by electrical elements, wherein the internal temperature is controlled.

## 6.2. FURNACE CONSTRUCTION

There are many ways to build a furnace. Basically, a furnace consists of a few essential parts, each of which can be varied according to the final operating requirements needed for the furnace. The following shows the essentials required for the proper design of a furnace:

### 6.2.1.- Elements of Furnace Design

a. a power source,
b. heating elements
c. thermal insulation,
d. a crucible to hold the melt
e. a sensor for temperature-feedback
f. a temperature set-point controller
g. a relay or other power-control device for the power source.

The power source, relay, set-point controller and sensor need not be discussed here. However, the insulation, heating elements, and crucibles involve materials, and need to be examined in more detail.

Heating elements can be metal wires or ceramic rods which become incandescent when an electrical current flows through them. Their compositions are critical and have been developed over many years to optimize their performance as heaters. Insulation is used to retain the heat generated and to disperse the heat uniformly throughout the heated space, namely the internal cavity of the furnace. Table 6-1 summarizes the thermal properties of Insulation and Heating Elements for furnace-construction.

**TABLE 6-1**

| INSULATION | | HEATING ELEMENTS | | | |
|---|---|---|---|---|---|
| | Max. Temp. Usable | **In Air** | Max. Temp. | Power Needed Voltage | Current |
| Glass Wool | 600 °C. | Nichrome Wire | 900 °C. | med. | med. |
| Fiberfrax | 1350 | "Globars"™(SiC) | 1475 | low | high |
| Quartz wool | 1100 | Kanthal Wire | 1300 | med. | med. |
| Fire Brick | 1100 to 1650°C | $MoSi_2$ | 1700 | low | high |
| Alumina(foam& beads) | 1850°C | Pt (40% Rh) | 1800 | low | high |
| Zirconia ($ZrO_2$) | 2400 | $ZrO_2$ :Y | 1900 | low | high |
| Magnesia | 2800 | $LaCrO_3$ | 1900 | low | high |

Those heating elements given in Table 6-1 are generally used in air atmosphere, up to about 1800 °C. If one needs to produce a melt above 1800 °C, it is necessary to use refractory metals which must be used in an inert atmosphere. Table 6-2 lists some of these heating elements:

Table 6-2

**Elements For NON- AIR Usage Only**

| | | Voltage | Current |
|---|---|---|---|
| Mo or W wire | 2400 to 2800°C | low | high |
| Iridium wire | 2400 | low | high |
| Graphite | 3400 | low | high |
| R.F. Current | 2800 | low | high |
| Oxy-hydrogen flame | 4000 | NA | NA |

The temperatures given are approximate, and are presented solely for comparison purposes. In furnace insulation, Fiberfrax™ (fibers of aluminum silicate) and "fire-brick" (bricks made from insulating silicate compounds) are the two most commonly used materials. For very high temperature work, Zirconia (= $ZrO_2$) in the form of beads, "wool", and boards pressed from fibers are used for temperatures above 1600 °C. Alumina, i.e.- $Al_2O_3$ , is much cheaper, but does not have the thermal shock, or very high temperature capability (i.e.- > 2000 °C.) of Zirconia.

Wire-wound elements are used most frequently. They are usable in air up to about 1200 °C. and consist of a heating wire or coil, wound upon an insulating- preform. They are cheap and will last for considerable lengths of time, especially if they are run at < 1200 °C. ( i.e.-Kanthal™ wire). On the other hand, wire made from precious metal, i.e.- 60% Pt - 40% Rh = {Pt(Rh)}, can be operated up to 1800 °C. in air, and will operate continuously at 1700 °C. for long periods. However, it is expensive and special care is required when wrapping a furnace core to form the furnace element. Another idiosyncrasy of this type of heating element is that the power-source needs to be especially designed. The Pt(Rh) wire has a very low resistance at room temperature. As it warms to a few hundred degrees, its resistivity changes considerably. Thus, a **current-limiter** is needed in the power control circuit during start-up, or the {Pt(Rh)} wire will melt.

"Globars"™ (silicon carbide rods) are the next most frequently used furnace heating elements. They will operate continuously at 1450 °C. and intermittently at 1500 °C. A newer type of element, Mo wire coated with silicide, i.e.- $MoSi_2$ (to protect the Mo wire against oxidation), has become common. Such heating elements have appeared as "hairpins" and will operate at 1750 °C. in air on a continuous basis. Another heating element coming into use is the defect conductor, $ZrO_2$ , doped with $Y_2O_3$ to make it conducting. However, this type of heating element has special operating characteristics which hinder its wide-spread usage. It does not become conductive until 600 °C. is reached. But, it can be operated continuously at 1800 ° C. in air for long periods of time, with a maximum of 1900 °C. It does require very high currents and low voltage to operate satisfactorily. Still another type of heating element is that of lanthanum chromite,

$LaCrO_3$. These are available in the form of rods and will operate at 1800 °C. in air for long periods.

Most heating elements fail due to the development of internal flaws in their structure. For example, a "globar" is formed by compressing SiC particles to form a rod, and then sintering it. During operation (especially if it is operated at the upper end of its temperature-operating range), it develops "hot-spots". These are due to oxidation and formation of localized resistive areas within the rod (due to diffusion and collection of vacancies at grain boundaries). These areas dissipate power locally, so that the rod eventually fails, i.e.- melts locally at the hot-spot and becomes non-conductive. Wire-wound elements fail in a similar manner, except that formation of vacancies within the metallic structure is the most prevalent mechanism which causes "hot-spots" in the operational heating coil. These vacancies also migrate to grain boundaries. The grain-boundary-junction decreases in conductivity, due to vacancy defect formation. Thereupon, hot-spots form within the wire during operation, causing ultimate failure of the heating element.

Among those heating elements which require the use of neutral, reducing atmospheres, or vacuum, Mo- wire or W-wire in the form of heating coils, graphite in the form of rods or semi-cylinders, are most often used. Iridium wire is also used but it is very expensive. Both Mo and W wire are usable up to 2800 °C while Ir can be used only to 2400 °C. Graphite heating elements can be used above 3000 °C.

We have also included R.F. (radio-frequency) current as a heating element, although it is only a heating **method** when employed with a suitable succeptor. Finally, one other method is listed for the sake of completeness, that of the oxy-hydrogen flame. It generates combustion products ($H_2O$) but the RF- method can be used in any atmosphere including vacuum.

The most critical element in melt-growth of single crystals is the container, or crucible. The first requirement for selection of a suitable crucible is that the crucible does not react with the melt. The second is that it be thermally shock- resistant. A third is that of operational-temperature capability. A fourth is that it be stable in the chosen atmosphere. These requirements eliminate many potential crucible materials for a given application. For use in air, the silica crucible ($SiO_2$) has no peer when used with oxide-based materials up to 1200 °C. It is almost non-reactive, thermally shock-resistant, and inexpensive. Mullite, i.e.- aluminum silicate, is more reactive, less shock-resistant but cheaper than silica. It is used in making glass-melts for the most part. Table 6-3 lists the crucible materials most often used and their temperature capabilities.

TABLE 6-3
COMPOSITIONS SUITABLE FOR CRUCIBLES

| **FOR USE IN AIR** | | **FOR USE IN NON-OXIDIZING ATMOSPHERE** | |
|---|---|---|---|
| MATERIAL | MAX. TEMP. | MATERIAL | MAX. TEMP. |
| Silica | 1200 °C. | Iridium | 2400 °C. |
| Alumina | 1700 | Zirconia* | 2400 |
| Platinum | 1750 | Magnesia* | 2800 |
| Mullite | 1400 | Carbon* | 2800 |
| Boron Nitride | 1400 | Platinum* | 1700 |

* Not in hydrogen

Most metals can be used as crucible materials, but only the precious metals seem to possess the non-reactivity required for melts at the higher temperatures. Alumina is less shock- resistant than silica but can be used at higher temperatures. Shock-resistance relates to how fast one can heat the crucible and its contents up to the melting point of the material without cracking the crucible. $ZrO_2$ and BN are most useful for metal melts. Pt is used for melting in air and remains chemically inert to most melts.

For preparation of melts in inert atmosphere, Ir stands alone. It is non-reactive, has a very high temperature limit, and is not subject to thermal shock or stress. Pt can also be used but at lower temperatures. Ir is about 4 times as expensive as Pt. A good rule of thumb is to use a **metal** crucible for oxide melts and an **oxide** crucible for metal melts, whenever possible.

Next, we need to determine exactly how one goes about obtaining a single crystal.

## 6.3.- STEPS IN GROWING A SINGLE CRYSTAL

Materials tend to grow polycrystalline. This behavior is related to the 2nd Law of Thermodynamics and the Entropy of the system. What happens is that a large number of nuclei begin to appear as the melt temperature approaches the freezing point (but before it freezes). All of these nuclei grow at about the same rate, and at the freezing point produce **a large number** of small crystals. If we wish to restrict the growth to just one crystal, we would like a single nucleus to grow preferentially at the expense of the others. But usually, it does not. However, if we use a "seed" crystal and set up conditions so that it will grow, then we can obtain our desired single crystal.

One method we might use is to cool the melt to incipient nuclei-formation, toss in the seed-crystal, and allow the melt to freeze into a single crystal. This is the KYROPOULOS method which we will discuss in detail later. Alas, this method only works for a few systems, notably alkali halides (cubic) and the like. We find that we can use a seed-crystal to grow single crystals, but only if we use it under carefully defined conditions. A modified Kyropoulos method has been used for many years to form single-crystal sapphire up to 13.0 inches in diameter. Plates cut from such crystals are used as windows and substrates for all sorts of integrated circuits, as well as watch "crystals".

The problems of obtaining a seed-crystal are not simple. We can freeze the melt to a polycrystalline state. When cool, we examine the boule (after first removing the crucible) to try to find a single crystal large enough for a seed (≈ 3 - 6 mm.). We could also cast the melt into a mold and then look for seeds. We could also freeze a polycrystalline rod by pulling it vertically from the melt. We would use a small loop of Pt wire to catch part of the melt by surface tension. By rotating the wire loop while pulling vertically, we find that a polycrystalline rod of small dimension builds up. Once we have the polycrystalline rod, we can reheat it next to the melt surface so that it remelts and "necks-in" as shown in the following diagram:

6.3.1.

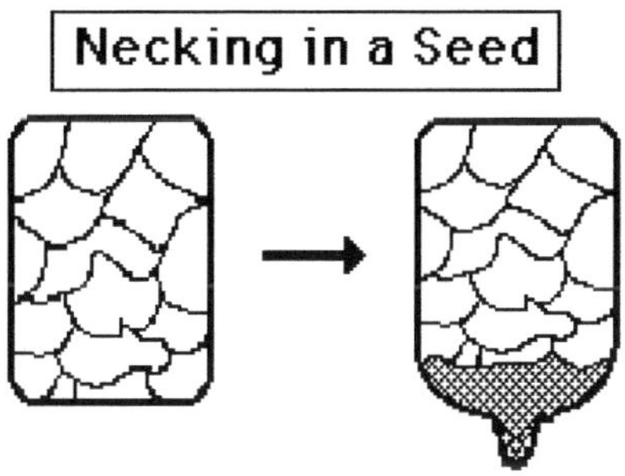


Fused Single Crystal Seed Tip

What happens is that the crystallites melt and fuse into a small tip. If we do this carefully, we will have our "seed". The tip's small size limits regrowth of the remelted part to that of a single crystal. Then, when we return the seed to the melt, we can initiate the growth of a much larger single crystal, provided that growth-conditions are suitable.

Another way to obtain a seed is to dip a capillary tube into the melt. Surface tension causes the tube to fill and when it freezes inside, a small seed results. The difficulty with this method is that it is difficult to obtain a tube of proper diameter, made of the proper material. Glass softens at too low a temperature and quartz melts around 1400-1500 °C. Usually, we are restricted to metals and even then, we must be able to cut the tube to obtain the seed, since it is confined within the tube. Once in a while, we can use the tube directly and obtain growth directly upon the seed, even though it has remained within the tube.

There is one other method that can be employed to obtain a seed. We use a metal crucible having a small tip and cause a melt to form. The tip acts in the same way to form the seed. Nevertheless, we have the same problem as when we use the metal capillary. We need to obtain the seed free from its holder.

6.3.2.- Crucible for Use in Obtaining A Seed from a Melt

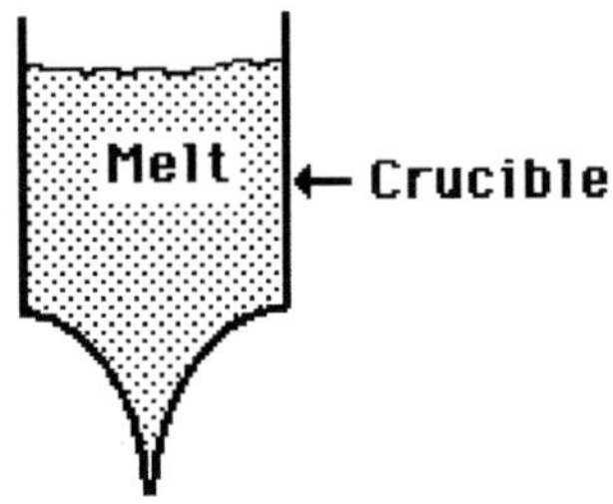


Thus, the crucible must be sacrificed, or else the whole mass must be extracted from the crucible with the seed in the tip intact. If there is a sufficient difference in contraction between the mass and the crucible, perhaps then we can obtain the whole mass intact, including our seed.

## 6.4.- CZOCHRALSKI GROWTH OF SINGLE CRYSTALS

The most common method for growing single crystals, whether they be a metal or a complex mixture of oxides, is to pull a single crystal from a melt. This is the so-called **Czochralski Method**, using the method invented in 1918 by the Polish scientist Jan Czochralski, which he called crystal pulling. Large crystals can be grown rapidly from the liquid formed by melting any given material (providing that the materials does not decompose upon melting). One attaches a seed crystal to the bottom of a vertical arm such that the seed is barely in contact with the material at the surface of the melt and allows the crystal to slowly form as the arm is lifted from the melt. Let us examine this method in more detail to illustrate its versatility in crystal growth. As an example, we will grow a crystal from a melt of oxides whose melting points exceed 1800 C. Automatically, we are limited to use of an iridium (Ir) crucible and we will use an R.F.- generator for the power source. In 6.4.1., (see next page), we illustrate the typical setup for the crucible in the Czochralski apparatus.

6.4.1.-

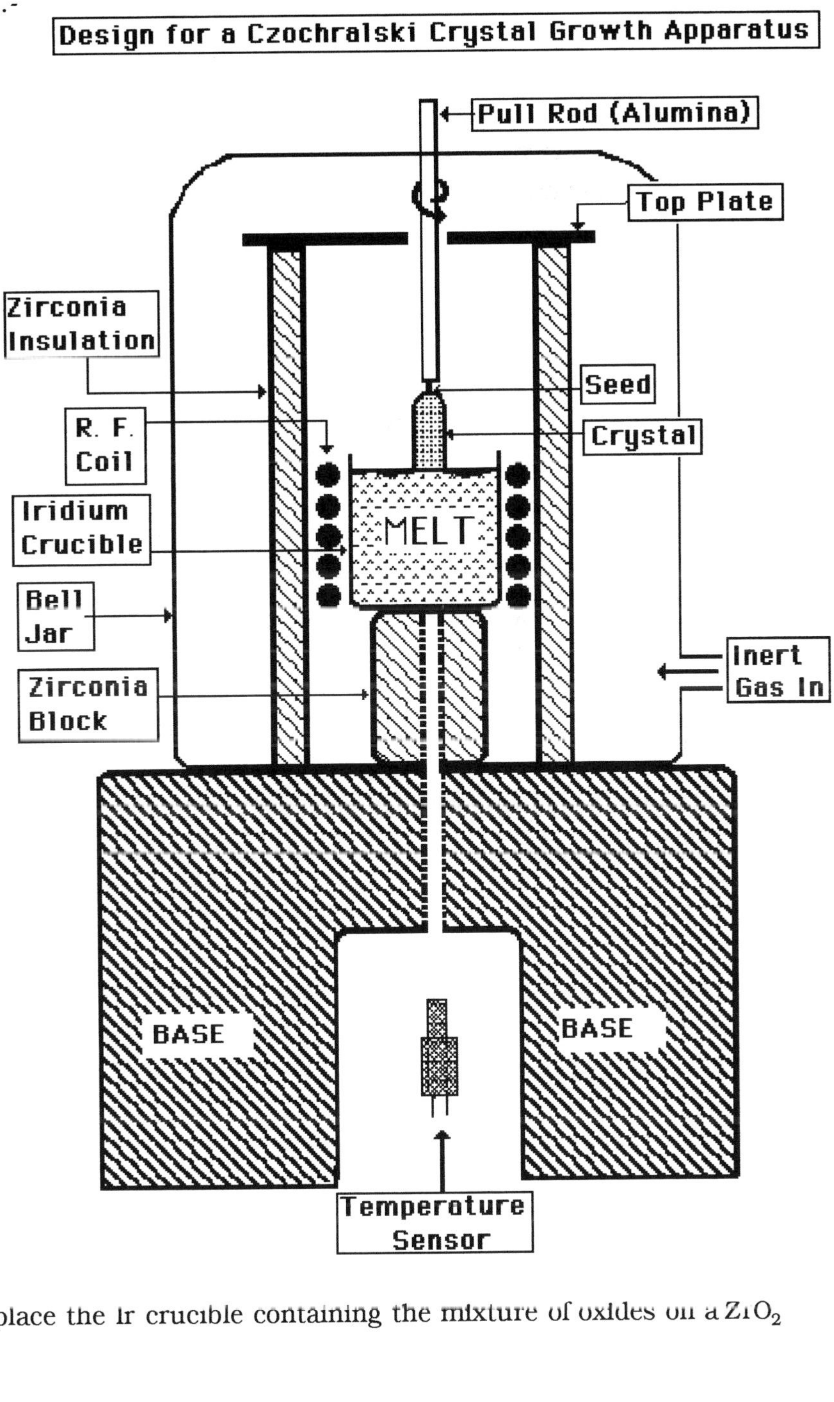


We place the Ir crucible containing the mixture of oxides on a $ZrO_2$

platform. This acts as a thermal barrier for the bottom of the apparatus. We may then place a larger $ZrO_2$ cylinder around the crucible for further insulation. An R.F. coil is placed in position around the outside of the insulation. Finally, an outside wall of insulation is put into place and a top cover plate is put into position. At 2000 °C., the outerwall thickness of $ZrO_2$ needs to be at least 2.5 - 3.0 cm. The whole is then covered by a bell-jar having a hole at the top. The dome and its contents are flushed for several minutes with an inert gas before the R.F.- generator is turned on. Gas flow is maintained throughout the entire operation. As the crucible heats ups, it's entire contents degas. As the materials melt, they contract so that it is usually necessary to add more material to the melt until the crucible is full before actual crystal growth is attempted. This may take 3 - 4 crucible volumes of material. Thus, the top-hole needs to be large enough to accommodate this operation.

**Pulling a good single crystal is not easy.** There are many factors involved, each of which imposes restraints upon the others. Some depend upon the crystal pulling system design, while others depend upon the nature of the material being grown. These factors, for a given system, become **parameters**, i.e. they are inter-related.

The parameters for CZOCHRALSKI GROWTH are listed in the order of their importance :

### 6.4.2.- CZOCHRALSKI CRYSTAL GROWTH PARAMETERS

1. Melt temperature and its thermal conductivity
2. Chemical stability of the melt
3. Degree that melt will "super-cool"
4. Temperature control and temperature gradient achievable
5. Interface angle which forms between crystal and melt
6. Degree of diameter control possible
7. Rotation rate used
8. Rate of pulling
9. Melt level maintained

All of these factors are dependent upon the nature of the crystal composition being grown including its thermal, physical and chemical properties. Because of the importance of R.F. generators to crystal growth, let us now examine them in more detail.

A radio-frequency generator is essentially a gigantic R.C.-tank circuit (R.C. = resistance-capacitance) which uses mercury-pool-diodes as oscillators. The frequency is fixed and the generator is self-regulating in power-output. As a result, change of power- output, which affects temperature regulation, is slow. The actual output of power occurs through the R.F. coil. The coil itself is a carefully-wound copper tube helix which has cooling water running through it, as shown in the following:

6.4.3.-

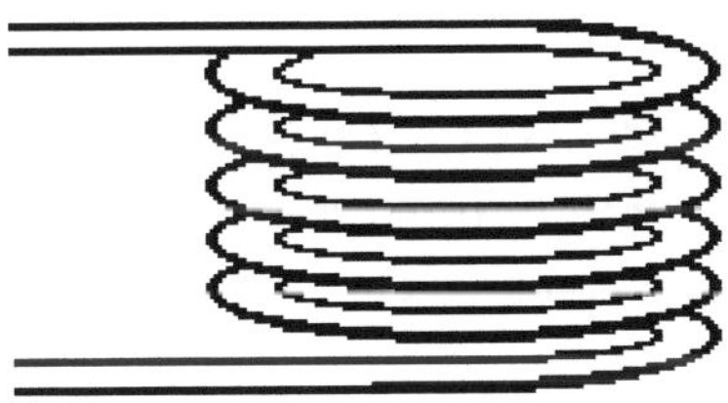

If a susceptor such as a metal crucible is placed within the coil, The R.F. power induces "eddy currents" in the crucible, causing it to heat up. Eddy-currents are circular electrical currents induced within the metal by the R.F. field of the coil. It is essentially a "skin" effect, and the depth of penetration, i.e.- depth of eddy-current generation within the crucible is defined by:

6.4.4.- $\text{depth} = 1/2\pi \, [\, \rho \,/\, \mu f \times 10^9 \,]^{1/2}$

where $\mu$ is a permeability , $\rho$ is a resistivity of the susceptor (crucible) and f is the frequency of the R. F. field. The spacing of the coils is extremely critical. One usually speaks of the "coupling ratio" for a given crucible- R.F. coil combination. This is the ratio of power transmitted to the crucible divided by the power delivered to the coil. R. F. generators come in various sizes, as follows on the next page:

| 6.4.5.- | 10 kc - 20 KW | 50 kc - 25 KW |
|---|---|---|
| | 100 kc - 50 KW | 2 meg. - 75 KW |

The size to use depends upon the susceptor being used, the temperature of operation desired and the heat losses within the system. For an Ir crucible, the 10 kc-20 KW generator works better than the 2 meg.- 75 KW generator.

Returning to Czochralski growth, of the 9 parameters given in 6.4.1. above, the first 3 are functions of the material whose single crystal we are trying to grow. #4 relates both to the material and the physical design of the melt-furnace plus that of the coil. Parameter #5 relates almost entirely to the operation of the system. As we have indicated, the steps in operating the apparatus are:

6.4.6.- OPERATION OF THE CZOCHRALSKI APPARATUS

1. Load crucible and start melt. Maintain inert atmosphere at all times.
2. Add material to melt until crucible is full.
3. When the crucible is full, gradually lower the melt temperature to achieve supercooling and incipient nucleation.
4. Allow system to equilibrate and dip seed into melt. Rotate seed at predetermined speed.
5. Allow seed to grow outward to desired diameter.
6. Begin pulling incipient crystal out of melt at a rate which maintains crystal size and growth.

We find that the rotation rate affects both the "interface angle", i.e.- the angle between crystal and melt, and control of crystal diameter, as shown in the following diagram, given as 6.4.7. on the next page.

Note that we have defined "interface angle" as the angle between the growing crystal and the residual melt. Rate of pulling also affects the quality of the crystal as well as the actual numbcr of intrinsic defects which may appear in the final crystal. In the upper left of 6.4.7., a flat-

6.4.7.-

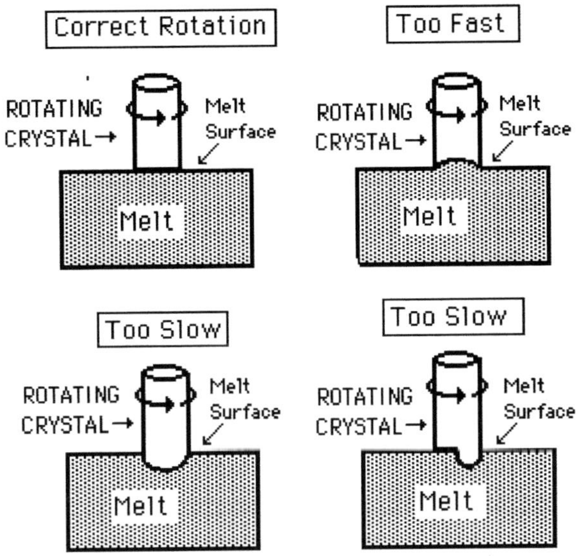


interface between the crystal and melt surface has been maintained by regulating the rate of rotation. If it is not maintained correctly, the edges of the crystal may grow too fast (resulting in a hollow crystal- upper right) or the edges may round (lower left) or facet (lower right). If faceting is allowed to continue, we find that the crystal so produced is not usable because the facets produce a polycrystalline rod rather than a single crystal.

Diameter control, Parameter #6, is important to the final quality of the obtained crystal. The sides of the crystal need to be straight because they reflect the regularity of the lattice planes within the crystal. Effects of deviation from "correct" growing conditions on the quality of the crystal so-produced are shown in the following diagram, given as 6.4.8. on the next page.

When growth conditions are not “optimum”, serious defects can appear in the growing crystal as a result of physical circumstances of the crystal growth apparatus. Because of this, we need to be very careful while the crystal is growing.

6.4.8.- Defects Produced in the Growing Crystal as a Function of Growth Conditions

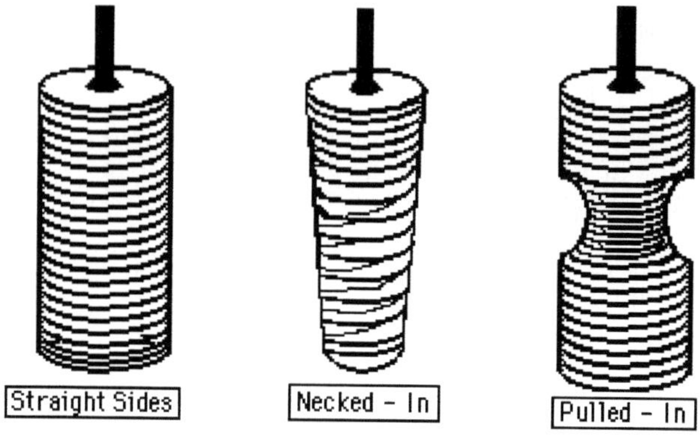


It is this factor which mandates the very careful design of the Czochralski Apparatus.

A proper design consists of a heavy base, to minimize effects of external vibrations on the surface of the melt, a precision screw driven by a controlled reversing-motor (so as to control rate of pulling precisely), and a precision motor controlling rate of rotation. Such a design is shown on the succeeding page as 6.4.9. Note that we have not shown the R.F. generator. However, we have shown one design of the pulling apparatus (there are many).

The next most important parameters in Czochralski growth of crystals are: the **heat flow** and **heat losses** in the system. Actually, all of the parameters (with the possible exception of #2 and #9) are strongly affected by the **heat flow** within the crystal-pulling system. A typical heat-flow pattern in a Czochralski system involves both the crucible and the melt. The pattern of heat-flow **is** important but we will not expand upon this topic here. Let it suffice to point out that heat-flow is set up in the melt by the direction of rotation of the crystal being pulled. It is also affected by the upper surface of the melt and how well it is thermally insulated from its surroundings. The circular heat flow pattern causes the surface to radiate heat. The crystal also absorbs heat and re-radiates it

6.4.9-

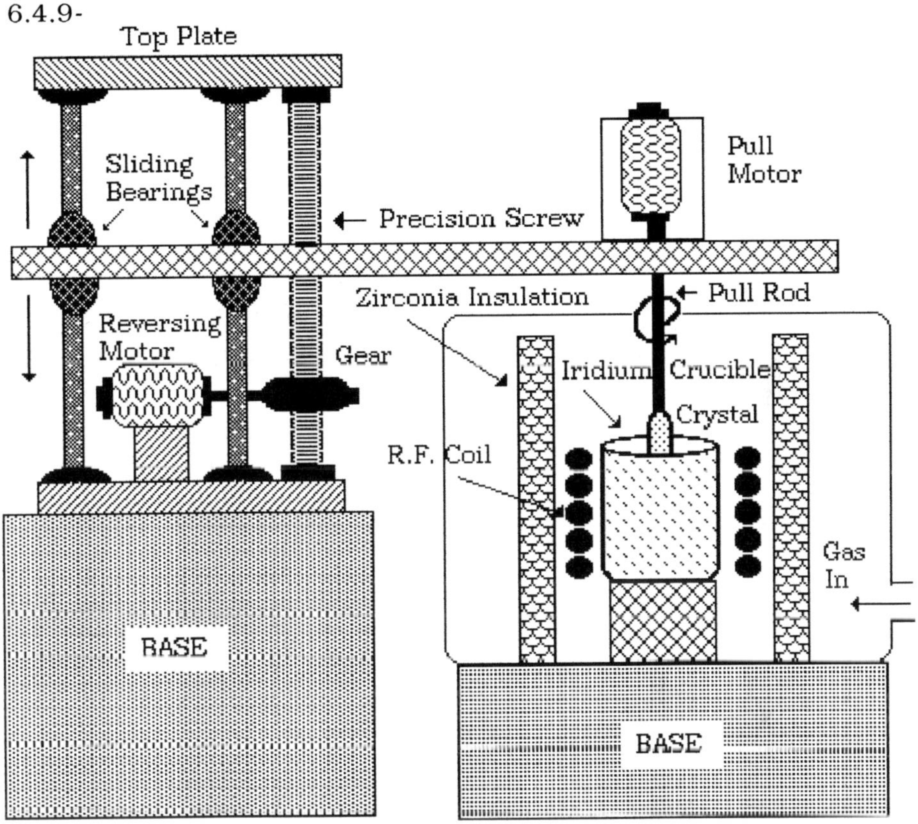


further up on the stem. If the crystal is kept stationary, we see that the heat flow pattern is uneven. Thus, it is mandatory that the crystal be rotated to control crystal diameter, so as to obtain a defect-free crystal. It should be clear that the most important part of the apparatus is the "pull-rod". In most cases, this is formed from a precision screw which is turned to raise the rod while the whole assembly is being rotated.

The above description applies to the system where only the growing crystal is rotated. There is at least one other way to "stir" the melt so as to control heat flow. This is illustrated as follows:

6.4.11.- $+$ crystal rotation $=$ $\pm$ crucible rotation
$\pm$ crystal rotation $\neq$ $\pm$ crucible rotation

We can rotate the crucible by itself, or in conjunction with the crystal. But another complexity arises, namely what direction of rotation and what relative speed of rotation should we use for both, or either? If we rotate the crystal clockwise and the crucible counter-clockwise. then the heat flow patterns become complex indeed. These complexities have been studied in detail but will not be enumerated here.

The best compromise seems to be fast- rotation for the crystal and slow or no rotation for the crucible. Of all the possible methods of stirring the melt, the static-crucible method seems to be the best, and this is the method used by most **crystal-growers.** The next best method seems to be rotating the crucible at a slow rate, counter to the direction of the crystal rotation. It is clear that crystal- rotation needs to dominate the stirring pattern so that mixing of the melt continues while the crystal is growing.

The Czochralski Method has remained the most frequently employed method for obtaining single crystals. The seemingly formidable complexities involved have been conquered by the use of computer-control of the crystal-growing Parameters. One such system, available commercially, is shown in 6.4.12. (on the next page).

Here, we show only a bare outline of the individual components in the overall system. This SYSTEM is capable of operation in inert atmosphere or vacuum. A slave-processor controls both crystal diameter and meniscus-contact of the growing crystal. As we have stated, this is most important if we wish to obtain a crystal essentially free from ingrown defects. This takes the form of a melt/seed contact monitor and a separate monitor is provided for the operator as a CCTV camera for observing the crystal diameter as it grows. There is also a crystal annealing furnace to remove any crystal strain that may have been induced by the crystal-growing conditions. A base heater helps to maintain a uniform temperature gradient in the melt-crucible during crystal growth. Note that both the crucible and crystal rotation can be controlled. In order to control the heat-convection patterns which normally appear in the melt, an external cryomagnet can be supplied. Its magnetic field helps control heat losses and maintains a better control of the crystal growth.

6.4.12.

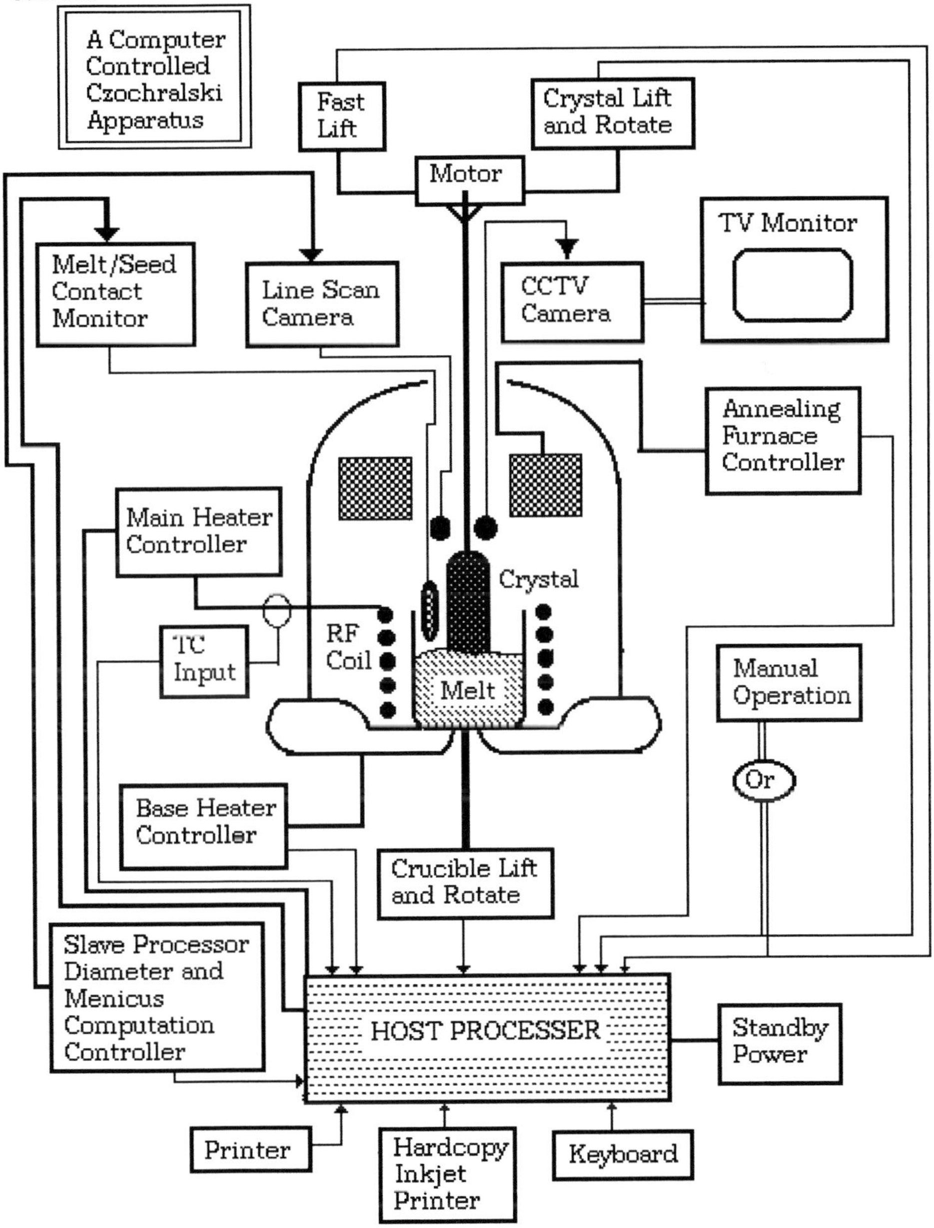


This has not been shown but includes a source of liquid nitrogen for the cryomagnet as well as a controller to maintain the correct magnetic field needed for defect-free growth of the crystal.

A final comment on the Czochralski Method: GaAs has become important in construction of integrated circuits for computers because of the promise of speed of response and density of components, compared to silicon wafers. Both As and Ga tend to oxidize in air as they approach the melt stage and $As_2O_3$ sublimes. Both elements are toxic to man. In the melt stage, they have high partial pressures as well, so that the use of an inert atmosphere is not sufficient to allow growth of a single crystal. One method that has been used for this case has employed a "liquid encapsulating agent". As shown in the following diagram, this consists of a lower melting agent used to form a liquid barrier, floating on the surface of the melt which serves as the source for the crystal (in this case, Ga-As).

6.4.13.-

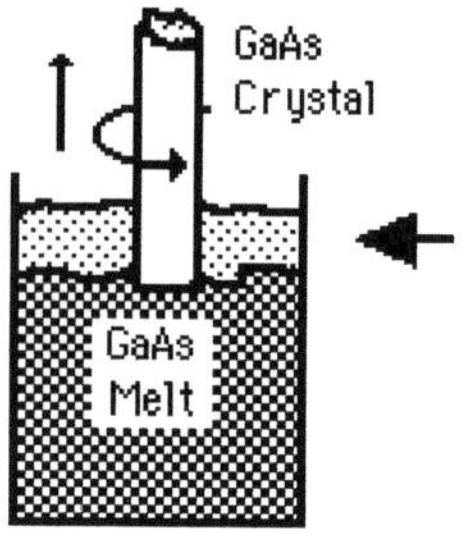


The arrow indicates the liquid barrier layer. This use of a barrier melt illustrates that there are several ways to grow crystals which would be difficult to obtain under "ordinary" means of crystal growth, i.e.- prevention of oxidation and evaporation of GaAs during crystal growth.

## 6-5: The Bridgeman-Stockbarger Method for Crystal Growth

Whereas the Czochralski method requires rather elaborate equipment to obtain single crystals of good quality, the Bridgeman (Stockbarger) method uses a fairly **simple** apparatus. This is shown in the following diagram, given on the next page as 6.5.1.

6.5.1.-

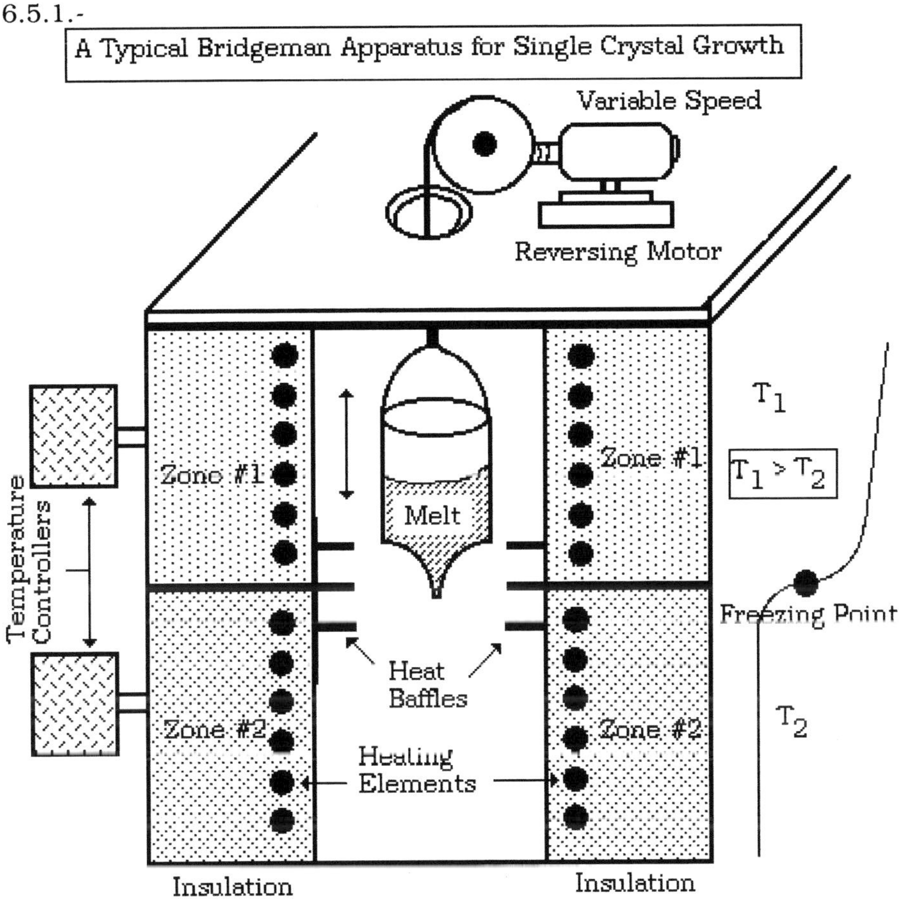


The major components needed to grow single crystals by this method are:

- ❋ A two (2) zone furnace
- ❋ Two (2) set point temperature- controllers
- ❋ A crucible with "seed" tip
- ❋ A constant-speed elevating and lowering device for the crucible.

The procedure involves obtaining a full crucible of melt whose composition mirrors that of the crystal we wish to grow. We then lower the crucible containing the melt through a baffled zone within the furnace

which are polished and used for UV transmitting windows, substrates for various electronic devices and even non-scratching faces on your watch. Note that sapphire is about as hard as diamond and has a heat-transmitting capability almost equal to many metals, while remaining essentially chemically inert and electrically non-conductive.

In other variations of the Kyropoulos method, we raise the melt through the freezing point of the melt, by raising it past the baffles of the furnace. Another possible method is one where we use a single- zone furnace, stabilize the melt, lower the furnace temperature to incipient nucleation and then cool to form the crystal. However, this works only for a few systems. The reason is probably supercooling. There are only a few melts where a large degree of supercooling occurs. Addition of a seed will then cause very rapid growth of the single crystal. But, it is usually strained and must be carefully annealed to obtain the mass intact. If the apparatus shown in 6.5.3. is used and the crucible is exactly positioned so that a vertical temperature gradient is maintained **exactly along the crucible height**, the melt, when seeded, will grow into a single crystal more slowly. Then, the degree of supercooling is not as important. As a matter of fact, melts with little or no degree of supercooling can be induced to form single crystals by following this procedure.

## 6.6.- ZONE MELTING AS A MEANS FOR FORMING SINGLE CRYSTALS

There is nothing to say that we must melt all of the material at one time. If we have a long rod of polycrystalline material, and melt a part of it, i.e.- a "zone", we can sweep this zone down the length and ultimately melt and freeze all of the rod. The major problem is, again, inducing a "seed" of single- crystal material to form (and to continue). In the forward direction of travel, the zone is melting whereas at the back of the zone, freezing occurs. If we can induce the back part to freeze as single- crystal, then we can transform the whole rod. This means that we can start with polycrystalline material and end up with a single- crystal:

6.6.1.- Powder $\Rightarrow$ Polycrystalline $\rightarrow$ Single Crystal

For the case of Zone-Melting to form a single crystal, we can distinguish between two separate cases, i.e.- horizontal and vertical zone melting. This is shown as follows:

6.6.2.-

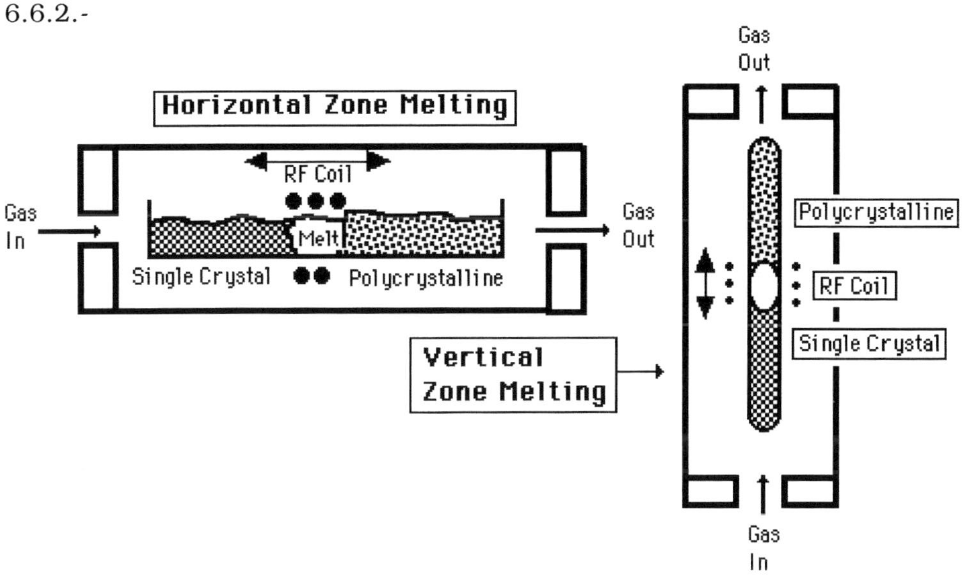


Our choice among heating elements is limited to the R.F.- coil, a hot-wire coil or perhaps the $MoSi_2$ "hairpin" element. The reason is that we must restrict the melt zone as much as possible, while moving the zone through the material. Heating by radiation is not very effective since it occurs in a $4\pi$ direction. One possible solution is to use a heat reflector around the wire heating element to limit the heating length of the melt zone. Another limitation of the method is that the power losses in the system are large because we must use a protective tube or other insulation around the material so as to be able to move the heat source. Unless we are willing to squander power, we cannot reach the very high temperatures of melting required for some materials. Therefore, the zone-melting method of single crystal growth has usually been confined to materials of moderate to low melting temperatures such as silicon.

The vertical method shown above has sometimes been called the "floating zone" method. The thickness of polycrystalline rod determines the maximum length of melt zone that can be used successfully. This relation can be calculated from:

6.6.3.- THIN RODS THICK RODS

$$Z_{Max} = 2\,(3r)^{1/2} \qquad Z_{Max} = 2.8\,[\,\gamma\,/\,\rho g\,]$$

where r is the radius in cm., $\gamma$ is the surface tension of the melt, $\rho$ is its density , g is the gravitational constant, and $Z_{Max}$ is the maximum length of the **floating zone**.

## 6.7.- ZONE REFINING

When a crystal freezes from the melt, it tends to reject impurities. If zone - refining is applied to such a crystal, the resulting single crystal **can be purer than the original.** If we make more than one pass, we can approach a high degree of purity. Such a procedure is called "zone- refining".

It has been determined that there is a distribution coefficient for the impurities between crystal and melt which favors the melt. We can see how this arises when we reflect that impurities tend to cause formation of intrinsic defects within the crystal and lattice strain as a result of their presence. In the melt, no such restriction applies. Actually, each impurity has its own distribution coefficient. However, one can apply an average value to better approximate the behavior of the majority of impurities.

One calculates the concentration of impurities left in the single crystal, as a function of a single pass, from:

6.7.1.- $$c_i = c_o\,[1 - (1-k_o)\exp\{-\,k_o L / Z\}]$$

where $c_o$ is the **original** concentration of impurities, L is the length the zone travels (in cm.), and Z is the melt-zone length. Note that we are taking a ratio of L to Z. We need to minimize Z in order to make the process efficient in segregation of impurities. $k_o$ is the distribution coefficient, defined by:

6.7.2.- $k_o = c_i\,(\text{solid}) \;/\; c_i\,(\text{melt})$

We plot $c_i$ / $c_o$ vs: L / Z, the number of zone-lengths solidified. This is illustrated in the following diagram:

6.7.3.-

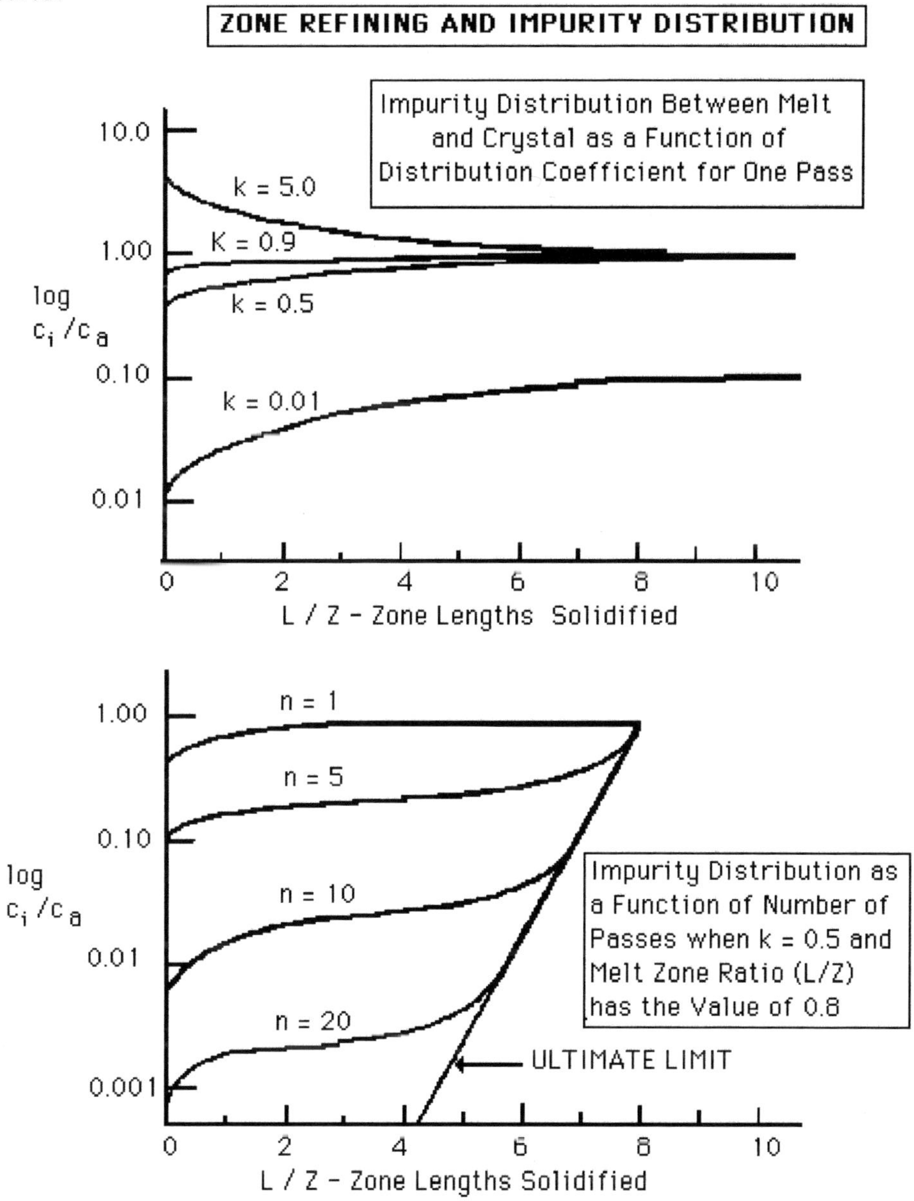

At the top of this diagram, we show the effects of **one pass** on various selected values of the distribution coefficient, $k_o$ . Note that at values of $k_o$ > 1, we do not obtain any significant improvement in impurity content. For example, we start at k = 5.0, but end up at 1.0. Even at $k_o$ = 0.5, the further we move the molten zone, the less we purify the material. This has been called the **impurity-leveling effect**. At the very low values of $k_o$, one still cannot improve the impurity level to any significant degree, using just one pass. We find that multiple passes are required to obtain purification. At the bottom of 6.7.3. is shown the effect of multiple passes upon the degree of purification obtained. Here, we have used:

6.7.4.- $k_o = 0.5$ and $L / Z = 8$

and have recalculated the data given before. Note that as the number of passes reaches 10, we obtain a change of some 90 times in the impurity content **at the front of the rod.** But, at the back, i.e.- L/Z = 8, we approach an ultimate distribution with 20 passes (the dotted line). If we could see the internal distribution of impurities after the rod had been zone-refined by 20 passes, we would see:

6.7.5.-

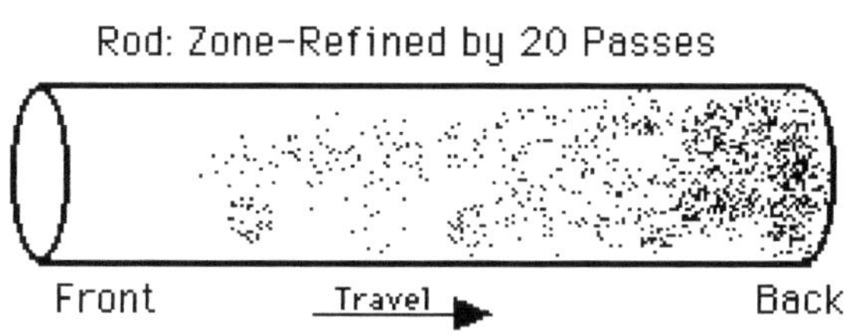


where the density of dots represents the concentration of impurities. What has happened is that we have moved many of the impurities from the front to the back, but not all. and we finally end up with an impurity distribution we cannot change.

To understand this, we must examine the impurity-leveling factor in more detail.

## 6.8. THE IMPURITY LEVELING FACTOR

When a melt-zone is moved through a long crystal, an impurity concentration builds up in the melt zone due to rejection by the crystal as it resolidifies. We can also say that the distribution coefficient favors a purification process, i.e.- k<< 1. Another reason (at least where metals are concerned) is that a solid-solution between impurity and host ions exists. It has been observed that the following situation, as shown in the following diagram, occurs:

6.8.1.-

The impurity, x , builds up at the solid- liquid interface as the liquid zone moves and the solid forms. We can write for the distribution coefficient:

6.8.2.- $$k_x = c_{xL} / c_{xS}$$

where $c_{xL}$ is the impurity- concentration in the liquid, and $c_{xS}$ is the impurity concentration in the solid. We then estimate $k_x$ from:

6.8.3.- $$k_x \, N_I / N_{MX} \exp - (E_M - E_X) / kT$$

where I is the impurity, MX is the compound under consideration, and $E_M$ , $E_X$ are the activation energies for formation of solid solution of MX · IX . Note that we have assumed that I is a cation impurity in the crystal, MX. We can differentiate between 2 separate cases, as shown in the following diagram, given as 6.8.4. on the next page.

Here, we show two cases for impurity segregation between melt and crystal as it grows in time. Note that an initial purification occurs in both cases but the distribution coefficient for the case on the right is such that the amount of impurity **actually incorporated** into the crystal, $k_i \cdot c_o$.

6.8.4.- Behavior of Impurities as a Function of Type of Solid Solution Formed

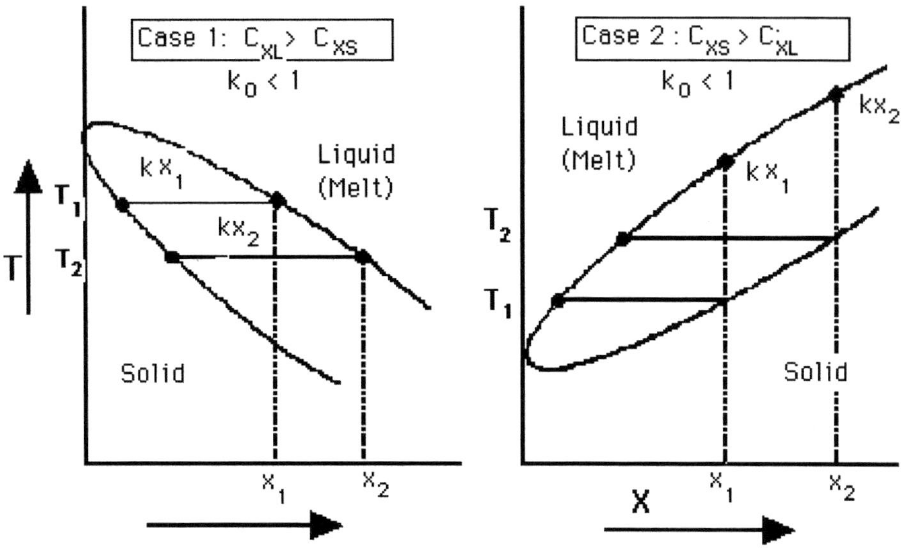


Note that it rapidly approaches the **original** impurity concentration, $c_0$ . At $t_1$, it actually **increases** over the original concentration at some point within the solid crystal. It should be clear that if the impurity freezes into the solid from the melt, there must be a certain amount of solid solution formation, even though the impurity content be less than 0.01%. Case I & II show simple solid solution behavior for $k < 1$. For $c_{xL} > c_{xS}$ , we will get an improvement in purity in the crystal. But for $c_{xL} < c_{xS}$ , the opposite effect is seen. Thus, two different impurities could manifest opposite behaviors, depending upon what host they were in.

The following diagram, given as 6.8.5. on the next page, illustrates further phenomena regarding zone refining. The same situation seen in 6.8.4. occurs for the case at the bottom right of 6.8.5. except that the distribution coefficient is such that the impurity buildup is slower. Nevertheless, simple solid-solution for impurity systems is rarely the norm. The most prevalent case is that of Case III of 6.8.5. Limited solid solution occurs, and we get a two-phase system.

6.8.5.-

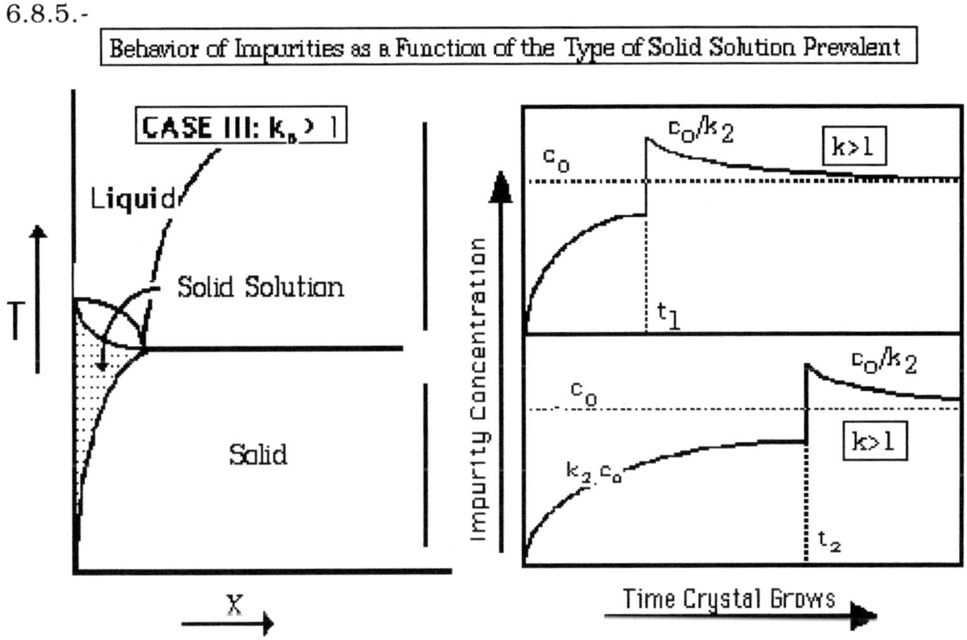


This illustrates the fact that impurity segregation and purification processes are dependent upon the type of impurity involved and its individual segregation coefficient. As we illustrated above in 6.8.1., the problem is that the impurity is initially rejected from the solid, but its concentration **builds up in front of the growing crystal.** The segregation coefficient, $k_i$ , then operates on that increased concentration and the product, $k_i\, c_o$ , increases.

If we use individual zone lengths, as shown in 6.7.3. , we would have:

6.8.6.-

| ZONE | IMPURITY CONCENTRATION |
|---|---|
| 1. | $(c_o - k\, c_o)$ |
| 2. | $(c_o - k\, c_o) + k(c_o - k\, c_o)$ |
| 3. | $(c_o - k\, c_o) + k(c_o - k\, c_o) + k^2(c_o - k\, c_o)$, etc. |

This is the reason for the behavior shown in 6.8.4. The impurity front builds up until its concentration surpasses the original concentration (sometimes by manyfold). Thereupon, the impurity concentration levels out (but not necessarily to the same $c_0$ we have used for illustration in 6.8.4.). It is for these reasons that zone-refining has found limited application, since actual purification by this method does not produce a crystal with a completely uniform impurity distribution.

The conclusion we must draw is that we should purify our raw materials **before** we attempt to grow our single crystal, not after. Zone refining has been found to be effective only in a few limited cases. It is not applicable to the general case or to high melting crystals. But if we wish to do **zone-leveling of an added impurity,** then we have a very useful technique. A good example would be to add a small amount of Ga to Ge . We can add a small volume of Ga to the front zone and will find that it has become evenly incorporated into the Ge crystal after several passes, using zone-refining.

6.9.- <u>THE VERNEUIL METHOD OF CRYSTAL GROWTH</u>

This method has also been called the "flame-fusion" method of crystal growth. If a powder is blown through a flame, it will melt if the flame is hot enough. The only flame hot enough to do this is the oxy-hydrogen flame. However, a specially designed burner is required so that the crystallites will melt during the short time that they pass through the flame-front. The original work was accomplished by Verneuil (1931) and the apparatus is named after this investigator. The tip of the torch is important. One design used for such a torch is given as follows:

6.9.1.-

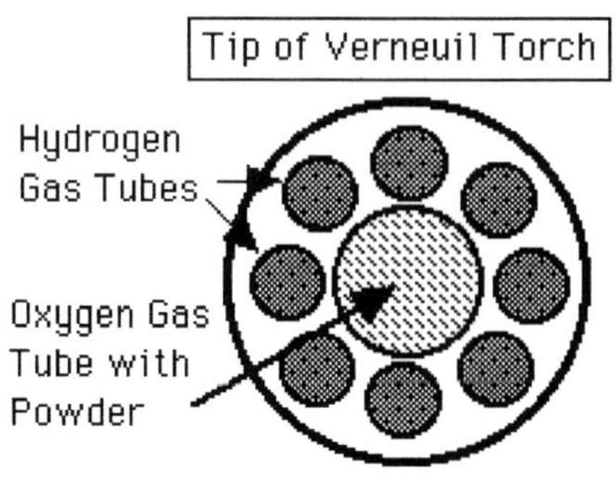

The center tube is used for oxygen gas which transports the powder. The burner is designed so that the outer tubes contain only hydrogen gas, with the **interstices** between the $H_2$ - tubes transporting additional oxygen gas. This design will melt $Y_2O_3$ powder which has a melting point of 2380 °C.

The overall Verneuil apparatus, shown in 6.9.2. on the next page, consists of a sealed hopper to contain the powder, the TORCH itself, a refractory pedestal to hold the growing crystal, and an after-furnace to anneal the crystal.

The procedure for using the apparatus shown in 6.9.2. is given as follows:

6.9.3.- Operating Instructions for Verneuil Apparatus

1. Fill powder hopper, close and attach to burner.
2. Light burner and adjust flame to oxidizing.
3. Adjust pedestal to proper height and begin rotation.
4. Slide after-furnace into position.
5. Turn on and adjust oxygen flow in powder hopper.
6. Begin tapping.
7. As melt begins to build up, adjust both height and speed of rotation so that crystal begins to form in a regular manner.
8. When the proper crystal size has been reached, stop powder flow by shutting off the oxygen flow to the hopper. Continue flame operation and pedestal rotation.
9. Lower the crystal into the after- furnace, and anneal crystal, Gradually lower furnace temperature and let cool to room temperature.

Thermal stability is important in this method because the high temperatures reached may be sufficient to cause decomposition of the material. The Verneuil Method of crystal growth is not generally applicable to all types of crystals. There are serious deficiencies in the method. For example, there is a large temperature drop of hundreds of degrees over a few millimeters within the crystal. This causes a large difference in thermal expansion within a limited space, and consequent

6.9.2.-

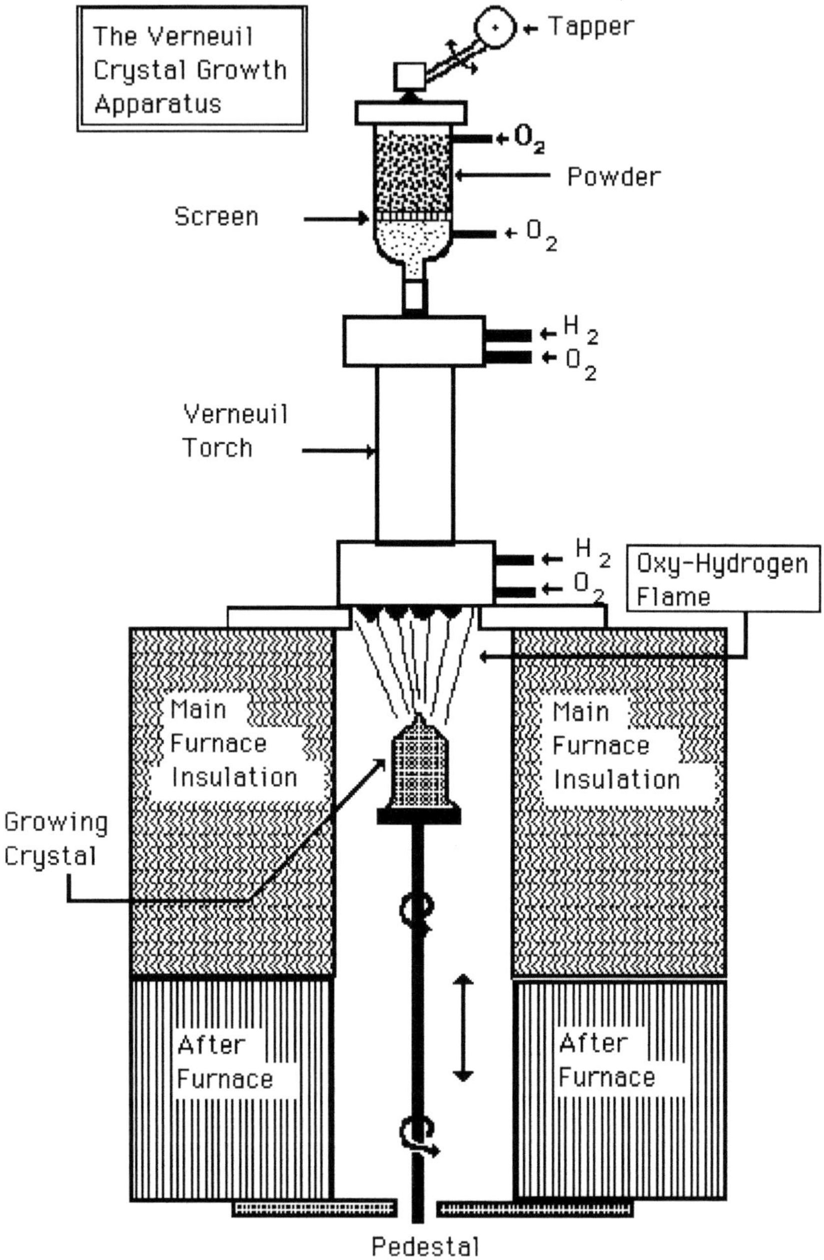


strain. Many crystals are not refractory enough to withstand the stress buildup and so crack into many smaller parts. This makes it very difficult to obtain a single crystal of any size. Thus, if a crystal is grown by this technique, it must be annealed carefully in order to obtain it intact. Generally, the method is restricted to crystals like $Al_2O_3$ whose thermal conductivity is high and whose refractive nature makes it possible to obtain a crystal. For the most part, one is restricted to growing simple oxides by this methods.

Complex oxides such as $Ca_2SiO_4$ or $Ca_2P_2O_7$ generally cannot be grown easily and the Czochralski Method becomes the method of choice.

The parameters involved in the Verneuil method are:

1. Melt Temperature
2. Thermal Stability at Melt Temperature
3. Temperature Gradient present in Furnace and after-furnace.
4. Sintering Volume Losses
5. Thermal Conductivity of Crystal
6. Annealing Temperature Required
7. Effects of Reducing Flame on Material
8. Chemical Stability of Matcrial bcing Uscd

6-10: <u>MOLTEN FLUX GROWTH OF CRYSTALS</u>

This method employs a molten flux which dissolves the material and re-deposits it upon a selected substrate. That is, the molten flux acts as a transport medium. The temperature of the flux can be varied to suit the material and to promote high solubility of the solute material in the molten solvent. One example is "YIG", yttrium iron garnet, i.e.- $Y_3Fe_5O_{12}$ . This material is used in the Electronics Industry as single crystals for microwave generating devices. It can be grown via the molten flux method.

A typical molten-flux apparatus is shown as 6.10.1. on the next page.

6.10.1.- A Molten-Flux Apparatus

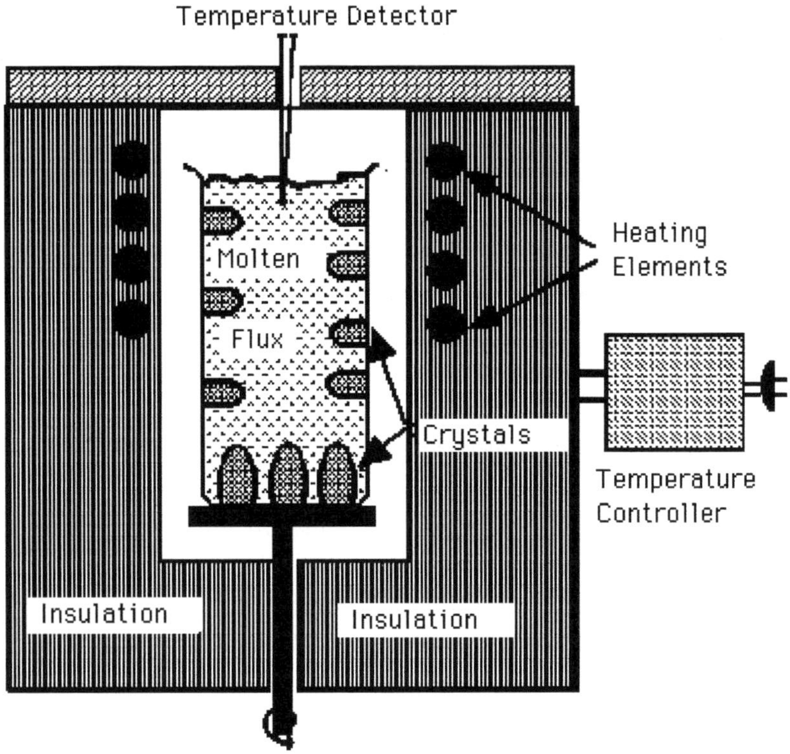


The six steps involved in using this method are as follows:

1. A flux such as lead borate is melted. $PbB_2O_4$ is useful in this method because it will undergo supercooling rather easily.
2. The material which is to form single crystals is dissolved in the molten flux to near saturation (Note that this requires prior knowledge regarding solubility of compound in molten flux).
3. The crucible, which is usually platinum, is rotated to obtain a uniform temperature distribution within the melt-compound solution.
4. The solution temperature is gradually lowered to incipient nucleation. At this point, because of the physical arrangement

of the heating elements, a temperature gradient will exist along the length of the crucible, from top to bottom.
5. Single crystals will begin growing along the bottom of the crucible, and sometime later along the edges.
6. The crucible is kept rotating to maintain a uniform mixing and heat flow while the crystals are growing.

This method has serious deficiencies for use as a general method. We find that if we dissolve MX in a $BX_2$ flux and grow an MX crystal, it is likely to be contaminated with B, or even $BX_2$.

The crystal-growing parameters for this method are:

6.10.2.- Molten Flux Growth Parameters

1. Flux melt temperature
2. Solubility of MX in $BX_2$
3. Degree of supercooling
4. Temperature gradient achieved
5. Rate of rotation used

If these are not within the correct range, one does not obtain single crystal growth. This method has been used in the past only because of its relative simplicity of apparatus and materials. However, the quality of crystals so-produced has been rather poor. Crystals produced by this method are suitable for structure determinations, but are poor in optical quality and are notat all suited for electronic applications.

There is one area where molten flux growth has been used to form an epitaxial growth of a single crystal film on a substrate. In this case, we can use molten lead tetraborate ($PbB_4O_7$), which melts at about 960 °C. The melt is raised to 980 °C. and the requisite oxides are dissolved therein. The temperature is lowered to near 960 °C. and supercooling begins. Since the substrate is cooler, a garnet film will grow on its surface under these conditions. The surface of the GGG slice will have been carefully polished to minimize surface defects.

6.10.3.- Epitaxial Growth on a Crystal Wafer

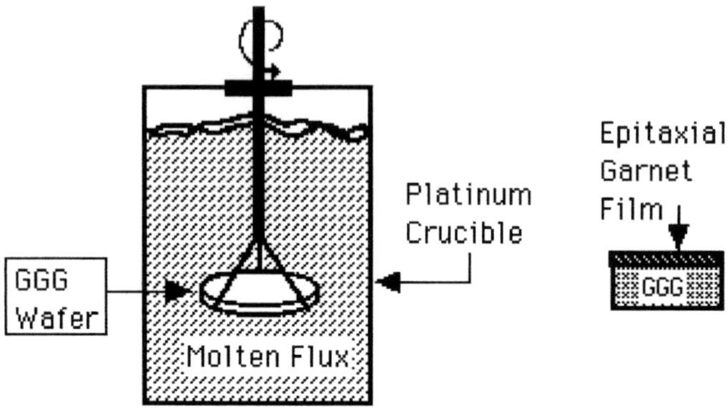


It takes about 6 minutes to grow a single-crystal film about 50 - 100 μ thick. The film grows in an epitaxial manner, that is- it builds up on the crystallographic planes of the substrate itself. It should be obvious that the degree of supercooling that the melt will undergo before the dissolved oxides precipitate out depends upon the amount of oxides dissolved. **This is generally determined by trial and error.** We lower the temperature to just above that point. A substrate slice of single- crystal GGG (gadolinium gallium garnet = $Gd_3Ga_5O_{12}$) is then submerged in the molten flux, while being rotated. Expitaxial growth was used in the past but has been supplanted by plasma deposition and the like.

6.11.- HYDROTHERMAL GROWTH

Single crystal growth by hydrothermal means utilizes water as the material-transport medium. The method is most often used for growing single crystal quartz. Quartz ($SiO_2$) is not very soluble in water, but its solubility increases considerably at higher temperatures. Thus, growth is accomplished at high pressures in a sealed autoclave. These quartz crystals are used as resonant frequency "tuning forks" for timing applications in digital watches. The following diagram, given as 6.11.1 on the next page, shows a typical apparatus.

Seed crystals are hung within an autoclave on a revolving hanger. Nutrient (high purity sand or natural crystal quartz) is contained at the

6.11.1.-

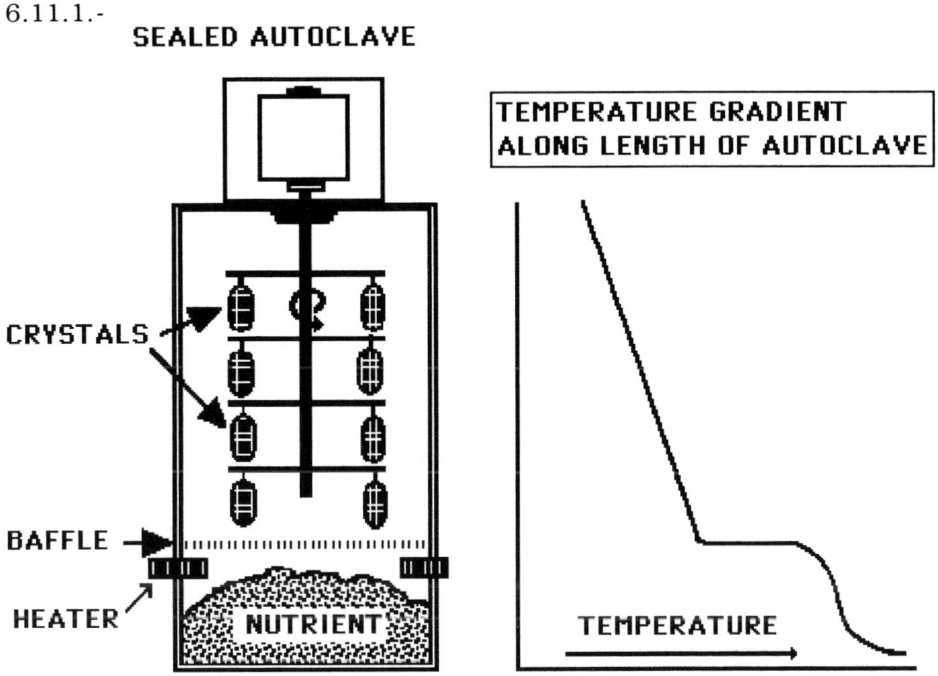


bottom and dissolves as the autoclave is heated. A temperature gradient is most often used so that the nutrient dissolves and is transported to the cooler area where it is redeposited as single crystal material. Note the temperature gradient at the baffle.

The solubility of the solute (in this case, quartz) is a function of both pressure and temperature. **Pressure** could be in theory be used as the controlling parameter rather than temperature. However, it is difficult to design an apparatus with a **pressure gradient**, whereas obtaining a temperature gradient is fairly easy.

Actually, what is controlled is the critical volume from the equilibrium:

6.11.2.- $$H_2O_{(g)} \Leftrightarrow H_2O_{(l)}$$

where l and g refer to the liquid and gaseous state, respectively. The phase diagram for water is shown at the lower left of the following diagram, given as 6.11.3. on the next page.

6.11.3.

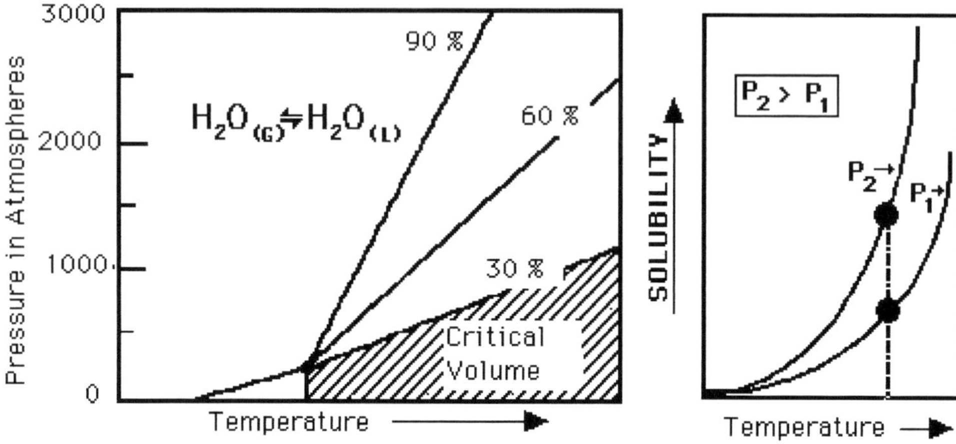


Above the critical temperature, water exists as a gas at all pressures. But, if we fix the volume of both gas and liquid (as in an autoclave), then no liquid is possible below the critical volume (~ 30 % by volume). By filling the available volume greater than 30% with water, we can go to very high pressures (i.e.- > 3000 atmospheres) and still maintain a liquid volume.

**In hydrothermal growth, the materials usually grown as single crystals are those classified as insoluble at standard temperature and pressure (STP).**

For quartz, growth is usually accomplished at 85% fill and 2000 atmospheres at 350 °C.

The parameters for hydrothermal growth of single crystals are:

6.11.4.- <u>Parameters Controlling Hydrothermal Growth of Single Crystals.</u>

1. Operating Pressure
2. Operating Temperature
3. Solubility in super-critical water
4. Degree of supersaturation
5. Degree of supercooling
6. Purity of Nutrient

The last three factors require some further consideration. We will discuss these in light of the growth of quartz, since it is this crystal for which the most experience has been gained. The rate of crystal growth is a function of seed-crystal orientation. The rate of growth, $r_{hkl}$ , may vary several orders of magnitude, depending upon the {hkl} plane orientation. However, a certain degree of supersaturation is mandatory as the nutrient dissolves, passes through the baffle, and moves to the seed area. Otherwise, the nutrient would precipitate before reaching the seed crystals. It has been determined that the rate of crystal growth is a direct function of the degree of supersaturation, Δ Sat:

6.11.5.- $$r_{hkl} = \alpha \cdot \mathbb{k}_{hkl} \cdot \Delta \text{Sat}$$

where $\mathbb{k}_{hkl}$ is the seed orientation factor and α is a constant dependent upon the crystal system. However, it has been found that the degree-of-supercooling factor is contra-indicative to the growth rate. To understand this, imagine the face of a growing crystal. Rejection of impurities at the growing interface causes a localized increase of impurity concentration. This is shown in the following diagram:

6.11.4.-

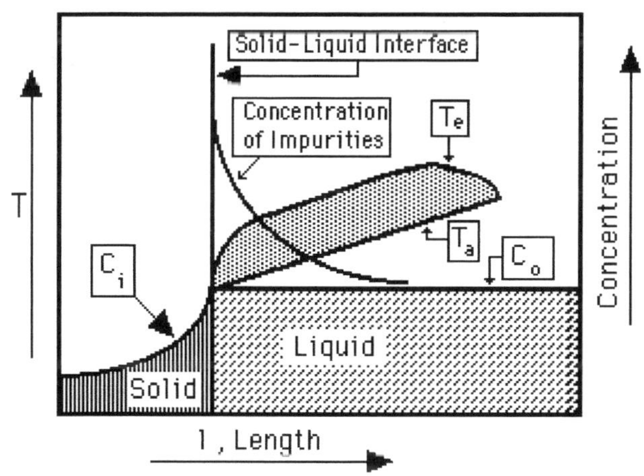


If the superimposed temperature gradient is $T_a$, i.e.- dT/dl is a constant, and $T_e$ is the equilibrium curve, i.e.- $dC/dT_e$ , then supercooling will occur. The rate of deposition thereby changes. If it becomes slower, crystal growth slows whereas if it speeds up, then dendritic growth or faceting occurs. Note that the impurity- rejection mechanism, observed for the case of Molten Flux growth, also occurs here.

For all these reasons, purity of materials is one of, **if not the**, most important of the parameters controlling hydrothermal growth.

## 6-12: VAPOR METHODS USED FOR SINGLE CRYSTAL GROWTH

If the material whose single crystal we want is volatile or sublimable, then we may choose a vapor-method of crystal growth. These methods have been used for a variety of crystals including ZnS and CdS. In this method, a carrier- gas is most often used for material transport and for the sulfides, $H_2S$ is the gas of choice. The following shows a simple apparatus:

6.12.1.- A Furnace for Growing Crystals from the Vapor Phase

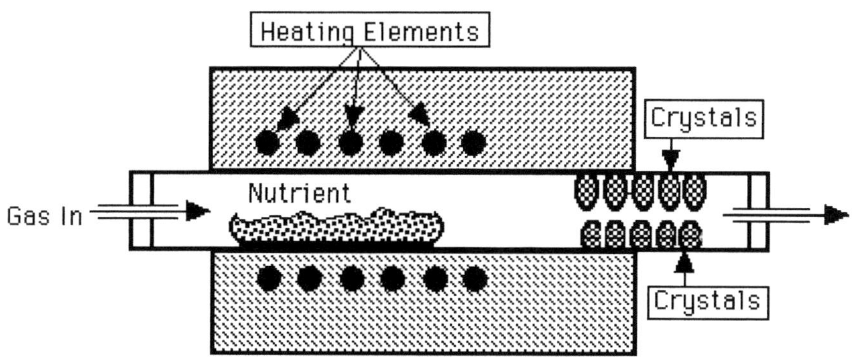


In this method, we start the gas flow and then begin heating. It is important to heat the material to just below the point of sublimation (volatilization) and let the system come to equilibrium. Gas flow need not be rapid but needs to be sufficient to carry the volatiles to the cooler part of the furnace. The tube used in the furnace is most often a silica-tube, although metal tubes have sometimes been used. The choice depends upon the nature of the material being sublimed and crystallized.

For ZnS and CdS, it is important to exclude all traces of oxygen since these materials are easily oxidized:

6.12.2.- $2\ ZnS + 3\ O_2 = 2\ ZnO + SO_2 \Uparrow$

The operating parameters for the vapor phase method of growing crystals are shown in the following:

6.12.3. - Operating Parameters for Vapor Phase Growth of Crystals

1. Temperature of sublimation (volatilization) of material
2. Amount of gas flow used
3. Degree of furnace- temperature set- point **above** material sublimation temperature.
4. Temperature gradient at crystal-growing junction of tube.

Obviously, whether the material has a low (high vapor pressure) or a high temperature of sublimation is important because of furnace- construction material considerations. In general, we cannot use this method for materials which are volatile above about 1200 °C. because most materials of construction cannot withstand the corrosive nature of the vapors at this high temperature. If we set the temperature of the furnace **above** the sublimation point of the material, it will volatize **all at once**. This would necessitate setting the gas- flow such that all of the material is transported by the gas. It would be better to set the temperature just below the vaporization (sublimation) point. There will be enough material transported to begin growth of single crystals. In fact, it has been determined that the slower the growth, the better are the crystals obtained.

There is another method that has been sometimes employed in the vapor phase growth of crystals. This method uses an **evacuated** capsule as shown in 6.12.4., given on the next page. The capsule is generally made from quartz, although platinum is sometimes used. **The capsule needs to be evacuated to remove any residual gas before heating is started.** Otherwise, the internal pressure would build until the capsule would explode.

6.12.4- Another Method Sometimes Used for Vapor Phase Growth

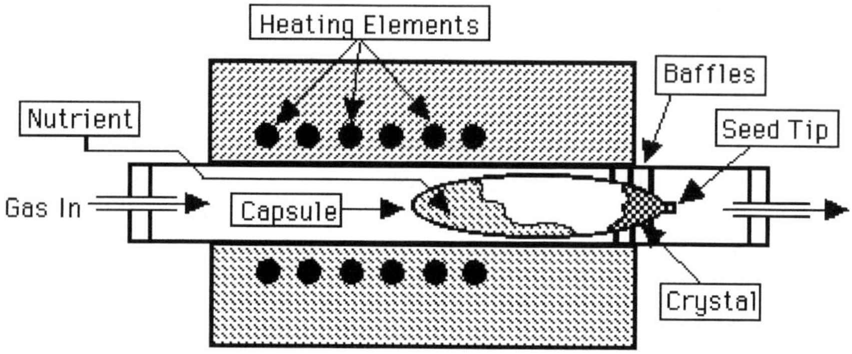


Even though it is evacuated, capsules have been known to explode because the quartz (metal) walls could not contain the internal vapor pressure of the material being grown as single crystal. Care must be exercised **not to handle the hot capsule** before and after crystal growth.

In addition to sublimation or vaporization, we can also use chemical transport as a method of single crystal growth. For example, we could use either of the apparati of 6.12.1. or 6.12.4. to grow a crystal of $ZnCl_2$ by the following reactions:

6.12.5.-
$$ZnO + 3\,HI = ZnI_2 \Uparrow + H_2O$$
$$Zn + I_2 = ZnI_2 \Uparrow$$

Metal carbonyls are also convenient for growth of certain **METAL** single crystals, vis-

6.12.6.-
$$M_{(s)} + x\,CO_{(g)} \Rightarrow M(CO_{(g)})_x$$
$$M(CO_{(g)})_x \Rightarrow M_{(s)} + x\,CO_{(g)}$$

In these cases, one is limited to the growth of single crystals of the transition metals since it is those metals which form volatile carbonyl compounds. The alkali metals, alkaline earth metals and certain of those elements which are allotropic in nature arc not at all suited for this type of crystal growth.

## 6.13.- EDGE DEFINED CRYSTAL GROWTH

Many times, the form of the crystal obtained is not suited for the end use contemplated. For example, a flat plate of $\alpha - Al_2O_3$ is often used as a base for integrated circuits (IC's) because of its high thermal conductivity and its low electrical conductivity. Si is vapor deposited on the surface of the Alumina plate, and through various photographic techniques and selective etching (with suitable additives diffused into the silicon layer), an integrated circuit is built up. The IC relies upon the high thermal conductivity of the Alumina base for its long life since "hot-spots" can quickly destroy the IC. One way to make these plates consists of pulling an $\alpha - Al_2O_3$ crystal by Czochralski means, slicing it into wafers, polishing both sides of the wafers and then cutting each one into smaller flat plates of the required size.

$\alpha - Al_2O_3$ (corundum) is extremely hard (9.5 on the Mohs scale) and is difficult to work with. It would be much easier if we could grow $\alpha - Al_2O_3$ directly as a flat plate. By using **edge defined growth**, one can do this. A typical apparatus is shown as follows:

6.13.1:

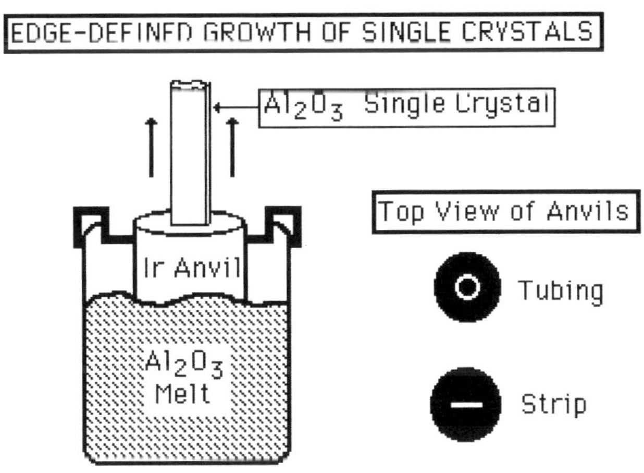


The apparatus consists of a normal Czochralski melt with an anvil at the surface of the melt. Once the crystal has started to grow, we pull it through the anvil, thus defining its size. Once it is in the form of a strip,

region of **partial- solid- solution.** In those regions, one solid freezes before the other, so that it is impossible to grow a crystal from a melt of these mixed oxides. Note that even at the eutectic composition (there are two of them), one still obtains a mixtures of oxides.

However, there is a congruently- melting composition and a range of compositions associated with it. The low temperature **stoichiometric** composition melts **incongruently** but it is not easily achieved from the melt. It shifts composition upon melting or freezing! It turns out that this behavior is more prevalent than one might realize. It is, in fact, typical for many systems and some of these are listed in Table 6-3, shown on the next page.

The maximum composition- existence range is given along with their **dystectic** composition, that is, the most usual composition.

A closer examination of Table 6-3 reveals a rather startling conclusion. It can be seen that most of these compounds are subject to structure-vacancy defects. Certainly all of the pnictides fall into this class. We are even given the defect concentration. Thus in GaP, we have:

6.14.2.- $$\delta\ p_P \leftrightharpoons p_{Ga} + (1+\delta)\ V_P + \alpha V_i$$

where $\delta$ **is usually** 0.00005. We can conclude that the melting and freezing behavior of a compound has a significant effect of its stoichiometry. However, it appears to manifest itself mainly as point defects in the crystal **produced.** The essence of this discussion is that although we melt the components, whether they be oxides or otherwise, to form a desired compound or composition, there are many cases where we do not obtain the stoichiometry we expect to get.

This is graphically illustrated when we consider the types of defects that are actually obtained in single crystals as they are being grown.

TABLE 6-3

Nonstoichiometric Nature of Some Binary Congruently Melting Compounds

| COMPOUND | MAX. EXISTENCE RANGE | DYSTECTIC COMPOSITION |
|---|---|---|
| $Ga_{0.5+x}P_{0.5-x}$ | + 0.00025 to - 0.00005 | + 0.00006 |
| $Ga_{0.5+x}As_{0.5-x}$ | + 0.00023 to - 0.00018 | + 0.00004 |
| $Pb_{0.5+x}S_{0.5-x}$ | + 0.0005 to - 0.00005 | + 0.0005 |
| $Pb_{0.5+x}Sc_{0.5-x}$ | NOT KNOWN | - 0.00005 |
| $Pb_{0.5+x}Te_{0.5-x}$ | + 0.0005 to - 0.00013 | - 0.00012 |
| $Sn_{0.5+x}S_{0.5-x}$ | + 0 to - 0.000008 | - 0.000005 |
| $Cd_{0.5+x}Te_{0.5-x}$ | + 0.000002 to - 0.000008 | - 0.000005 |
| $Mg_{1+x}Al_{2-2x}O_{4-2x}$ | + 0.47 to - 0.42 | NOT KNOWN |
| $Gd_{3+x}Ga_{5-x}O_{12}$ | + 0.30 to - 0.002 | + 0.05 |
| $Li_{1-5x}Nb_{1+x}O_3$ | + 0.031 to O- | + 0.0092 |
| $Li_{1-5x}Ta_{1+x}O_3$ | + 0.025 to O- | + 0.0066 |

## 6.-15: ACTUAL IMPERFECTIONS IN SINGLE CRYSTALS

Now we can summarize **all** of the imperfections likely to appear in single crystals. Some of these, particularly **stacking- faults**, were discovered only when single crystals were grown large enough so that the deviation from long range order became apparent.

The intrinsic defects found in single crystals include vacancies, interstitials, impurities and impurity compensations. reverse order, and combinations such as V- S and I- S, etc. Their numbers are well described by:

6.15.1.- $N_i = N_o \exp - \Delta E_i / kT$

where i refers to the intrinsic defect. In addition, we have dislocations, both edge (line) and spiral (screw), both of which are three- dimensional. Their numbers are well described by:

6.15.2.- $\Delta G_i = n \Delta H_i - T (\Delta S_{config} + n \Delta S_{vib})$

This equation arises because both of these **extrinsic** defects affect the energy of the crystal. We can also have grain boundaries which may be: clustering of line defects or mosaic blocks. The latter may be regarded as very large grains in a crystallite.

Another imperfection in crystals is called "twinning". This usually happens when **enantiomorphs** are present, or possible. A good example is quartz, i.e.-

6.15.3.- $\alpha$– quartz $\underset{573\ ^\circ C}{\leftrightarrows}$ $\beta$– quartz

In this case, **two** crystals grow and are joined at a given plane, each being a mirror image of the other.

The other crystal imperfection we have not covered is "stacking faults". A good example is SiC. Here, we have **two sublattices,** one based on Si and the other on C, each of which is hexagonal. In stacking **alternate** layers of identical atoms, we first stack Si and then C. If we refer to "A" as the 1st layer, "B" as the 2nd layer, and "C" as the 3rd layer, then the normal stacking sequence for the hexagonal lattice is:

6.15.4.- **ABC - ABC - ABC**

where we have shown three sequences. For SiC, "A" in the first sequence is Si, "B" is C, and "C" is Si. We still have the same packing since only "A" atoms are over "A" atoms, as shown in the following diagram:

6.15.5.-

**STACKING SEQUENCES IN THE HEXAGONAL LATTICE**

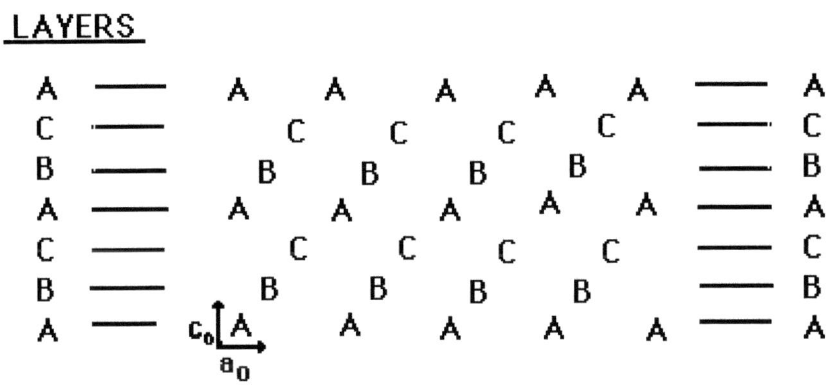


Here, we have arranged the layers on a two-dimensional structure, even though the layers are arranged in three dimensional order. Note that only two crystallographic axes are indicated, We call this the natural stacking sequence because of the nature of the hexagonal close- packed lattice.

However, SiC also exhibits other stacking sequences, as shown in 6.15.6., given on the next page. These arranged layers are called "polytypes" and are prevalent where simple compounds such as SiC and SiN are involved. In many cases, the properties of such compounds depend, to a large extent, upon the specific stacked layers obtained during formation.

| 6.15.6. | β - SiC | = | ABC ABC |
|---|---|---|---|
| | SiC - 4H | = | ABCA ABCA |
| | SIC - 6H | = | ABCACB ABCACB |
| | SiC - 15R | = | ABCBACABACBCACBABCBACABACBCACB |

For SiC, we can also have "polytypes" where two stacking sequences like 4H 6H can combine to form a unit. Another "polymorph" is 4H -15R.

This phenomenon has been thoroughly studied and polymorphs of 87R and 270R have been reported.

Another type of stacking fault is called "**polystructure**". A good example is ZnS, which is dimorphic (has two forms). The cubic form of ZnS is called sphalerite, whereas the hexagonal form is called wurtzite (These are their mineral names, after the first geologist who discovered them). The stacking sequence for sphalerite is: AB or ABBA; that for wurtzite is: ABC. A polystructure sometimes results when sphalerite is converted to wurtzite:

$$6.15.7.\text{-} \quad ZnS_{cubic} \overset{\sim 1200\,°C.}{=} ZnS_{hexagonal}$$

The polystructure sequence is: ABC ······ AB·······ABC·····AB······ABC. This may be regarded as cubic- close- packing within a hexagonal- close- packed structure. We may also note that ZnS is subject to the same "polytype" stacking faults as those given for SiC above. Thus, the stacking fault patterns noted are a function of the type of lattice involved, not on the chemical composition of the material. For the most part, these stacking faults are found only in the high symmetry lattices such as hexagonal close-packed and cubic close-packed structures.

## 6.16- ELECTRONIC PROPERTIES OF CRYSTALS

Although we have described the growth of crystals in some detail, we have not considered the behavior of electrons in either crystallites or single crystals. It is the electronic properties of such materials that are useful in industrial applications. Therefore, it would behoove us to consider the factors that cause change in electronic properties of solids as a function of structure and bonding. It is evident that this aspect is very important, since the electrical nature of solids varies according to both composition and arrangement of atoms in the structure. Some materials are semi-conducting and others are dielectric insulators. Still others are good conductors of charge (metals). Still others have unique optical properties, due to arrangement of electrons and atoms within the solid state.

One example of the latter is the crystal, β- $BaB_2O_4$. This crystal belongs to the trigonal system with point symmetry of $C_{3v}$ . It is a negative uniaxial crystal and its main use is **to double the frequency of a laser beam**. There is a high temperature form, α–$BaB_2O_4$ (transition temperature of the β to α–form = 925 °C.), but the α-form is not optically active. It is natural to ask why this should be so. In the case of β- $BaB_2O_4$ , we know that when photons from a laser beam traverse the crystal, many are absorbed, causing local atoms to become energized (i.e.- go into an excited state). Because of the prevalence of a dense phonon spectrum (i.e.- vibrational states of the lattice), the photon emitted by each excited atom is enhanced by extra energy added by the phonon states. These phonon states are coupled to the atomic excited states in the β–form but not in the α- form (probably due to the difference in lattice symmetry and structure). The emitted photon thus is twice as energetic as evinced by the fact that its frequency is twice (2X) that of the original photon absorbed by the lattice while the wavelength is half (1/2X). This phenomenon occurs only with certain classes of lattice symmetry and certain compositions of matter. Among these are: $LiNbO_3$ , $KTiPO_4$ , $LiBO_2$ and $K_2TiO_3$ .

In semi-conducting compounds, we know that some of the electrons form bonds between the cation and the anion, either as covalent or ionic bonds (or somewhere in between). What happens to the rest? Do they remain around the parent atom? Why are some solids conductive while others are not? The following discussion addresses these questions. Obviously, we cannot be exhaustive but we can examine the main features of each phenomenon to show what happens in the solid. We will not derive the equations associated with each subject. This aspect is left to more advanced studies.

### A. Conductivity in Ionic Compounds

We have already discussed diffusion in solids to some degree. While bulk properties such as heat capacity are not sensitive to defect concentration, many other properties such as conductivity are. Thus, the method of preparation becomes important if one wishes to obtain a conductive or even a semi-conducting solid. Most materials in Industry are polycrystalline rather than single-crystal because of cost and difficulty of manufacture. However, only the single-crystal state suffices for some applications. The degree of conductivity of solids is dependent upon at least two factors, structure and ionicity. Ceramic materials are notoriously low in conductivity because their band gaps between the valence and conduction band are relatively large. This translates to conductivities of $10^{-6}$ or lower. Contrast this to metals where conductivities of $10^2$ are common.

Conductivity in solids depends upon movement of charge through the lattice and includes both types of charge. In oxygen-dominated materials such as solid ceramics, the diffusion of charge can only involve electrons and the cation through the structure. We have already considered this aspect in Chapter 3. In certain structures, the electrical conductivity associated with ion motion can exceed conductivity from electrons by several orders of magnitude. Diffusion and ionic conductivity are related through the Nearst-Einstein equation, vis:

6.16.1.- $$\sigma = \frac{Dnq^2}{kT}$$

where D is the diffusion coefficient, n is the number of charge carriers per unit volume, q is the charge per ion, k is Boltzmann's constant and T is the absolute temperature. D generally refers to the more rapidly moving species. However, there are many ambiguities in the use of this equation to characterize the degree of conductivity, especially when a direct current is involved. Diffusion usually occurs by the movement of ions to a neighboring crystal defect. In a salt crystal, the measured current may be induced by movement of interstitial ions (Frenkel) or vacancies (Schottky). The position of a defect changes when a neighboring ion moves in to fill it. Schottky defects predominate in KCl where both cation and anion vacancies occur. In AgCl, some Ag ions occupy interstitial sites producing positive Frenkel effects and negative Schottky defects. Nonetheless, for stoichiometric compounds, the vacancy concentration and ionic conductivity is very low. If one adds certain "dopants", the conductivity will change and that is what is done to produce solids with suitable conductivity for certain electronic devices. To illustrate exactly what we are discussing, consider the following diagram, which is based on a solid array of water molecules, i.e.- "ice":

6.16.2.-

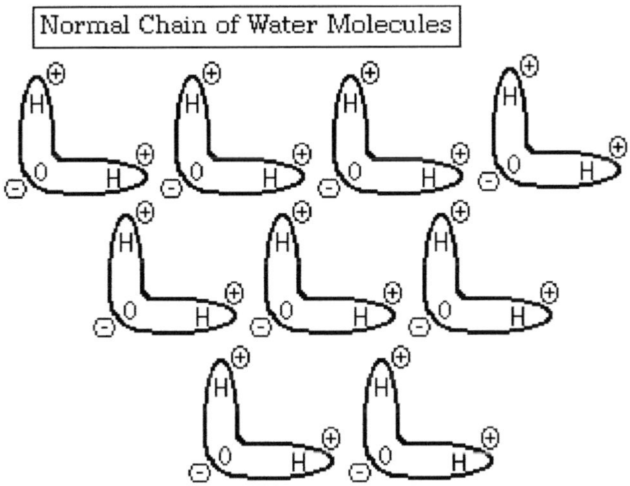


Here, we have a two-dimensional chain of water molecules which are bonded by van der Waals forces between positive and negative charges of the individual water dipoles. If an electrical field (via a voltage) is applied to ice, charge can move by diffusion of protons. Each water molecule in ice is surrounded by four (4) neighboring molecules. Hydrogen atoms lie near lines joining adjacent oxygen atoms and are closer to one than the other. Two neutrality conditions are satisfied:

1) Water molecules are intact and electrically neutral

2) The integrity of the hydrogen-bonding network remains intact.

One and only one hydrogen directly connects to two neighboring oxygen ions. Violation of either neutrality rule produces an electrically-active defect. A positive bonding defect would look like this, as shown in the following diagram as 6.16.3. on the next page.

6.16.3.-

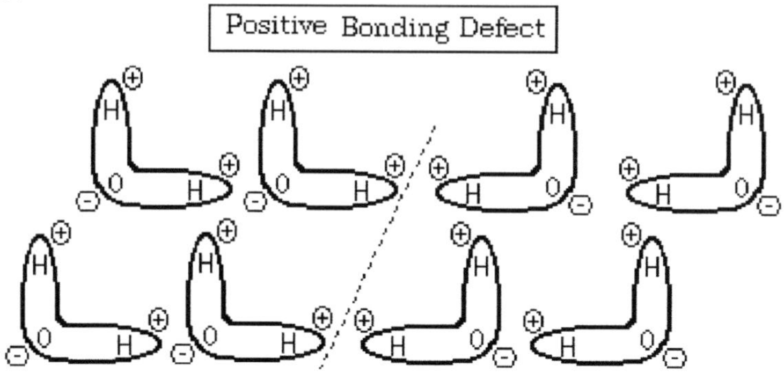


If a proton is transferred to another molecule, a positive hydronium ion is created along with a negative hydroxyl ion, i.e.:

6.16.4.- $$2\ H_2O \Leftrightarrow H_3O^+ + OH^-$$

The dotted line shows the extent of the positive defect due to packing of the water molecules in a lattice structure which violate the hydrogen-bonding network rule.

The same can be seen for the negative bonding defect:

6.16.5.-

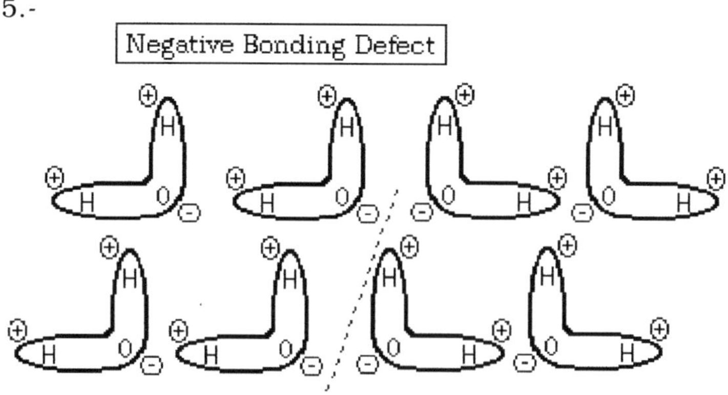


Note that the actual bonding condition is three-dimensional. In both of these cases, rotation of a water molecule results in either a positive or negative bonding defect which is electrically active. In the first case, a positive defect results when 2H are on the same bond and a negative defect when no hydrogens are present. Note that this type of defect disrupts the structure to a much greater degree than the bonding type of defect. It is possible for the two types of ions, i.e.- hydronium and hydroxyl, to be nearest neighbors in the lattice structure (although we have not shown them as such here). This gives rise to both positive and negative ionic defects in the structure, as shown in the following diagram:

6.16.6.-

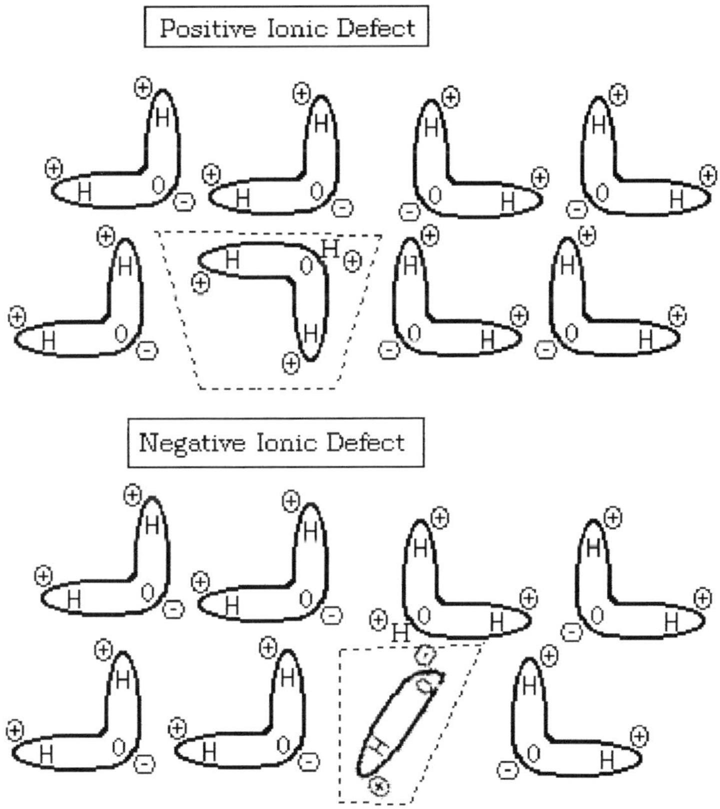


The motion of hydronium and hydroxyl ions leaves molecules oriented against the applied electrical field. The molecular dipole moment is then anti-parallel to the field. If bonding defects are passed by either of the hydronium and hydroxyl ions, the molecular dipoles are left parallel to the field. Bonding defects dominate the polarization in pure ice because they are more numerous than ionic defects. Concentrations of bonding defects average about $10^{-6}$ per mole of ice. Although ionic defects are fewer in number than bonding defects, they have higher mobility (compared to the diffusion mechanism of bonding defects). Thus, the transport of charge due to ionic defects is about equal to that actually transported by bonding defects, when a direct current is applied to ice.

Our motive in presenting this analysis of the motion of charge in ice when an electrical voltage is applied has been designed to acquaint you with the complexities of motion of charge in a crystalline solid. Motion of charge in an amorphous solid such as glass is even slower than most crystal solids. Solids are divided into three groups based upon their conductivity, as we have already discussed. They are: metals, semi-conductors and insulators.Aqueous solutions are the other class of conductors which can be used to carry a current of electrons. Metals carry charge because their valence and conduction bands overlap, allowing electrons to be easily transported. Semi-conductors have a narrow band gap between the two energy bands in the solid and a slight electrical field is all that is necessary for them to transport current (electrons).

## 6.17.- Silicon Single Crystals and Integrated Circuits in Commerce

We have already referred to the use of quartz crystals as used in "quartz-crystal" watches and timepieces. However, the single crystal grown in largest quantity is silicon (Si). Single-crystal silicon is endemic throughout our culture and is responsible for the current "Information-Age" engendered by the use of computers and other electronic equipment. However, growing defect-free silicon crystals is not an easy task.

### A. Silicon

Silicon (Si) is a nonmetallic chemical element of the carbon family (Group IVa of the periodic table) and makes up 27.7 percent of the Earth's crust. It is the second most abundant element in thc crust, being surpassed only by oxygen.

Its electronic properties are:

Atomic number = 14; atomic weight = 28.09
Melting point = 1410 °C; boiling point = 2355 °C
Density = 2.33 g/cm$^3$; oxidation state = -4 & +4
Electron configuration = $1s^22s^22p^23s^23p^2$.

Silicon was first isolated and described as an element in 1824 by Jöns Jacob Berzelius, a Swedish chemist. Silicon does not occur uncombined in nature, i.e.- as an element. It is found in practically all rocks as well as in sand, clays, and soils, combined either with oxygen as silica ($SiO_2$= silicon dioxide) or with oxygen plus other elements (e.g., aluminum, magnesium, calcium, sodium, potassium, or iron) as silicates. Its compounds also occur in all natural waters, in the atmosphere (as siliceous dust), in many plants, and in the skeletons, tissues, and body fluids of some animals.

Pure elemental silicon is a hard, dark gray solid with a metallic luster and with a crystalline structure the same as that of the diamond form of carbon. For this reason, silicon shows many chemical and physical similarities. There is also a brown, powdery form of silicon having a microcrystalline form. The element is prepared commercially by reducing the oxide by reacting it with carbon (as coke) in electric furnaces. On a small scale, silicon has been obtained from the oxide by reduction with aluminum metal.

Silicon, like carbon, is relatively inactive at ordinary temperatures. But, when heated, it reacts vigorously with the halogens (fluorine, chlorine, bromine, and iodine) to form halides and with certain metals to form silicides. It is unaffected by all acids except hydrofluoric. At red heat, silicon is attacked by water vapor or by oxygen, forming a surface layer of silicon dioxide. When silicon and carbon are combined at electric furnace temperatures of 2,000 to 2,600 °C (3,600 to 4700 °F), they form silicon carbide (Carborundum™ = SiC), which is an important abrasive. When reacted with hydrogen, silicon forms a series of hydrides, the silanes. Silicon also forms a series of organic silicon compounds called silicones, when reacted with various organic compounds.

Silicon's atomic structure makes it an extremely important semiconductor. Highly purified silicon, doped with such elements as boron, phosphorus, and arsenic, is the basic material used in computer chips, transistors, silicon diodes, and various other electronic circuits and electrical-current switching devices. Silicon of lesser purity is used in metallurgy as a reducing agent and as an alloying element in steel, brass, and bronze.

The most important compounds of silicon are the dioxide (silica) and the various silicates. Silica in the form of sand and clay is used to make concrete and bricks as well as refractory materials for high-temperature applications. The mineral quartz, i.e.- $SiO_2$, may be softened by heating and shaped into glassware. Silicates, most of which are insoluble in water, are employed in making glass as well as in the fabrication of enamels, pottery, china, and other ceramic materials. Sodium silicates, commonly known as water glass, or silicate of soda, are used in soaps, in the treatment of wood to prevent decay, for the preservation of eggs, as a cement, and in dyeing. Silicones are synthetic organo-silicon oxides composed of the elements silicon, oxygen, carbon, and hydrogen. They are used as lubricants, hydraulic fluids, waterproofing compounds, varnishes, and enamels because, as a class, they are chemically inert and unusually stable at high temperatures. Three stable isotopes of silicon are known: silicon-28, which makes up 92.21 percent of the element in nature; silicon-29 =4.70%; and silicon-30 = 3.09 %. Five radioactive isotopes are known.

### B. Silicon as a Semi-Conductor

In semiconductors such as silicon, each atom in the structural lattice has four outer electrons, each of which covalently pairs with an electron from one of the four neighboring atoms to form the interatomic bonds, i.e.- the "diamond" structure. Completely **pure silicon** thus has essentially no electrons available at room temperature for electron conduction, making it a **very poor conductor**. However, the key is getting the silicon pure enough. Originally, silicon was thought to be a natural semi-conductor until really pure silicon became available.

Growth of silicon crystals begins with the preparation of extremely pure polycrystalline silicon having fewer than 1 dopant atom per 10 billion silicon atoms. This silicon is melted in a quartz-lined furnace. The temperature of the molten silicon is reduced to just above the melting point, and a small bar (the seed) of silicon in single-crystal form is introduced into the surface of the melt. The molten silicon freezes slowly onto the seed with a crystalline structure that is contiguous with the structure of the seed (For more details, see the Czochralski method given above). The seed is slowly withdrawn, while rotating under carefully controlled conditions, to form a single-crystal cylindrical ingot of silicon. This ingot is greater than 200 millimeters in diameter and weighs up to 100 kilograms (220 pounds). After growth, the silicon crystal is ground to a smooth cylindrical shape and sliced into thin wafers, using diamond tools. These wafers are usually about 0.6 millimeter thick. The surfaces of the wafers are polished flat by a series of successively finer abrasives until one side has a perfect mirror finish. For the manufacture of integrated circuits, the methods used to achieve planarity and smoothness of the individual discs are extremely important. We will discuss these below.

If an atom from column V of the periodic table, such as elemental phosphorus, is substituted for an atom of silicon, four of its five outer electrons are used for bonding in the tetrahedral lattice, but the fifth is free to move within the crystal (which then becomes a "negative-type" semi-conductor). If the replacement atom comes from column III of the periodic table- such as boron, it will have only three outer electrons, i.e.- one too few to complete the four interatomic bonds. Since the crystal attempts to become electrically neutral were this bond complete, the vacancy acquires a positive charge because an electron is missing in the overall electronic structure. A neighboring electron can move into the vacancy, leaving another vacancy in the space previously occupied by that electron. This vacancy, with its positive charge, is thus mobile and is called a “hole.” Holes in semiconductors move about as readily as electrons do, but because they are positively charged, they obviously move in a direction opposite to the motion of electrons. This form of silicon is called a "positive-type" semiconductor.

The process of substituting elements for the silicon is called doping, while the elements are referred to as dopants. The amount of dopant that is required in practical devices is very small, ranging from about 100 dopant atoms per million silicon atoms downward to 1 per billion. Dopants are usually added to the silicon after the crystal growth process, when an integrated circuit is being formed on the surface of the wafer.

C. Semiconductor Devices

The fact that pure silicon is not a conductor but is "doped" to make it a good semi-conductor might seem odd, particularly since considerable effort goes into obtaining very pure silicon. Nonetheless, the undoped silicon is not conductive. This allows one to form discrete areas of n-type and p-type silicon on the same wafer and to position these in conjunction with each other in a manner so that a current amplifying device results. An example of such a device is shown in the following diagram:

6.17.1.-

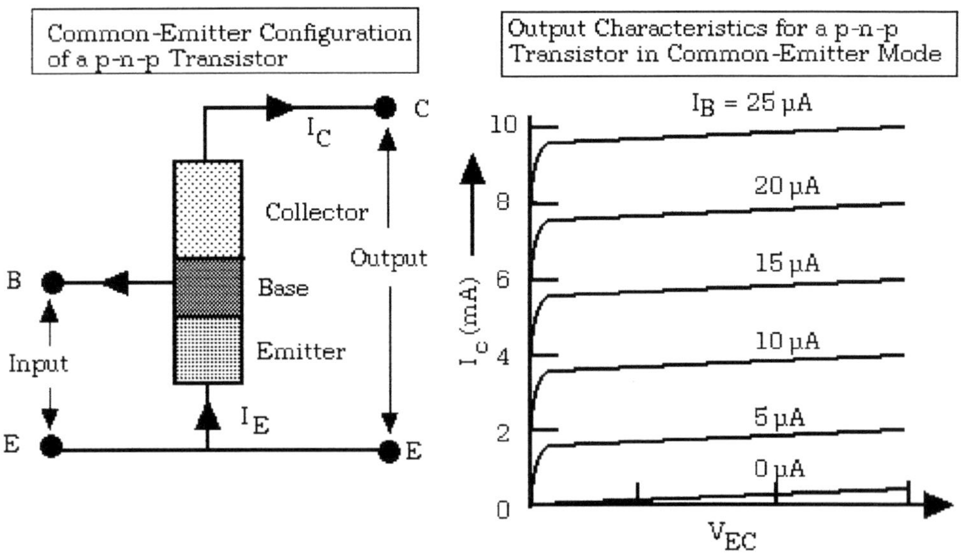


Here, we show a so-called "common-emitter" configuration of a p-n-p transistor, i.e.- the emitter states are coupled together. In general, such a configuration would be fabricated by forming the p- and n-types of silicon through injection of the proper impurity at neighboring sites on the surface of the silicon wafer. As shown in the diagram, a voltage is applied between the points E & B (where E is electrically connected to the emitter portion of the p-type areas and B is connected to the base (or n-type) area). The connections are usually conductive lines printed on the surface of the non-conductive Si-wafer.

One p-type area is used as the emitter (hole-conduction) while the other is used as the collector (of current) when the input voltage is applied between points B & E at the transistor. This results in a current, $I_B$, from the base as well as a current, $I_C$, from the collector. Thus two local circuits are at work, one through the emitter-base and the other through the base-collector. The important part to notice is that although the current, $I_B$, is in micro-amps, the current, $I_c$, is in milliamps, an amplification of 1000 x. Thus, the p-n-p configuration on the surface of a silicon wafer results in an amplifier of current (also known as a transistor).

A number of discrete steps are required to make such a transistor and passive parts such as resisters and condensors are also added to make what is called an "integrated circuit" or IC.

## D. Integrated Circuits

An integrated circuit (IC) is a monolithic assembly of electrically isolated circuit elements. What this means is that each circuit element is formed on top of , or beneath, other circuit elements to form a compact assembly. Each conductive layer is separated by a non-conducting layer, usually composed of an oxide such as silicon dioxide, $SiO_2$. The assembly includes both active semiconductor devices (transistors and diodes) and passive components (capacitors and resistors). These are connected together by means of electrically conducting interconnections separated by insulating layers with holes for layer-to-layer interconnection. All elements are fabricated in situ by an iterative process of lithographic definition, deposition, and etching on a common substrate (in most cases, silicon) in such a manner that the resulting interconnected elements perform the desired electrical circuit function. Lithographic refers to the photographic process used where a photoresist (a material which forms either a water-soluble or water-insoluble area upon exposure to the proper wavelength of light) is used to define the areas exposed. It is these areas which are then subjected to definition, deposition and etching. These are among the many processes to which each wafer is subjected during the process to make IC's. Many ICs, typically about 5.5 mm long or up to 33 square millimetres in area, are fabricated concurrently on silicon wafers up to 30 centimeters (12 inches) in diameter and subsequently sawed into individual chips (dies) prior to packaging. An IC chip thus produced is usually sealed in a plastic package with electrical leads that are internally connected by fine wires to output pads on the silicon die that permit the packaged IC to be attached to a circuit card.

The integration of a large number of semiconductor devices on a single die of silicon is made possible by the high operational efficiency of the individual devices. Because of this, power dissipation is minimal as are the requirements for heat removal. The result is an IC of high dependability, which is further enhanced by the process of integration itself. It is the method of manufacturing the electrical interconnections between the devices and circuit elements by metal deposition and etching that yields an extremely dependable circuit structure. The number of devices integrated on a single tiny chip has increased from an initial few to nearly 10,000,000 individual circuit elements per die with ever-smaller features in the development stage. This has, in turn, led to a progressive increase in complexity, as exemplified by the metal-oxide-semiconductor (MOS), the dynamic random-access memory (DRAM), and the MOS microprocessor.

Since manufacturing cost is proportional to chip area, the use of smaller features has not only led to an increase in complexity and resulting functionality but also has produced a decrease in cost per transistor and an increase in circuit performance.

There are two reasons for the latter:

(1) the performance of the individual active devices improves as their internal dimensions become smaller.

(2) the quality of the performance of the entire circuit improves as the active devices are positioned closer to one another.

Large-volume applications for IC dies have further lowered the manufacturing costs per IC. The cumulative result is that the integrated circuit has become the most pervasive technology of the 20th century. It has provided the cornerstone of modern microelectronics and has promoted the development of the so-called information society. Applications of ICs include their use in computers as well as other instruments. It is these applications that are bringing about revolutionary advances in medical diagnosis, biotechnology, aeronautical and space engineering of our age. Other technologies affected (and vastly improved) are tele-communications, and military defense systems. IC's have fueled the development of new consumer products capable of bringing new types of services and information to the home and office environment that otherwise would not have been possible.

Our next step will be to describe the steps required in the manufacture of Ics. As you will see, they have become quite complicated.

## E. Manufacture of Integrated Circuits

The process of fabricating semiconductor devices is a complex series of more than 600 sequential steps, all of which must be done with utmost precision in an environment cleaner than a hospital operating room. If particles are present, they can cause electrical shorting among the conductive lines etched on the surface of the silicon wafer. Thus, no impurities which might affect the wafer surface either physically or chemically can be tolerated. A series of layers is thus formed. This reality mandates the manufacture of IC's in a closed "clean-room". Since a human being emits around 5000 particles per minute (notably skin particles), it is necessary that all operating personnel involved in the manufacturing process wear special clothing and masks to prevent contamination of the wafer during the various 600 operations required to make a finished IC.

other becomes soluble (de-polymerization) upon exposure to light of the proper wavelength. The use of one or the other depends upon the actual step in IC manufacture being accomplished. One can be removed by a simple washing procedure, whereas the other is removed by either washing or more strenuous means such as baking or ashing. Before removal of the non-exposed parts of the photoresist, the next step is to expose the silicon surface to whatever reactant is required to form the layer under construction on each IC wafer. This is done reiteratively until the basic IC circuit is formed. Once this is done, the layers are covered to form the final form of the wafer.

Then the portion containing each circuit is cut into a small square called a "die", many being called "dice". The scale of line dimensions, i.e.-width, in integrated circuits in 1997 was 0.25 microns (2.5 x $10^{-7}$ meters). It continues to decrease year by year. Approximately 1,850 dice can be obtained from a 220 mm wafer, if all of the dice test "good". Thus, the diagram given above is not accurate because the details of my drawing cannot be made precise enough to accurately detail all of the characteristics of dice formation on the surface of the silicon wafer and still illustrate the point. A high-power semiconductor device for industrial use, on the other hand, may be so large as to require a slice of silicon measuring well over 125 millimeters in diameter. Of late, the current size (or cross-section) for a separate line or conductor in an integrated circuit is 18 microns. Research has decreased this size to 13 microns in 2003 with 8 microns (μm) in the offing. Obviously, this increases the yield in the number of dice from one wafer by decreasing the actual size of each individual die being formed. This year, the first 300 millimeter wafer came on stream in production units making integrated circuits. This enlargement in size was designed to increase the yield of dice by 145% over wafers only 200 mm in diameter. A major change in equipment was needed to handle the larger wafers and the total cost to do so was in the billions of dollars.

F. Steps in the Manufacture of Integrated Circuits

The fabrication of ICs begins with the preparation of silicon of very high purity. Single crystal boules with a diameter as large as 30 centimeters are produced. The boules are sliced into wafers having a specified crystal orientation. This means that the direction of single-crystal growth must be controlled. When their surfaces have been polished flat to a mirrorlike finish that is free of defects, the silicon wafers are ready for fabrication into IC devices. The methods of planarization and polishing are of supreme importance to the overall quality of the chips or dice being formed. For example, the following lists some of the factors that are specified for acceptable silicon wafers which are used to form 0.13 μm lines in the IC:

| | Critical Process Step |
|---|---|
| Oxygen content > 30 ppm | Crystal growth |
| Flatness = 1.5 μm (side to side) | Lapping,Grinding,Polishing |
| Warp = 25 μm | Slicing |
| Particles = < 100/wafer | Polishing, Cleaning, Crystal growth |
| Non-critical metals = < 16 ppm | Cleaning |
| Critical metals = < 45 ppm | Cleaning |

Note that the oxygen content is controlled in the crystal growing process. The presence of metals on the wafers is due to the slicing and polishing steps where the cleaning process did not remove all of these **external** metal particles. The difference between critical and non-critical metals is whether they would affect the conductivity of the silicon if they were accidentally caused to combine with the silicon comprising the wafer. The other factor not yet addressed by us is the damage caused by the various processes used to form the finished wafer. Among the damage capable of being evoked on the surface of the silicon is the formation of vacancies, scratches and dislocations in the silicon structure. For these reasons, each manufacturer of wafers has developed its own proprietary process. Some of these processes are slow and, of course, contribute to the overall cost of the so-produced wafers.

This is shown in the following table, presented on the next page.

Table 5-4
Comparison of Some Wafer Manufacturing Operations

| Operation | Depth of Damage | Flatness of Processed Wafer | Relative Speed of Process | Ease of Automation |
|---|---|---|---|---|
| Slicing (ID saw) | **high** | **poor** | **slow** | **good** |
| Slicing (wire saw) | **high** | **poor** | **medium** | **medium** |
| Lapping(multiple wafer) | **medium** | **good** | **fast** | **poor** |
| Grinding(single wafer) | **low-med** | **good** | **fast** | **good** |
| Wet chemical etching | **zero** | **medium** | **medium** | **good** |
| Rough, one-side polish | **zero** | **good** | **slow** | **good** |
| Rough, two-side polish | **zero** | **very good** | **slow** | **poor** |
| Finish polishing | **zero** | **good** | **very slow** | **good** |
| Plasma chemical etching | **zero** | **excellent** | **slow** | **good** |
| Plasma torch & ADP* | **zero** | **good** | **medium** | **good** |

ADP = Atmospheric Downstream Plasma

Note that some processes produce little or no surface damage but that the process is slow. What this means is that the manufacturer must choose between finishing processes and acceptable damage during processing. When some damage is left on the surface of the wafer or if the planarity is not good, then the IC processes used result in defects in the IC circuit and/or in the individual transistors formed. This results in a "bad" die or chip, i.e.- it does not function properly. It is for this reason that the number of "good" dice or chips vary from wafer to wafer. Note that electronic testing is required to determine if a die is good or not. The typical IC manufacturing process involves a sequence of more than 600 operations. These operations may be categorized according to function:

1. Deposition of thin films
2. Introduction of impurities (doping)
3. Lithographic patterning of IC features corresponding to those of the physical layout
4. Etching to dcfine the features of individual circuit elements

5. Annealing to cause a chemical reaction or to form the desired microstructure of deposited films or interfaces between films
6. Polishing to level and smooth the surface
7. Cleaning to prevent contamination and the consequent introduction of defects into circuit elements by particulate matter.

These operations are repeated again and again in different sequential order until the IC fabrication is completed. At the same time, they are not the actual steps needed to form the individual dice. We will address these later on after we have described the individual operations used in performing each step in the IC manufacturing process. It is very important to distinguish between manufacturing operations and manufacturing steps.

The actual manufacturing steps involved include:

1. Wafer cleaning
2. Lithography
3. Planarization
4. Ion implantation
5. Sputtering
6. CVD- chemical vapor deposition
7. Resist strip
8. Wafer cleaning
9. Thermal processing
10. Metrology

The following diagram, given as 6.17.3 on the next page, shows the cyclic nature of the steps used to form the intermediate layers of an electronic device. In this case, we have shown at least 8 individual steps needed to fabricate this structure. Note that the use of photoresist is not indicated in the CVD steps, although it plays a significant role in the overall process. Note that these same 8 steps can be repeated so as to build up a series of electrical and isolated layers which comprise the overall IC design

6.17.3.-

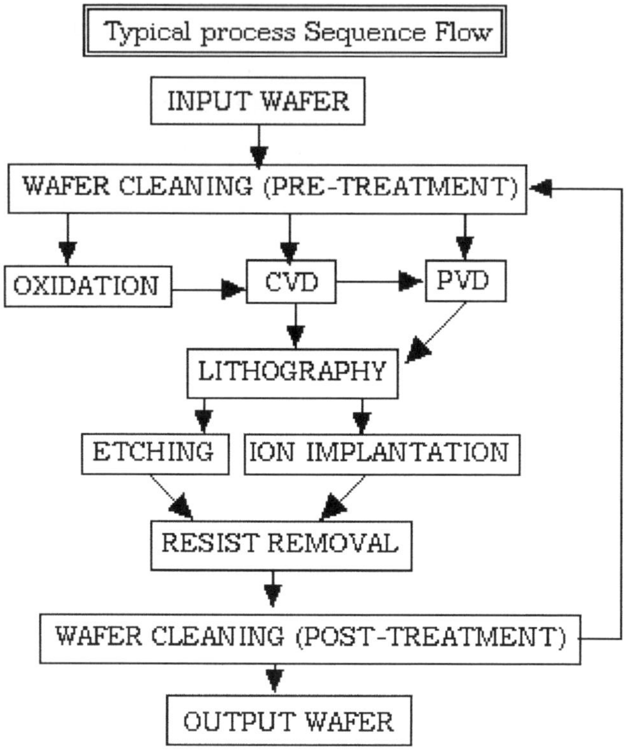


circuits. However, even this diagram does not show **all** of the cyclic steps and machinery needed to accomplish the 600 individual steps referred to above in the manufacture of an integrated circuit on a silicon wafer.

As shown in 6.17.4. given on the next page, the manufacture begins with a polished wafer. Study this outline well. It gives most of the 600 steps involved in IC manufacture.

The first step is usually formation of an oxidized film layer, followed by microlithography steps. We have indicated the individual steps involved in lithography, as well as those involved in mask making. Note that the next step involves etching and stripping steps. Inspection is the next major step, followed by doping and then diffusion and oxidation (depending upon the specific layer being formed). Reinspection then takes place. If

6.17.4.- SEMICONDUCTOR FABRICATION STEPS

Equipment Includes: Wafer Processing - Test - Assembly

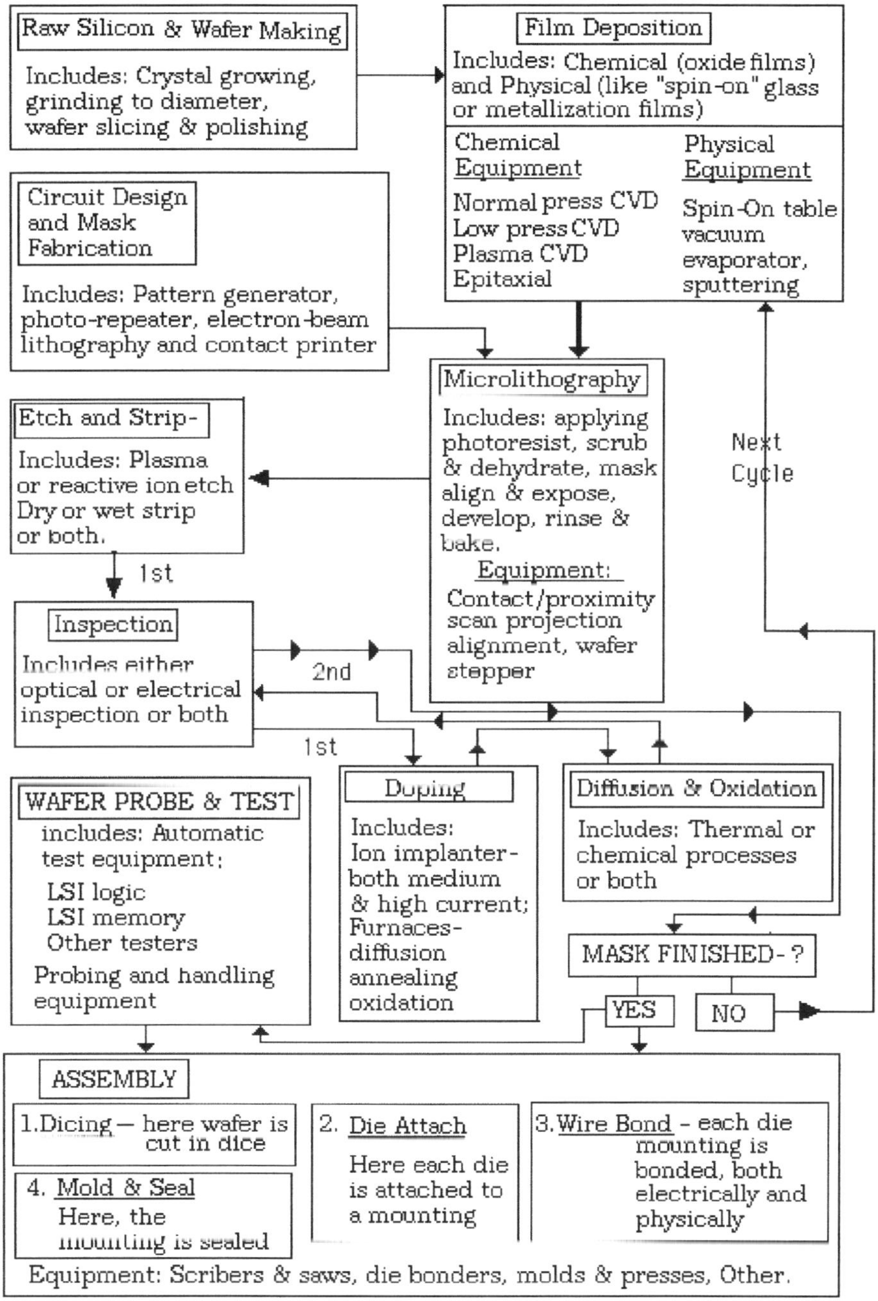

the mask lithography is finished, then wafer-probe testing then takes place. If not, then the cycle begins anew at the film deposition stage. After wafer probe and testing, assembly of the cut dice into a mounting takes place by attaching wires to appropriate points of the integrated circuit, the wires are then bonded, and the whole is sealed before final testing and shipment.

In order to illustrate exactly what we are discussing, the following diagram of a CMOS design used in 1997 is presented on the following diagram. That is, a p-type silicon source is formed on an n-type well, and vice-versa.

6.17.5.-

A Double Wall CMOS Structure Used in 1997

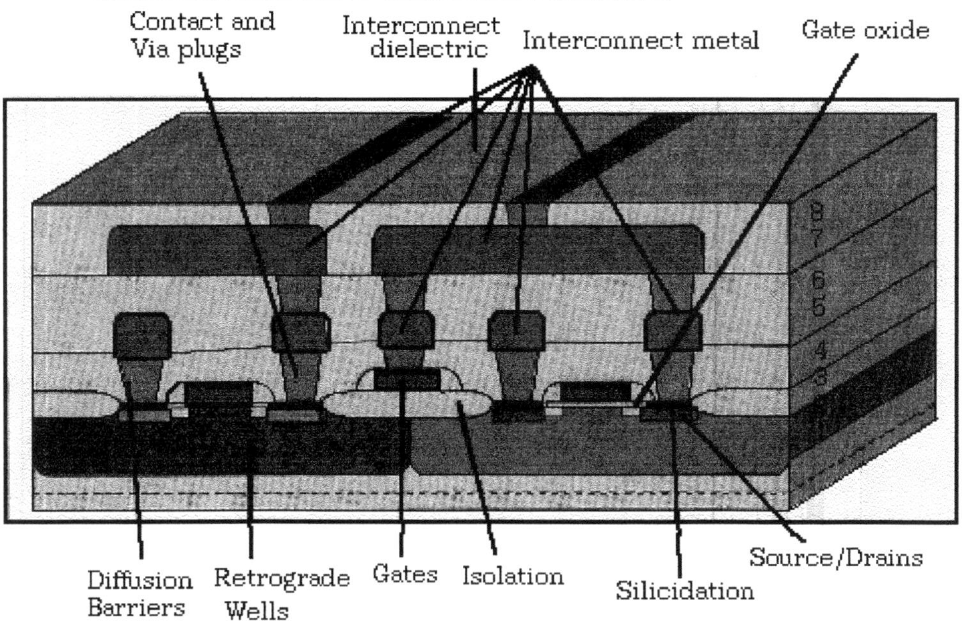


"Sources" are formed in the first step and then "Gates" are formed. A silicide is then used to lower the contact resistance between the silicon of the gate, source and drain and the contacting plug. The thickness of gate oxides is only about 40-50 Å. They are thermally grown rather than deposited by CVD. Nitrides, sometimes combined with oxides, are being used to suppress boron penetration from the gate. Silicides are formed by depositing a metal such as titanium, tungsten or cobalt, followed by rapid thermal processing to form the silicide. The silicide in greatest usage was that of Ti, where the silicide was essentially a $TiO_xSi$ type of compound. Ti/TiN is mostly used now. The depth of the source/drains for 0.25 μm line CMOS designs is about 60 nm., i.e.- quite shallow. Elemental boron is implanted by plasma implantation, using about 5 kV acceleration of $B^{3+}$ ions. Polysilicon gate transistors are formed separately as p-type and n-type. Diffusion barriers are used to prevent diffusion of materials to or from the retrograde wells already formed. The smoothness and thickness of these films are of particular concern. Smooth films help subsequent via-hole filling by hot or warm aluminum deposition. Thin films are desirable to keep the resistance of the plug low.

Tungsten is used for electrical contact and via-plugs. Interconnect metals used in 1997 were aluminum, sometimes combined with copper. This was called "damascene", but in 2003, copper is being used almost exclusively. Once the metal connect is deposited by evaporation or CVD, metal etching is used to remove those parts of the metal film not needed. Interconnect dielectrics include the use of "spin-on-glass" (which is a slurry of low-temperature-softening glass particles) or silicon oxide films formed by high-density plasma processes. Note that at least 8 steps, shown in the above figure, are needed to form the finished CMOS transistor structure. The outline given above is the fabrication technique used to manufacture this **simple** CMOS device on a silicon wafer. Remember that the structure given and described above was fabricated and duplicated about 1800 times on a specific silicon wafer. Following the final step, the wafer was cleaned and then cut into dice. Each die is then mounted and electrical connections made prior to its use in a given device. What we have not described adequately is the complexity of achieving just one of the features, i.e.- layers, shown in 6.17.5. Consider the actual fabrication steps needed to make just one of the features shown therein, i.e.- the "gates" of the CMOS structure. At level 3 of 6.17.5., the gates are shown. The separate steps are illustrated in the following diagram, presented as 6.17.6. on the next page.

6.17.6.-

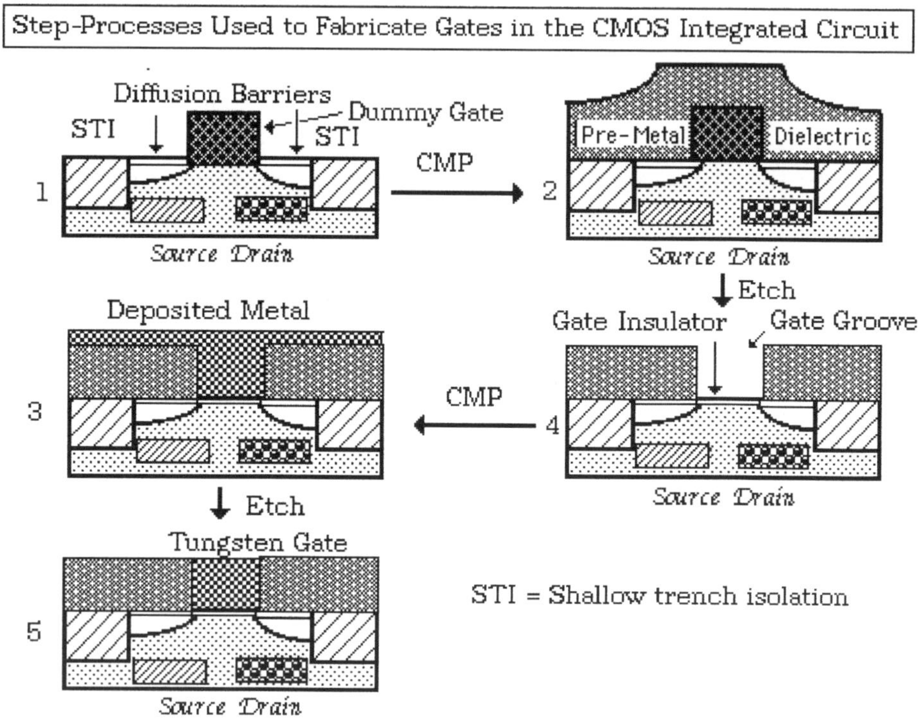


The major steps are labeled 1 to 5. However, there are several intermediate steps also involved. At "1", the shallow trenches are formed, source and drain implants were accomplished, diffusion barriers were formed from Ti/TiN and a dummy gate made from $Si_3N_4$ was deposited on a dummy oxide gate. Five separate photomasks were used to accomplish these intermediate steps. At "2", a pre-metal dielectric film, $SiO_2$ , was deposited by low pressure CVD and planarized by CMP ("Chemical Mechanical Planarization") in order to uncover the top of the dummy gate (2 steps). At "3", wet and dry etching was used to remove the dummy gate and the gate groove was fabricated. A gate oxide of $Ta_2O_5$ was then grown in the groove. (4 steps). At "4", a W/TiN layer was deposited (2 steps) and then planarized to its final form in "5". Note that a total of 14 intermediate steps were required in the formation of the gates.

Metallization of contacts, vias and interconnect metals still need to formed and connected.

A significant number of different machines are needed to accomplish the above fabrication steps. Such **machines** include:

1. Lithographic- Optical wafer steppers are the most expensive and critical machines used in the Fabrication Laboratory.

2. Planarization- Chemical mechanical planarization machines are used to level the topology of the films deposited on the wafer. Chemical-mechanical smoothing is the process used to improve process uniformity, repeatability, removal rate, planarity and defect reduction. Thus, planarization is requisite to producing defect-free CMOS circuits.

3. Ion Implantation- these plasma machines are used to form the various parts of the coupled transistors at voltages of 5 kV to 2 MV. These include retrograde well formation, gates, drains and sources.

4. Sputtering- Deposition of copper doped (damascene) aluminum interconnects, as well as diffusion barriers, and anti-reflective stacks of Ti/TiN is done by this technique. The metal is heated by an electron beam and "sputters" onto the target. Nitrogen gas is used when TiN is required.

5. Chemical Vapor Deposition- Deposition of silicon oxide films is accomplished by CVD equipment. Either plasma CVD or ozone oxidation is used. Blanket tungsten films are also deposited by CVD equipment to create contact and via plugs. Polysilicon and silicon nitride films are deposited in hot-wall furnaces. TiN diffusion barrier films are deposited by either sputtering or CVD, the latter giving superior step coverage.

6. Etching- metal and oxide etch equipment are used to remove excess parts of the deposited film. High density plasma sources

provide a more rapid removal of material and produce cleaner etching.

7. Resist stripping- Both wet and dry removal processes are used at this step in processing. Dry ashing removes the bulk of the photoresist and wet stripping removes remaining residues.

8. Wafer cleaning- wafers are cleaned in automated wet benches using the RCA wet clean method (which uses a combination of dilute acids, peroxide and bases in a particular order). Environmental concerns has prompted the search for a new method of cleaning.

9. Thermal processing- Diffusion furnaces are used not only for the anneal of implanted dopants but for growing high quality thermal oxides, depositing polysilicon nitride films ($SiN_x$) and for rapid thermal processing of deposited films.

10. Metrology - The goal of most metrology machine efforts is to keep the process under control, whether it involves making measurements of physical size of individual features and film thickness, or making electrical measurements of parametric test structures. Defects are also measured and estimated, including excess particles and misplaced features in the composite.

Let us now discuss in more detail several of the operations used to fabricate IC's. These include:

G. Film Deposition.

Several kinds of thin films are deposited on a silicon wafer by different methods (see above) during various stages of the fabrication process. The initial step following cleaning is the formation of a silicon dioxide film. This film is grown by placing the silicon wafer in an oxidizing environment at high temperature as, for example, in a quartz-walled furnace tube. This operation may be followed by the deposition of a film of silicon nitride. Later in the fabrication process, metal (e.g.- tungsten metal) and dielectric (e.g.- glass) films are deposited by means of sputtering. In this technique, a cathode made of the material to be deposited is bombarded with electrons, resulting in the ejection of atoms and molecules from the cathode. This material is deposited on the wafer surface and forms the desired film.

Metal and polysilicon films are formed by a chemical-vapor deposition process using organometallic gases that react at the surface of the IC structure. Various metal silicide films may also be deposited in this manner by reaction with the surface of the silicon wafer to form metal silicides. Glass and polymer films are deposited or spin cast or both, as are photoresist films (those of a photosensitive material). This process is accomplished by applying a liquid polymer onto a rapidly rotating wafer. The exact method used varies from manufacturer to manufacturer and usually remains proprietary.

### H. Impurity Doping.

Selected impurities (e.g.- boron and phosphorus) are introduced into the silicon substrate to control its conductivity in a selective manner. This is accomplished by the combination of two methods: ion implantation and thermal diffusion. In the ion implantation process, the silicon wafer is exposed to a beam of energetic particles (i.e.- ions accelerated through a potential of several hundred kilovolts) of the substances that are to be incorporated into the silicon. These impurities are driven into the silicon wafer via kinetic energy to a depth ranging from a few hundred angstroms to several micrometers, depending on the energy and the mass of the ions in the beam. The wafer is annealed after implantation of the impurity ions in order to remove damage caused during the ion implantation process and to diffuse the impurities to the desired positions in the silicon wafer.

### I. Lithographic patterning.

The process of lithographic patterning determines the geometric features specified by the layout and patterning as the integrated circuit is fabricated layer by layer. A photomask, containing pattern information in the form of an etched chromium layer on glass, is prepared for each layer. Note that a specific photomask is required for each layer of the IC. Thus, many differing photomasks are required for the overall process to produce all of the layers comprising the integrated circuit and the individual components.

An image of the photomask is projected onto the surface of the silicon wafer. This is accomplished with a very high-precision optical projection printer after the wafer has been coated with a thin layer of photoresist. A positive photoresist is a material which is monomeric to begin with and which is caused to polymerize when a light beam from the photomask strikes selected portions of the photoresist so as to become insoluble in either water or a selected solvent. There are both positive and negative photoresists, so-called because the major portion of the photoresist is caused to become either insoluble or soluble by the action of being exposed to selected wavelengths of light. The light is focused down to a point. Thus, to produce a 18 micron line in the photoresist, the focus cannot exceed this value. At this point in time, the focus has approached the focal-limit of the natural wavelength of visible light and even ultraviolet light cannot be focused beyond a certain point. Thus, x-ray sources are being developed to produce the 0.13 micron lines required in advanced IC circuits. We will not fully explore the chemical nature of the photoresists required in this technology except to say that considerable effort has gone into their development.

The exposure is done one die at a time, stepping from each die to the next to complete the exposure of the wafer. So-called "steppers" are an important part of the overall process for making IC circuits on a silicon wafer. The resist containing the pattern image is developed (usually by exposure to light), and the exposed photosensitive material is removed by water-washing from the areas that have not been exposed to light (in the case of a positive photoresist). The desired pattern is thereby transferred from the photomask to the photoresist on the surface of the partially fabricated IC. This photoresist pattern serves to define the areas of the wafer where, during a subsequent process step, film is to be deposited, i.e.- in a via (passage), material is to be removed by etching, or impurities are to be introduced. Once the selective deposition or removal of material has been accomplished, the remaining photoresist is cleared off the wafer.

J. Etching.

During this process, material is selectively removed from the wafer surface as defined by the patterned photoresist in order to define the structure of the previously deposited layer. The etching process is accomplished by exposing the wafer to a gas plasma, which both chemically reacts with the material to be removed and physically ablates it. At the completion of etching, the remaining photoresist is cleared from the wafer.

K. A Final Look at the IC Manufacturing Process

Although the basic fabrication process of an CMOS circuit has been illustrated above for the process used in 1997, the exact materials and sequence of steps has changed considerably in just two years. The following illustrates how the process steps described above have been modified. The details of the components used in manufacturing an CMOS circuit, having a line width of 0.18 μm, is shown in the following diagram, given as 6.17.7. on the next page. Contrast this diagram to 6.17.5.

In this case, we have shown a number of layers comprising the CMOS device and have numbered some of the layers. At "1", the metal interconnects are shown. These are composed of tungsten or copper metal and are either evaporated or applied by electrolysis in an inert gas atmosphere. The use of "damascene" has decreased but is still used for some types of DRAM's. The use of "diffusion barriers" continues, connecting the Cu metal to the active semi-conductors. Inter-level dielectrics being used are fluorocarbon polymers and low-k glasses, both in the form of spin-on suspensions. Siloxane glasses (which are a form of silico-organic materials) are also in use. For purely electrical "connects", tungsten is still used since its tendency to diffuse through the dielectric layers is almost nil.

6.17.7.-

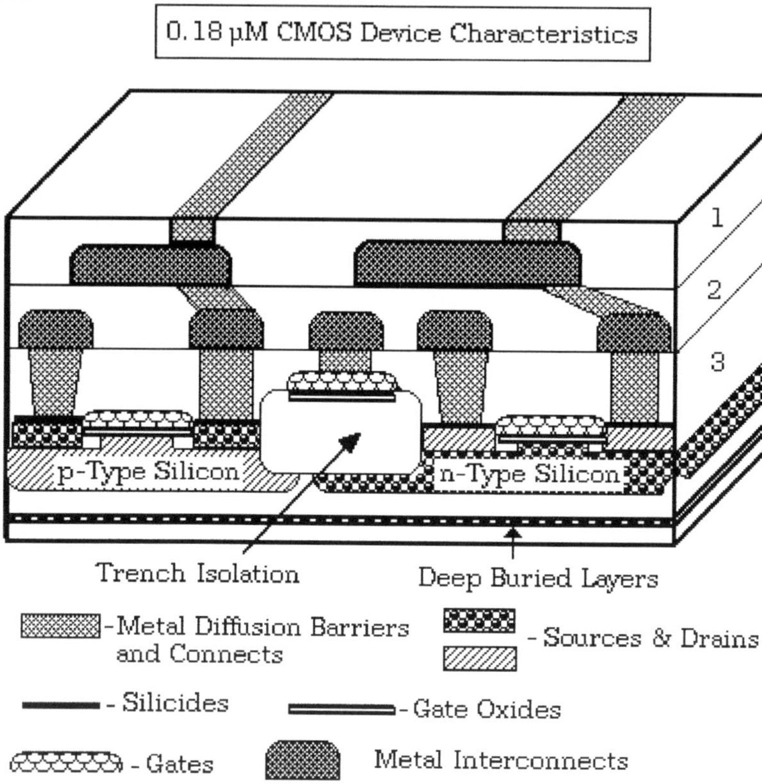


If a diffusion barrier is required, then a titanium/nitride (Ti/TiN) is used. This counteracts the tendency of most metals to diffuse through a given structure, particularly if the layers are composed of several different types of metals. connections.

After the photoresist has been developed and the isolation oxide/nitride mask defined by etching, the wafer is ion-implanted to electrically isolate the areas where the MOSFETs are to be formed, and oxide is grown over the implanted regions. The oxide/nitride film is then removed, and a thin gate oxide is regrown in these regions, which are the sites for the MOS transistors. Polysilicon may be deposited and patterned in the next photolithography process using a second photomask to define the gate structure of the NMOS transistors. A subsequent ion-implantation process introduces the required impurities into the source/drain regions of the

MOSFET's, and silicon oxide is deposited. The source/drain positions where electrical contact is to be made to the MOSFET's are defined, using the oxide-removal mask and an etch process. For shallow trench isolation, anisotropic silicon etch, thermal oxidation, oxide fill and chemical mechanical leveling are the processes employed. For shallow source/drains formation, ion implantation techniques are still be used. For raised source/drains (as shown in the above diagram) cobalt silicide is being used instead of Ti/TiN silicides. Cobalt metal is deposited and reacted by a rapid thermal treatment to form the silicide. Capacitors were made in 1997 from various oxides and nitrides. The use of tantalum pentoxide in 1999 has proven superior. Platinum is used as the plate material.

Metal is then deposited into the opened vias (openings) in the oxide layer and over its surface. During the subsequent photolithography process, it is patterned to form the desired electrical interconnections. These two steps are repeated for each succeeding level to produce additional levels of interconnections. Finally, a protective overcoat of oxide/nitride is applied (passivation), and vias are opened so that the wires connecting the IC chip to its carrier package can be bonded to output pads.

The most important part of the overall process for building or layering an integrated circuit involves the use of dielectrics to isolate the electrically conducting transistors built into the structure. Diagram 6.17.8., given on the next page, shows how these insulating layers are used.

Here, it is easy to see the various layers and steps necessary to form the IC. We have already emphasized the formation of the n- and p-wells and the individual process steps needed for their formation. Note that an epitaxial layer is used in the above model. There are isolation barriers present which we have already discussed. However, once the polysilicon gate transistors are formed, then metal interconnects must then be placed in proper position with proper electrical isolation. This is the function of the dielectric layers put into place as succeeding layers on the IC dicc. Once this is done, then the wafer is tested.

6.17.8.-

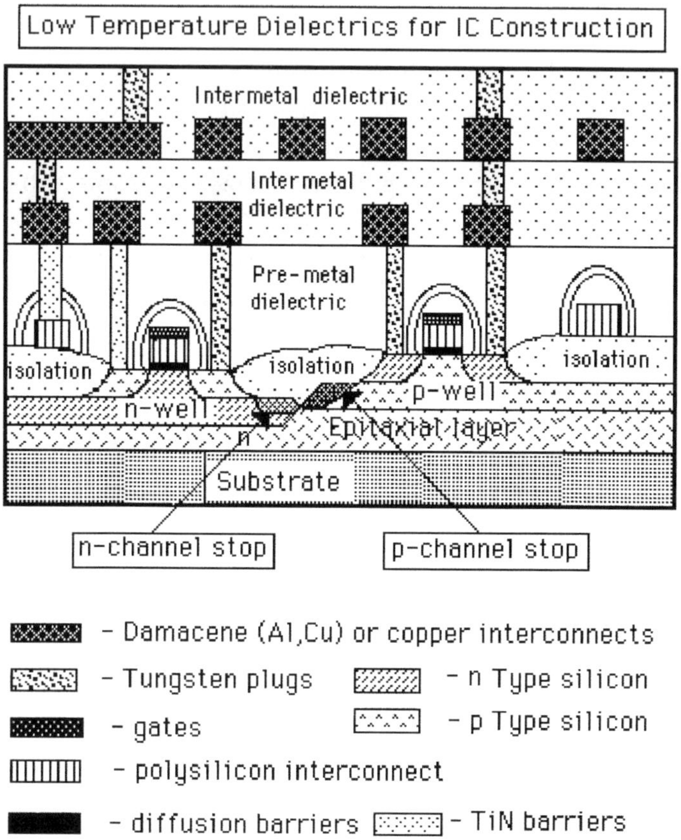


The assembly process begins with an electrical test of the ICs die by die on each wafer prior to separation. This locates and enumerates the number and location of the good units. The wafer is then sawed into individual dies, and each good unit is installed in a plastic or ceramic package by attaching the die to a lead frame and soldering wires between the output pads on the die and the internal leads of the package. A final test is performed on the packaged die to determine whether the unit operates within the specified standards.

The following diagram, shown as 6.17.9. on the next page, shows an IC chip mounted upon a circuit board, with attendant resistors, condensers

6.17.9.-

Mounting and Packaging of Integrated Circuits

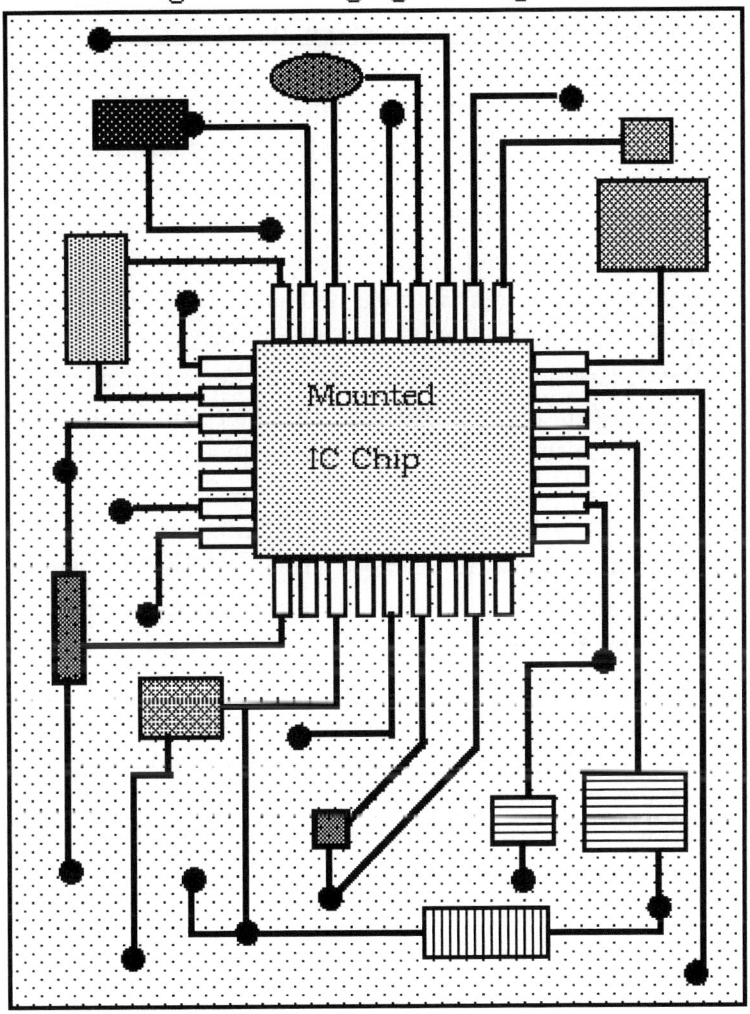


and solder "bumps" (the black spots). The interconnections (shown as black lines on the diagram) are usually copper. They are also formed by a photo-micro-lithographic process similar to the lithography process used in manufacturing IC circuits. Here, the lines are in mils, i.e.- thousands of an inch, rather than in microns, i.e.- millions of a meter. Note that a number of connections are hidden within the device mounting and cannot be seen directly.

## L. Crystal Growth and Crystal Defects Affecting IC's

Under conventional crystal growth conditions of silicon boules, vacancies agglomerate to form small, low-density octahedral **voids**, commonly called "D-defects", i.e.- dislocation defects. Interstitials can form distributed **dislocation** clusters. Both have serious effects upon the process of forming integrated circuits. The void defects have been clearly associated with dielectric breakdown failures in integrated circuits, while the dislocation defects have been associated with controlling the rate of the electrical defect producing reactions due to integrating the management of defects throughout the entire growth process. In other words, the dislocation defects tend to cause the formation of defects in the integrated circuit at the point of the dislocation defect. Additionally, such a defect may not appear until the IC is being operated. Such defects are known to be related to certain classes of leakage current. The void defects have recently become a particular concern throughout the industry in spite of their extremely low density.

In the growth of the single-crystal silicon, the crystal boule is pulled from the melt and is allowed to cool as it is being pulled. By annealing the upper part of the boule and finally subjecting it to a controlled cooling rate, this "zone engineering" ensures that the reactions that produce either the void or dislocation cluster defect are completely suppressed. The result is crystals that are completely free of both void (vacancies in our terminology) and dislocation cluster defects. Eliminating one or the other can be readily achieved. Crystals have been made that are completely free of both classes of defects and have earned this material the name "perfect silicon,".

At present, defect-free silicon crystals have been achieved at only at diameters of 200 mm. Comparisons of crystal quality were made among three techniques: a typical conventional Czrochralski crystal growth technique, a slow-cooled controlled reaction and the "perfect silicon" process. The quality levels achieved in D-defect levels of the material is mirrored in the gate oxide integrity of these materials.

This technique is extendible to 300 mm growth where the propagation of in-grown defects is a major challenge. Engineering work on newly developed 300 mm crystal pulling systems has already begun and is expected to be complete in 2003. As device geometries shrink and active gate areas increase, the probability of one of these defects occurring in a sensitive area increases. This can result in a crystal-related device yield limiting factor that can become critical in certain advanced IC structures. Thus various processes that reduce the density of these void defects are being developed to improve yield potential. This has been achieved primarily by controlling the crystal cooling rate through certain important temperature ranges that have the effect of controlling the defect generating reaction. The effect is fewer, but not zero, voids.

"Blow-cool" silicon is a common expression used for this general type of approach to the problem. But as device generations progress, it is less clear that such an approach will be successful in producing sufficiently low void densities to produce acceptably profitable yields. Epitaxial silicon provides an alternative but is problematic in certain cost sensitive applications, particularly DRAMs.

## 6.18 - Future Methods for Manufacture of Integrated Circuits

At this point, we have described how integrated circuits have been designed and manufactured. At the dawn of the DIGITAL AGE (which we as a society have already entered), Gordon Moore (who founded Intel Corporation which manufactures nearly all of the chips used in computers) predicted that the number of transistors on an integrated circuit, and hence its speed, would double every 18 months. He has been accurate and this has become known as "Moore's Law". Since 1965, this law has been the guiding light of the Semi-conductor Industry and has been good for the Industry as well as the consumer since computers have dropped drastically in price while rising dramatically in speed of computing, i.e - rate of handling bytes. Speed is measured now in megabytes per second whereas in 1965, it was in terms of kilobytes per second. However, Moore's Law will soon collide with a much less flexible set of laws, those of the laws of physics. By 2004, the optical lithography used to make integrated circuits will reach its limits because of the diffraction limit of visible light. Although we have described current dimensions in terms of 0.18 μ, lines less than 0.10 μ will not be possible. However, the solution seems to be that either very short wavelength ultraviolet will be used or that a beam of electrons will be employed to etch the silicon circuits in a wafer. Both methods will employ a much different approach to the same problem.

For far-UV at 193 nm., quartz optics can be used. However, the capacity for finer lines is not as great as for 157 nm. Light sources for these wavelengths do exist but the optics required involve $CaF_2$. This material is not stable in air and is attacked by water vapor. One solution for this optical problem has been found to be the use of reflecting mirrors. The following diagram shows a conventional system for lithography:

6.18.1.-

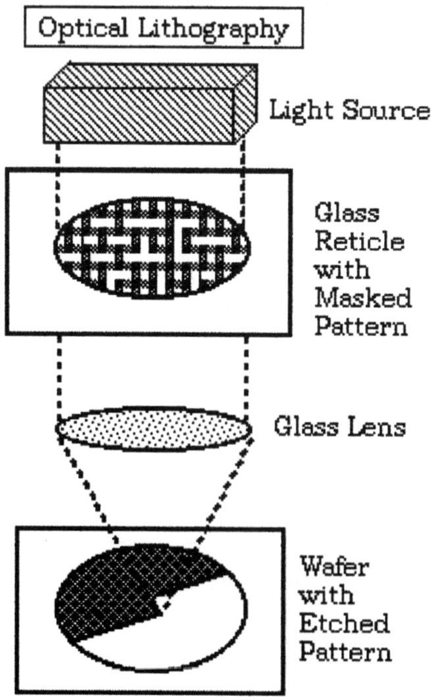

Here, we shown a "conventional" system. It consists of a light source, an etched, transparent circuit diagram and a lens to focus the pattern upon the wafer which has a photoresist layer. Upon exposure, the unexposed part of the photoresist is washed off. Sometimes, the exposed area is removed, depending upon weather the photoresist is positive or negative. Contrast this simple system with a method under development is the "Extreme Ultraviolet (EUV) system, using radiation from excited Xe gas, i.e.- 157 nm. The EUV system is shown in the following:

6.18.2.-

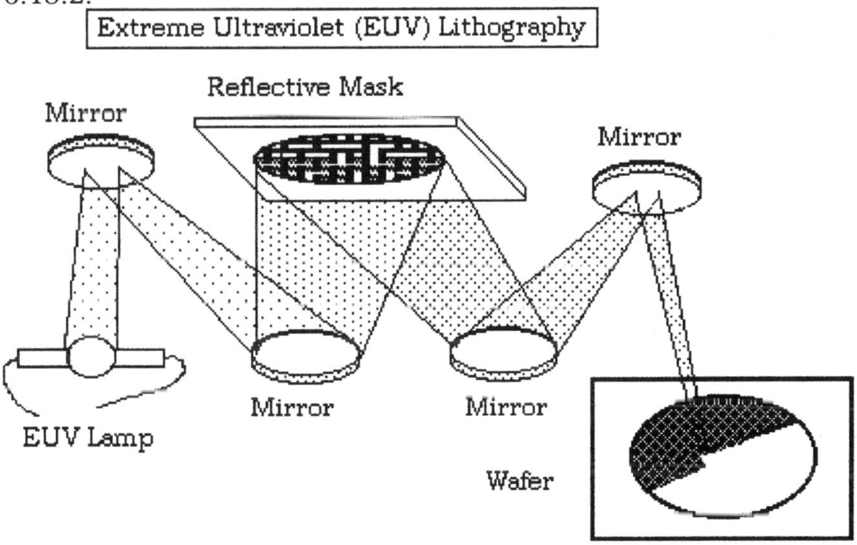


It has similar characteristics to the above systems shown in 6.17.2. and 6.18.1. But, if the EUV method is successful, it will require a near vacuum since oxygen and water vapor strongly absorb 157 nm radiation.

## 6.19.- Pushing the Limits of Semi-Conductor Technology

For the past 30 years, the semiconductor industry has followed Moore's law, which states that transistor performance and density double every 3 years (1). Although not truly a law, Gordon Moore's statement has yet to be violated. But now it seems to be in serious danger. Fundamental thermodynamic limits are being reached in critical areas, and unless new, innovative solutions are found, the current rate of improvement cannot bemaintained. The dominant electronic device used today in integrated circuits is the silicon-based metal oxide semiconductor (MOS) transistor, which consists of a source, drain, channel, and gate region (see the figures given above). The source region provides a supply of mobile charge carriers, enabling current to flow from the source to the drain when the transistor is turned "on". The source and drain regions are electrically isolated from one another by an oppositely charged channel region. A controlling gate electrode is separated from the channel by an insulating oxide material. By applying a voltage across the insulating material, an electric field is created. If the applied voltage repels the channel charge and attracts the source and drain charge, a conducting layer is formed, and current can flow from the source to the drain. In contrast, if the applied voltage attracts the channel charge and repels the source and drain charge, no conducting layer between the source and drain can form, and the transistor is "off"' The transistor thus acts as a digital switch that is turned on or off by applying a voltage to the gate. Individual switches are combined to form the building blocks for microprocessor and memory chips.

For more than 30 years, the switching speed of the MOS transistor has been increased by reducing the size of the device. For proper operation, MOS transistor scaling theory (i.e.- size requirements) requires all vertical and lateral dimensions to be scaled simultaneously (2). At the same time, the total amount of charge in the source, drain, and channel regions must not decrease in order to maintain low device resistance. Charge in the source, drain, and channel regions is created by locally adding dopant atoms to the silicon lattice. In silicon, each silicon atom is covalently bonded to the four nearest neighbors in the lattice. Adding a dopant atom with five valence electrons increases the mobile charge concentration by donating an unbound electron. A dopant atom with only three valence electrons needs to accept an extra electron to substitute a silicon atom; the localized trapping of the extra electron effectively creates a positive free charge referred to as a "hole." Transistor scaling requires an increase in the concentration of these donor and acceptor atoms to maintain a constant total charge in the source and drain regions.

The dopant concentrations in MOS transistors have increased more than 100-fold over the past 20 years and are on the order of 1% of the silicon lattice density for current device technologies. The maximum thermodynamically stable concentration of atoms in silicon, or solid solubility, varies for different dopant atoms. It is a fundamental thermodynamic property and is not dependent on the method of incorporation of the dopant atoms. Above the solid solubility limit, the dopant atoms begin to interact with each other as a result of electrochemical interactions and the strain fields caused by the atomic size mismatch of the dopant atoms and the silicon lattice. This leads to the formation of clusters of dopant atoms, which are not located on silicon lattice sites and do not increase the mobile charge density (3). Unfortunately, the charge concentrations needed for current process technologies are at the solid solubility limit for the dopant atoms currently in use. New dopant atoms have been evaluated, but none have yet been found to create higher concentrations of mobile charge. Thus, unless new methods are developed, future scaling of the transistor will result in a loss of total charge, an increase in resistance, and a potential decrease in performance.

Scaling of the gate oxide insulating material is facing an equally critical fundamental limit. The gate oxide-silicon system can be thought of as a parallel plate capacitor. By applying a voltage on the gate electrode charge is attracted or repelled at the silicon interface. Thinning the insulating oxide material increases the electric field and results in a stronger coulombic force, which increases' the charge density in the silicon and leads to lowered resistance and improved device characteristics. However, increasing the electric field in the insulating oxide can cause the material to break down, resulting in device failure. For previous technology generations, this has determined the minimum oxide thickness that could be used.

As supply voltages are scaled with each generation, the oxide thickness has been scaled to maintain the same maximum electric field. However, within the last few technology generations, a new fundamental limit to the scaling of the insulating oxide has emerged. The oxide layer has become so thin that quantum mechanical tunneling of electrons from the silicon substrate to the gate electrode is now possible (4). The probability of an electron tunneling through a potential barrier depends exponentially on the thickness and potential energy of the barrier. State-of-the-art gate oxide thicknesses are currently between 1.5 and 2.0 nm (5). This represents 3 to 4 atomic layers of oxide. At these dimensions, current flow through the gate oxide as a result of electron tunneling becomes substantial. The tunneling process does not appear to damage the oxide, but the resulting gate leakage can close circuit failures because circuit designs assume no appreciable gate current. Even if circuit techniques can be designed to deal with the leakage, the amount of power consumed will become unacceptably large.

If the insulating gate dielectric cannot be scaled, MOS device performance will be severely degraded. Scaling of the gate dielectric is required not only for the capacitive coupling of the gate to the channel that decreases device resistance; it is also critical for scaling the transistor length. Increased coupling of the gate to the channel allows a higher doping density to be used in the channel while maintaining a low resistance when the transistor is switched into the conducting state. This increase in the channel doping density increases the channel barrier, thereby improving the isolation between source and drain when the transistor is turned off. This permits the lateral distance between the source and drain regions to be scaled. Thus, decreased capacitive coupling and inability to scale lateral dimensions may result if oxide thickness cannot be scaled.

Statistical fluctuation is also a potentially fundamental limit for continued transistor scaling (6). The transistor dimensions have become so small that only about a hundred dopant atoms are needed to control the electrical characteristics. As a result, small changes in the exact number and distribution of the atoms can cause appreciable changes in the device behavior. The statistical nature of the dopant distribution is inherent to the fabrication process and cannot easily be changed. For very large integrated circuits that can use more than 10 million transistors, this statistical variation can cause serious design problems. Unless ways for reducing statistical variation are found, it may not be possible to scale dimensions to the point where tens of atoms determine the device characteristics.

Solutions for these problems have not yet been found (7). It has been proposed that the semiconductor material must be changed to continue transistor scaling. Alternate semiconductor materials such as GaAs and SiGe have been evaluated for more than 20 years, but although these materials have found a niche for certain applications, neither has been able to solve the problems of silicon without causing even more complex problems. Alternate insulating materials for the gate dielectric are also under evaluation. By using an insulating material with a dielectric constant much larger than that of $SiO_2$, the thickness of the material can be increased while still increasing the capacitive coupling. The increase in thickness would strongly decrease the electron tunneling current and would permit continued scaling of the transistor. Unfortunately, no material with a substantially increased dielectric constant that is also compatible with MOS transistors has yet been found.

A substantial effort is being made to increase charge concentrations by creating dynamic equilibrium. Processes such as laser annealing and epitaxial growth have been proposed for creating ultrahigh mobile charge concentrations. Unfortunately, these carrier densities are fragile, and the metastable states are extremely difficult to maintain during processing of the device.

These fundamental issues have not previously limited the scaling of transistors and represent a considerable challenge for the semiconductor industry. There are currently no known solutions to these problems. To continue the performance trends of the past 20 years and maintain Moore's law of improvement will be the most difficult challenge the semiconductor industry has ever faced.

## 6.20.- The Solar Cell

In one way or the other, the Sun is responsible for most of the Earth's energy. Green plant (chlorophyll) photosynthesis has provided the basis of conversion of light energy into coal and crude oil in past eons. Heat from the Sun is also responsible indirectly for other forms of energy used by Man. Examples are wave and wind power (Here, energy is converted into electrical current by use of generators that are rotated by either form of energy). Thus, solar power usage by Man has evolved into two distinct technologies:

1) Direct use of the Sun's heat energy for supplying hot water for homes by special designs of rooftop units (Also known as Solar Thermal)
2) Photovoltaics, defined as the conversion of light energy into electrical energy.

It is the latter with which we will be concerned, a technology which has been touted as the answer to pollution abatement. Additionally, it has been recognized that World crude oil supplies will dwindle to zero within the next 100 years, and that a suitable energy replacement will be required.

The Solar Energy industry traces its origin back about 150 years. Development of solar-cell technology stems from the work of the French physicist Antoine-Cesar Becquerel in 1839. Becquerel discovered the photovoltaic effect while experimenting with a solid electrode in an electrolyte solution. He observed that voltage developed when light tell upon the electrode. About 50 years later, Charles Fritts constructed the first true solar cell using junctions formed by coating the semiconductor, selenium, with an ultrathin, nearly transparent, layer of gold. Fritts' devices were very inefficient converters of energy as they transformed less than 1% of the absorbed light-energy into electrical energy. Though inefficient by today's standards, these early solar cells fostered among some a view of abundant, clean power. In 1891, R. Appleyard wrote of "The blessed vision of the Sun, no longer pouring his energies unrequited into space, but by means of photoelectric cells which would cause the total extinction of steam engines and smoke"........

By 1927, another metal semiconductor junction solar-cell (In this case made of copper and the semiconductor copper oxide), had been demonstrated. By 1935, both the selenium cell and the copper oxide cell were being employed in light-sensitive devices such as photometers for use in photography. However, these early solar cells still had an energy conversion efficiency of less than 1 percent. This impasse was finally overcome with the development of the silicon solar cell by Russell Ohl in 1941. Thirteen years later three other American researchers, G.L. Pearson, Daryl Chapin, and Calvin Fuller, demonstrated a silicon solar cell capable of a 6% energy-conversion efficiency when used in direct sunlight. By the late 1980's, silicon cells, as well as those made of gallium-arsenide, with efficiencies of more than 15% had been fabricated. In 1989, a concentrator solar-cell, a type of device in which sunlight is concentrated onto the cell surface by means of lenses achieved an efficiency of 37%, because of the increased intensity of the collected energy. In general, solar cells of widely varying efficiencies and cost are now available.

At present, solar-energy voltaics are based upon monocrystalline silicon (Si) discs, similar to the integrated-circuit industry. As a matter of fact, reject-discs from that industry are used to make solar-cells wherein the discs are cleaned and the reject IC circuits are removed before the silicon-disc is processed to make a solar-cell. However, solar cells based on the III-V transition elements (such as GaAs alloys, where the valence state of the two metals is $Ga^{3+}$ and $As^{5+}$) can also be fabricated. But they are, at present, too costly. Cost is a major consideration in solar-cell technology since any electricity generated by solar power must be priced similar to the actual cost of using coal or oil to generate electrical power (what we call "electricity"). At present, solar-power electricity generated from silicon discs costs approximately twice that of coal- or oil-based generating plants. (Factored into this is the cost of preventing pollution from the generating plants). Since solar power is non-polluting, it has, potentially, a large advantage if the actual energy efficiency can be raised from about 15% to higher values. Thus, the major factors that must be addressed by solar cell manufacturers are the efficiency and cost of conversion of light energy into electrical energy. Although other types of solar cells based upon transition elements such as GaAs and the like exhibit higher efficiencies, the much lower cost and availability of Si discs make it the material of choice when cost considerations are taken into account.

A solar cell requires a p-n junction (positive and negative) which is activated when photons come into contact with it. The p-n junction then acts as a "diode" to generate a voltage and resulting electrical current. In solar cells based on crystalline silicon, the p-state is achieved by diffusing a trivalent ion such as $Ga^{3+}$ into tetravalent silicon, $Si^{4+}$, while $As^{5+}$ can be used to form the n-layer. Conduction in the p-layer involves the movement of "holes" through the electrically neutral silicon while the n-layer conductivity involves electrons. Because $Si^{4+}$ has four bonds, it has a low conductivity because of its diamond-like structure, i.e.- it forms tetrahedral bonds throughout the structure. If a trivalent ion is diffused into the structure, an electrical defect results in formation of a "hole", i.e.- lack of an electron. Likewise, the diffusion of a pentavalent ion results in an "extra" electron, resulting in a n-layer within the bulk of the silicon.

### A. Crystalline, Polycrystalline and Amorphous Solar Cells

Solar cells based on crystalline silicon discs generally have conversion efficiencies of about 15%, or less, depending upon the manner of fabrication. The efficiency of silicon solar-cells is limited to about 20% because the long-wave emission, i.e.- near infra-red and infra-red radiation, from the Sun is not absorbed in the silicon matrix, or is not absorbed in the silicon close enough to the p-n junction for the photo-generated charge carriers to be collected by the junction. To decrease costs, some manufacturers have turned to poly-crystalline silicon where melted silicon is cast in a mold and cooled. If this is accomplished correctly, a large-grained polycrystalline silicon results. Another less obvious benefit results. The silicon can be cast in a square mold and the resulting square discs will fit together more tightly that the round monocrystalline discs, grown by the Czochralski crystal-pulling technique. Efficiencies of about 16% have been achieved. Another innovation has been the development of amorphous, i.e.- glassy (not crystalline), silicon solar-cells. These arc madc by evaporating (or sputtering) a thin-film of silicon upon a suitable substrate. Both p- and n-silicon can be evaporated sequentially to form a suitable p-n junction. The advantage of thin-film solar-cells lies in efficient use of silicon material. But, the disadvantage lies in lower energy conversion efficiency and most amorphous silicon solar cells only have an efficiency of 11-12%. However, if the silicon is subjected to hydrogen gas during manufacture, a hydrogenated film results, said to have improved efficiencies of up to 16%. Thus, three (3) types of silicon have been used to fabricate solar cells: monocrystalline, polycrystalline and amorphous films.

Silicon solar-cells are fabricated using silicon discs in a manner similar to that described previously for manufacture of integrated circuits. The basic external structure used in solar cells is shown in the following diagram:

6.20.1.

External Construction of a Silicon Solar Energy Cell

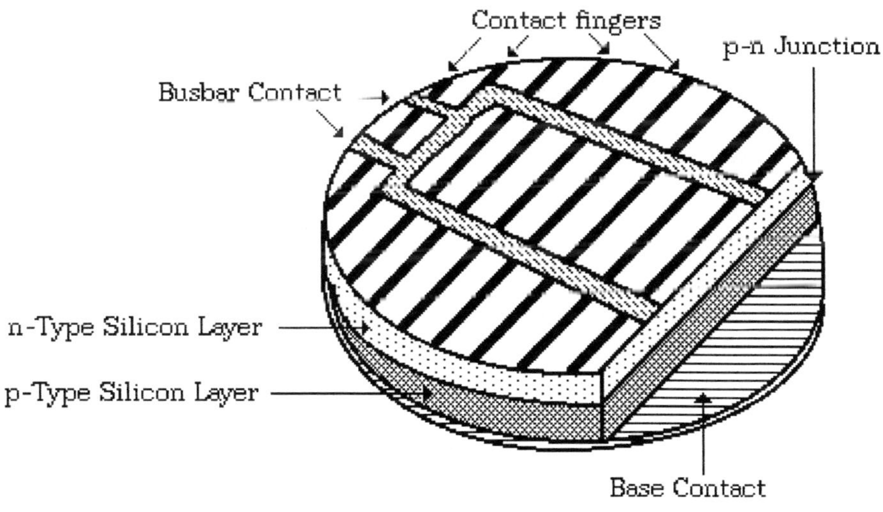


Several differences from that of an integrated circuit can be noted. First of all, two (2) electrical contacts must be established across the bulk of the silicon wafer. When light strikes the surface of the solar cell, its absorption within the silicon bulk releases electrons which are collected as a current. Also shown is the p-n junction. However, the actual silicon disc is only about 350 μm. in thickness. Diffusion processes are used, as a matter of practicality, to form both the p-layer and the n-layer. Thus, the actual physical and electrical configuration is better visualized as shown in the following diagram:

6.20.2-

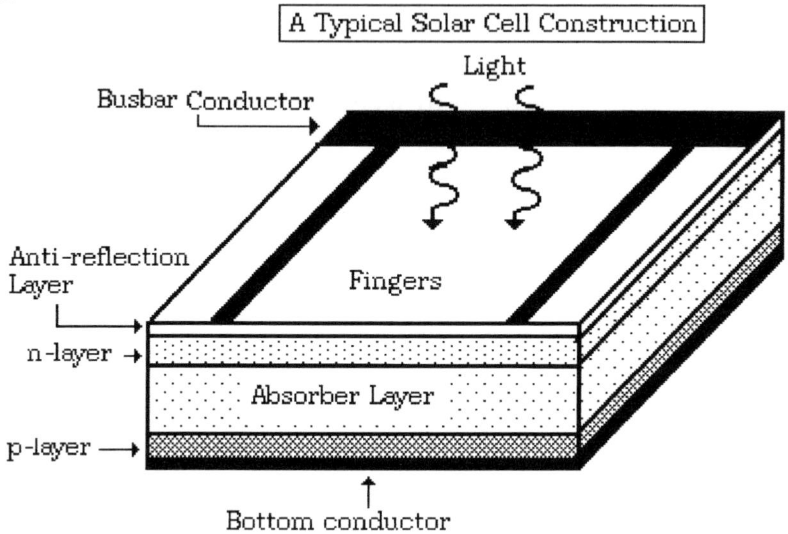


Once the silicon disc is cleaned, the first step is diffuse ions into either side of the silicon disc to first form either the p-layer or the n-layer. Some manufacturers like to have the n-layer closer to the light source, as shown in the above diagram, while others prefer the opposite. At any rate, ions like $B^{3+}$ and $P^{5+}$ are generally used to form the active electrical layers. A number of differing processes have been developed to do this, the exact nature of which depending upon the specific manufacturer of solar cells. Sputtering, vapor-phase and evaporation are used. The most common process uses a volatile boron or phosphorous compound to contact the surface.

Obviously, the other side **must be protected** during the process to form one or the other active layer. Clearly, the silicon disc needs to be heated as well during the process to aid the diffusion process. Note that the surface will be rich in diffusing species and that the density of species declines within the interior, forming a diffuse layer which is dense near the top and thinner in the interior of the silicon. What happens is that once the ion contacts the silicon surface, it "hops" from site to site into the interior of the bulk of the silicon matrix. Since diffusion processes are notoriously slow, the electrical-layers will be only a few microns (μM) thick. The silicon disc is about 350 μM thick so that the bulk of the silicon thickness is not electrically (but is optically) active. This has a beneficial effect on the solar-cell performance since the thickness labeled "Absorber layer" is actually the p-n junction. The photons striking the surface of the solar cell are absorbed within the interior layer, having traversed the "anti-reflection" layer. The function of this layer is to trap the light falling upon the solar cell and to promote the transmission of photons into the energy conversion layers below. Materials such as silicon oxides and titanium oxide are employed as the anti-reflection layer in most silicon-based solar cells. The trick is to form an anti-reflection layer whose refractive index is equal or just slightly above the refractive index of the matrix, here the silicon disc. The base contact is almost always that of a conductive metal like silver or aluminum. The top contact is made via metal stripes or contact fingers which are patterned across the top of the silicon solar cell. A busbar is used to connect all of the collection fingers to the electrical output. The collection fingers allow most of the surface area of the disc to become active. Only about 0.5% of the top surface is usually covered by the metal stripes. The rear aluminum contact also serves as a highly reflective layer as well.

In a conventional solar cell based on a mono-crystalline disc, the stripes or contact fingers are **screen-printed**, using a silver-based alloy, in a resin-based ink. After firing to remove the organic resin, the busbar contacts are screen-printed and then fired. However, some of the current-generating surface is still shaded by the busbar and attendant stripes. The most recent innovation has been the use of a laser to scribe these grid-lines as grooves. The collection grid is thus buried in the surface of the disc, reducing cell shading and making possible energy conversion efficiencies of up to 18%.

The top conductor is almost always in a grid-pattern to allow as much area as possible open to contact with light photons. Note that metals are not light-transparent. The bottom contact does not need to be patterned.

Hence the top grid pattern is usually widely spaced but not the extent that the electrical contact layer will have difficulty in collecting the current produced by the cell's other active layer. Clearly, the silicon disc needs to be heated as well during the process to aid the diffusion process. Note that the surface will be rich in diffusing species and that the density of species declines within the interior What happens is that once the ion contacts the silicon surface, it "hops" from site to site into the interior of the bulk of the silicon matrix.

When different materials are placed in contact with another, an electrical field exists at the interface, or junction, between the materials. It is this fact that allows the formation of an optical-diode such as a solar cell. The role of the n- and p-layers is to establish this built-in electrical field. This field is needed since it dictates the direction of electrical current. When light is absorbed within the absorber layer, free electrons result. In other words, specific Si atoms within the bulk layer are excited and electrons can move. The absorption of light within the interior absorber layer results in energetic free-electrons, causing a current which is collected by the conductors for use in an external circuit to do useful work. Silicon is particularly advantageous since its absorption spectrum lies is strong in the red and near infrared portion of the Sun's spectrum, a region where much of the Sun's energy lies. The latest innovation has been the use of "layered" solar-cells in which thin-transparent films are transposed upon the same substrate. The following shows the types of semi-conductor being used:

6.20.3-

| Cell Type | Spectral Region |
|---|---|
| Germanium, Ge | Near Infrared |
| Silicon, Si | Red or Near Infrared |
| Gallium Arsenide, GaAs | Orange and Red |
| Gallium Indium Phosphide, $GaInP_2$ | Green and Yellow |
| Gallium Nitride, GaN<br>or Gallium Phosphide, GaP | Ultraviolet and Blue |
| Projected Efficiency | 48.8% |

In this Chart, the individual semi-conducting layers are presented as well as the portion of the solar spectrum each one is supposed to absorb. Obviously, if each layer is not transparent, then part of the Sun's energy is wasted. Thus, the 48% efficient of such solar cells may not be achievable.

Note that a five-layer solar-cell is proposed. However, cost-considerations may make this type of solar cell impractical since the cost of electricity will be too much. A three-layer type of solar-cell has been built which had an efficiency of about 39% (consisting of Si, GaAS and GaN). But until the price of other energy sources mount, or until the cost of gallium-based materials drops, a layered solar-cell will not be economically feasible.

## B.-Thin Film Solar Cells

We have already mentioned amorphous silicon solar cells. New processes have been developed to manufacture solar cells based upon deposition of very thin films of photosensitive materials. Such processes have a distinct cost advantage since once the films are deposited, little further processing is needed to form the final solar cell module.

Amorphous silicon is the only thin film process used for mass production of solar cells. Initially, such cells possessed very low efficiencies of less than 10%, due to their inherent degradation problems. To overcome this deficiency, extra layers have been added. In some cases, a triple junction is formed to increase the internal electrical field and hence the collection efficiency. Additionally, extra layers which capture light at other wavelengths because of change in their bandgap energy have been employed. All of this increases costs but improves conversion efficiency. As mentioned above, one of these modifications has included formation of the silicon hydride, $SiH_X$ , which has a slightly modified absorption spectrum.

A prime contender for leading thin film technology as applied to solar cells is cadmium telluride (CdTe). Its bandgap is almost ideal for use as a solar cell for energy conversion from the Sun's spectrum. Here, CdTe and cadmium sulfide (CdS) are used to produce a low cost thin film solar cell without the inherent stability and degradation problems of amorphous silicon. The process uses electrodeposition upon a substrate like glass. Obviously, a conductive coating like tin oxide is needed. Efficiencies over 25% are common for the CdTe solar cell. In 1998, a full scale production effort was started by British Petroleum Corp., using their "Apollo" process.

6.21.- Piezo-Electric Materials and Their Uses

Piezoelectricity involves the creation of an electric charge (or voltage) by applying pressure to a material. Discovered by Pierre and Jacques Curie in 1880, this phenomenon occurs in many non-metallic crystals with non-centrosymetric crystal structures and some degree of ionic bonding. Upon the application of stress, a polarization charge (P) per unit area is created which equals d$\sigma$, where $\sigma$ is an applied stress and d is the piezoelectric modulus. The constant, d, which defines piezoelectric behavior, is a third rank tensor (which means that the applied force generates voltage in varying degrees, depending upon the direction in the piezoelectric crystal structure)..

Thus, piezoelectric behavior is highly dependent on the direction(s) along which a crystal is stressed. For example, in a quartz crystal, stressing along the [100] direction will cause polarization (a voltage), while stressing along the [001] will not. While some single crystal ceramics, such as quartz, are still used in piezoelectric devices, the emergence of polycrystalline barium titanates ($BaTiO_3$) in the 1940s, and lead zirconate titanates ($PbZrTiO_3$), known as PZT, in the 1950s, led to most modern applications of piezoceramics. Both materials (and modifications of them) have high values of d, and thus can create a relatively large voltage for a given Stress— called a *direct* piezoelectric effect. Alternatively, the materials produce a "large" strain if an electric field is applied—called a *con-verse* piezoelectric effect.

For polycrystalline piezoceramics to work, the electrically charged dipoles within the entire piezoelectric component must be aligned by placing the piezoceramic within a high electric field—a process known as poling. The ability of piezoceramics to almost instantaneously convert electrical currant to mechanical displacement and vice versa makes them useful transducers. The efficiency of conversion between mechanical and electrical energy (or the converse) is measured by a parameter called the coupling coefficient. PZT based materials are widely used because of their high coupling coefficients.

## I. Applications

Sonar: The use of piezoceramics (converse mode) to generate a pulsed underwater acoustic wave and then receive a reflect(~l echo (direct mode)—sonar—was the first practical application of piezoelectrics. By the end of World War I, French engineers could measure the depth of submcrgcd objccts using quartz transducers. Contemporary sonar devices generally have a pulse emitter of PZT based material optimized for generating strong pulses. The "receiver" consists of arrays of hydrophones will ceramic compositions optimized to produce the maximum electrical signal from a minimal echo. Modern hydrophone transducers are composite structures with piezoceramic rods mountcd in a polymer. This enhances the output voltage per unit under pressure. Key to these systems are very sophisticated electronics that can separate physical and instrumental noise from the actual returned signal.

Medical Ultrasound: Many of us have first seen our unborn children or grandchildren thanks to medical ultrasound technology. This application is somewhat related to sonar in that both use pulse echo methods to send and receive signals. However, in this short-range application, transducer arrays can be designed to form images.

Micropositioning and Micro-motors. The converse piezoelectric effect produces solid state motion. Piezoceramics are used as actuators in many micro-positioning applications, such as fiber optic alignment, mirror tilting for optical communication and imaging, wafer mask alignment in the semi-conductor industry, and hydraulic servo valves. In exceptional cases, positioning can be controlled to a nanometer or less.

Piezoelectric Transformers: If a two-layer transducer is designed so that a low voltage applied to the first layer to produce a strain, a larger strain is imposed on the second layer. This causes the second layer to generate a higher voltage than the one applied to the first layer. Thus, we have a step-up transformer. When operated at the appropriate frequencies, voltage increases from 3 V to 1500 V are attainable. Such transformers (inverters) are used to power cold-cathode fluorescent lamps for notebook computer displays.

Active noise and vibration damping: Noise and vibration are produced by waves pulsing through the air or through structural materials. Creating an "anti-wave" of the same frequency spectrum, but opposite in amplitude, will cancel out or at least decrease the objectionable noise or vibration. Piezoceramic transducers coupled with sophisticated signal detection and generation electronics are used for such damping applications.

## Suggested Reading

1. "The Growth of Single Crystals" - R.A. Laudise, Prentice-Hall, New York (1970).

2. "The Art and Science of Growing Crystals"- Ed. by J.J. Gilman, Wiley, New York (1963).

3. "Crystal Growth"- Ed. by S. Peiser, Pergamon, New York (1967).

4. "Phase Diagrams-Materials Science & Technology"- Ed. by A.M. Alper, Academic Press, New York (1970).

5. "The Metal-Insulator Transition in Selected Oxides" - Honig & Van Zandt, *Ann. Rev. Mat. Sci.*, **5** , 225-278 (1975).

**References on Silicon Devices**

1. Gordon Moore, ***IEDM Tech. Dig.*** (1975), p.11.
2. RH Denard et al, IEEE J. Solid-State Circuits **SC-9** 256 (1974)
3. KS Jones, S Prussin & ER Weber, Appl. Phys., **A45** 1 (1988)

4. J Maserjian & GP Peterson, Appl. Phys. Lett., **25** 50 (1974)
5. HS Momose et al, ***IEDM Tech, Dig***. (1994), p. 593
6. B Hoeneisen & CA Mead, Solid-State Electron., **15** 819 (1972)
7. S Thompson, P Packan & M Bohr, Intel. Tech., **Q3** 1 (1998)

## Problems for Chapter 6

1. From 6.2.1. "Elements of Furnace Design", prepare a drawing showing how these components are connected together, both physically and electrically, to form a furnace suitable for heating a crucible containing a powder for solid state reaction in air, up to 1200 °C. Be sure to use the right insulation, heater and crucible.

2. Do the same for a furnace designed to heat a crucible in a reducing atmosphere, up to 1600 °C.
3. Design a furnace capable of melting $Al_2O_3$ in air; do the same for melting YAG. Be sure to use the right crucible.

4. List the ways that one can obtain a "seed" crystal. Is it possible to use one type of crystal-growth method to obtain a seed crystal for Czochralski growth?

5. We wish to grow crystals of CaS and SrS, activated by 0.01 mol of $Ce_2S_0$. These crystals will be used to evaluate the luminescent processes which take place in the $Ce^{3+}$ activator center, particularly the polarized emission properties. Determine which one (or more) crystal growing methods is most suitable for obtaining single crystals in this cubic system. Note that we wish to grow the crystal along the {1,1,1} plane since this is the direction of highest atom-density and will give the best crystals for our optical study. Note that a trivalent cation is being situated upon a divalent lattice site.

6. MgO has a wide infra-red transmission spectrum, spanning 0.2 μm to 9.0 μm of usable range. Outline a crystal growing method to obtain single crystal material suitable for this purpose. What methods are not suitable?

7. $TiO_2$ is a material which has far-reaching uses as a UV modifier. Coatings on steel have been shown to inhibit bacterial growth and even kill bacteria on its surface. We wish to grow a single crystal in order to clarify and define this anti-bacterial action. Outline one or more methods for obtaining a single crystal of $TiO_2$. Since this action may be one of a defect-structure, i.e.- oxygen vacancies, include at least one method for obtaining a controlled defect structure of $TiO_{2-X}$, where x can be controlled.

8. We wish to test a new type of ceramic tube to the $Al_2O_3$ tube normally used to fabricate high-pressure sodium lamps in order to compare lamp qualities and life-time operation. Select a method which would produce the desired results and describe how this would be accomplished. Note that the ceramic tube requires both strength and a high melting point.

# Chapter 7

## Measurement of Solid State Phenomena

When solids react, we would like to know at what temperature the solid state reaction takes place. If the solid decomposes to a different composition, or phase, we would like to have this knowledge so that we can predict and use that knowledge in preparation of desired materials. Sometimes, intermediate compounds form before the final phase. In this chapter, we will detail some of the measurements used to characterize the solid state and methods used to follow solid state reactions. This will consist of various types of thermal analysis (TA), including differential thermal analysis (DTA), thermogravimetric analysis (TGA) and measurements of optical properties.

### 7.1.- Methods of Measurement of Solid State Reactions

If we wish to characterize a solid state reaction, from initial compound(s) to final product(s), there are only a few methods we can use. Of primary importance is x-ray identification since we must know what we started with, and what we end up with. We have already discussed the x-ray method in some detail in the second chapter and how one goes about using that method.

If a solid is stable at room temperature, it will remain in that state until some form of energy is applied. In general, it is the application of heat that causes such change. We find that two effects can occur simultaneously, a thermal change and a weight change (but not always). As an example, consider $CaCO_3$. When it is heated to about 840 °C., it forms calcium oxide, CaO, by solid state reaction, vis:

7.1.1.- $$CaCO_3 + heat \Rightarrow CaO + CO_2 \Uparrow$$

The arrow indicates that a reaction has taken place. The gas formed, $CO_2$, is volatile and the final product has a lower molecular weight than the starting material. Thus, a **weight loss** occurs. The **orthorhombic** structure of $CaCO_3$

changes to the **cubic** form of CaO. Thermal energy is required to rearrange the atoms. What has actually happened is that we have exceeded the bonding energy of one compound ($CaCO_3$) by increasing the vibrational energy of the atoms to the point where chemical bonds are broken (by adding heat). This occurs at about 840 °C. and $CO_2$ gas is formed which is stable (and volatile). When we cool the product, we find that we have CaO. Because this change requires heat to be absorbed, the overall process is called **endo**thermic. If it had released heat during the change, it would be called **exo**thermic.

Measurement of the weight change is called thermogravimetric analysis (TGA) whereas measurement of the thermal change accompanying a structural metamorphosis is called thermal analysis (TA). Some compounds do not lose weight but merely change their structural form. A good example of this involves zinc sulfide, ZnS. This material is obtained by precipitation from solution and is cubic at room temperature. When heated to about 1100 °C, cubic ZnS (sphalerite) changes to a hexagonal (wurtzite) form. If wurtzite is then cooled to room temperature, it remains hexagonal. Only if wurtzite is cooled very slowly through the transition temperature does it revert to the cubic sphalerite form.

ALL CHANGES IN PHASE involve a release or absorption of calories. One reason for this is **that each solid has its own heat capacity.** That is, there is a characteristic heat content for each material which depends upon the atoms composing the solid, the nature of the lattice vibrations within it, and its structure. The total heat content, or enthalpy, of each solid is defined by:

7.1.2.- $\Delta H = \int C_p \, dT$ , where : $C_p \equiv (\partial q/\partial T)_p$

where q is a heat quantity. Thus, as we go from one solid to another, we see a change in caloric content. One way to measure this involves the use of DTA.

A. DIFFERENTIAL THERMAL ANALYSIS (DTA)

In 1821, Seebeck discovered that by joining two different metal wires together to form a loop (two junctions), a direct current (DC) would flow in the circuit. A representation of this is given in the following:

7.1.3.-

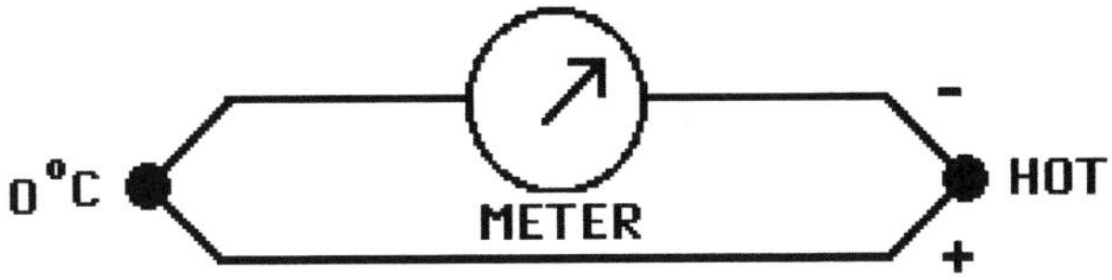


Seebeck used antimony and copper wires and found the current to be affected by the measuring instrument (ammeter). But, he also found that the **voltage generated** (EMF) was directly proportional to the **difference** in temperature of the two junctions. Peltier, in 1834, then demonstrated that if a current was induced in the circuit of 7.1.3., it generated **heat** at the junctions. In other words, the SEEBECK EFFECT was found to be reversible. Further work led to the development of the thermocouple, which today remains the primary method for measurement of temperature. Nowadays, we know that the SEEBECK EFFECT arises because of a difference in the electronic band structure of the two metals at the junction. This is illustrated as follows:

7.1.4.- The Fermi Level at a Junction of Two Dissimilar Metals

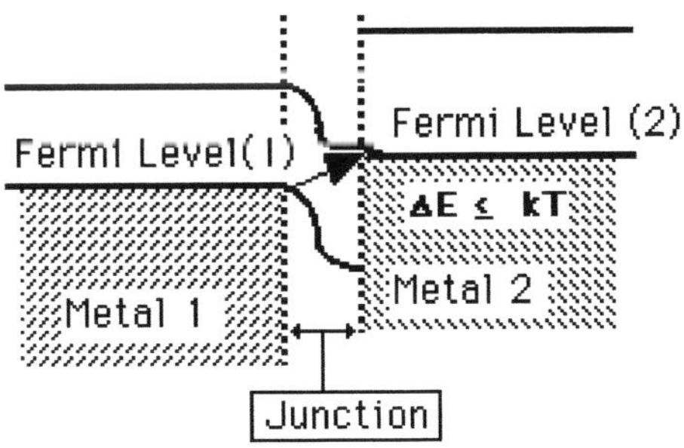


In this diagram, we show the band model structure at the juncture of two metals, each of which has its own Fermi Level. The Fermi Level is the energy level of the electrons contained in the metal. That is- when metal atoms (each

having its own set of electrons) assemble and/or condense to form a metal, the electrons form a "cloud" around **all** of the atoms. This cloud of electrons has an average energy and the top of the energy level is called the Fermi Level. Using such a concept helps one to mentally conceive how electrons form energy bands in any given solid. In contrast to a metal, the electrons in most inorganic solids form well defined energy bands which are determined by the atomic positions of the atoms composing the structure. The main difference between inorganic solids and metals is that the constituent electrons are confined to allowed energy zones, i.e.- Brillouin zones, in the former, but not in the latter. It is for this reason that most metals are conductive, whereas most inorganic solids are not.

Referring to the above diagram, flow of electrons is indicated by the arrow. Since the height of the Fermi Level is proportional to temperature, then the EMF generated is a function of temperature also. It is thus apparent that a thermocouple (TC) will consist of a negative and a positive "leg". The common thermocouples in use today are listed in Table 5-1 along with the temperature range over which they are useful. Also listed is the approximate EMF generated over this range, as well as the nature of each "leg", i.e.- positive or negative.

TABLE 7-1

USEFUL TEMPERATURE RANGES FOR COMMON THERMOCOUPLE

| Composition | Code | Output Range (millivolts) | Useful Temp. Range, °F. | Useful Atm.* |
|---|---|---|---|---|
| (+)Copper-Constantan(-) | T | -5.28to 20.81 | - 300 to 750 | A,N, R |
| (+)Iron-Constantan(-) | J | -7.52to 50.05 | -300 to 1600 | R |
| (+)Chromel-Alumel(-) | K | -5.51 to 56.05 | -300 to 2300 | A, N |
| (+)Chromel-Constantan(-) | E | 0 to 75.12 | 32 to 1800 | A,N, R |
| (-)Platinum-Pt(10%Rh)(+) | S | 0 to 15.979 | 32 to 2900 | A , N |
| (-) Pt- Pt (13% Rh)(+) | R | 0 to 18.636 | 32 to 3100 | A , N |
| (-)Tungsten(5%Re) - W(26%Re) (+) | C | 0 to 38.45 | 32 to 5000 | N, R |

* A = air or oxidizing; N = neutral ; R = reducing

The compositions of the alloys used to make the thermocouples listed in Table 7-1 are given as follows:

7.1.5.- Compositions Used to Make Thermocouples

| | |
|---|---|
| CHROMEL: | 90% Ni - 10% Cr |
| ALUMEL: | 95% Ni - 5% Al , Si , Mn |
| CONSTANTAN: | 57% Cu - 43% Ni |

In differential thermal analysis, i.e.- DTA, we use one thermocouple "bucked" against the voltage output of another of the same composition to produce a "net" EMF. What this means is that either the positive (or negative) legs of both thermocouples are electrically connected so that the net EMF at any given temperature of the two is zero. Only if one thermocouple temperature differs from that of the other does one obtain an EMF response.

7.1.6.- A Thermocouple Used for Differential Thermal Analysis

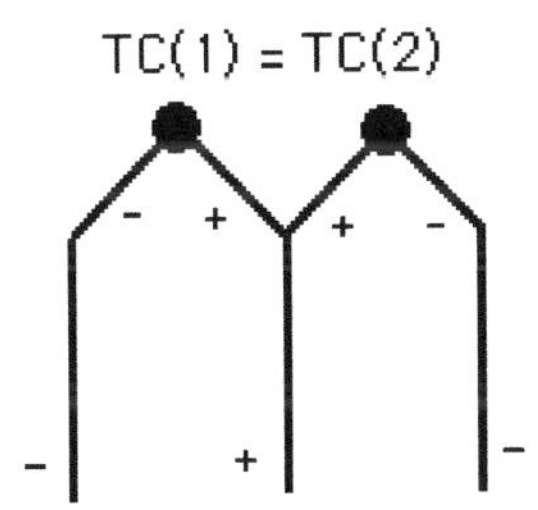


If we put a sample next to one thermocouple and a "standard" or “reference” next to the other, we can follow any thermal changes that may take place as both are heated since each TC generates its own EMF as the temperature changes. Thus, if we put a reference material, R, directly in contact with the "TC(1)" thermocouple junction (hereinafter, we will refer to this thermocouple junction as "R") and a sample, S, at TC(2), i.e.- “S”, then we can detect any thermal change that may occur if either R or S undergoes a transformation as it is heated.

In this configuration, we can detect any thermal changes that occur in the sample as compared to the reference. Note that if both TC-1 and TC-2 of 7.1.6. are at the same temperature, **no EMF is generated.** Actually what we are measuring are changes in heat flow as related to $C_p$ (see 7.1.2.).

For inorganic materials, the best reference material to use is $\alpha - Al_2O_3$. Its heat capacity remains **constant** even up to its melting point (1930 °C.). What this means is that no thermal changes occur in R so that any change detected will be that of the sample, S. In DTA, we want to measure $\Delta C_p$, but find that this is actually:

7.1.7.- $$[C_{p(S_f)} - C_{p(S_i)}]dT_S \pm [C_p]dT_R$$

where $S_i$ is the initial state and $S_f$ is final state for a given solid state reaction of the sample, S (No reaction occurs for R). It should be apparent that we must maintain an equal **heat flow into both R & S simultaneously**, at a uniform rate. If we raised the temperature by steps, we would find that the actual heat flow in both R & S lags behind the furnace temperature considerably, as shown in the following diagram:

7.1.8.-

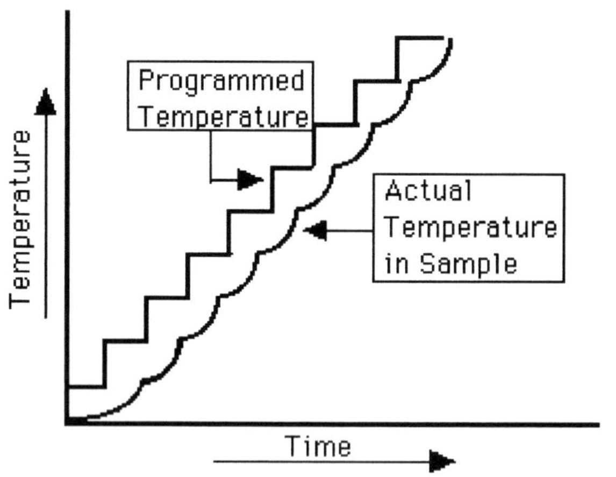

The critical parameters associated with the DTA Method have been determined to be :

7.1.9.- FIVE PARAMETERS ASSOCIATED WITH THE DTA METHOD

1. dT/dt = k
2. Sample Size
3. Rate of Heating
4. Degree of Crystallinity of Sample
5. Effects of External Atmosphere

The rate of heating generally used for most inorganic materials ranges between about 2° to 20°/min. while that for organic compounds lies between about 15° to 100°/min. In the following diagram, a typical thermogram is shown:

7.1.10.-

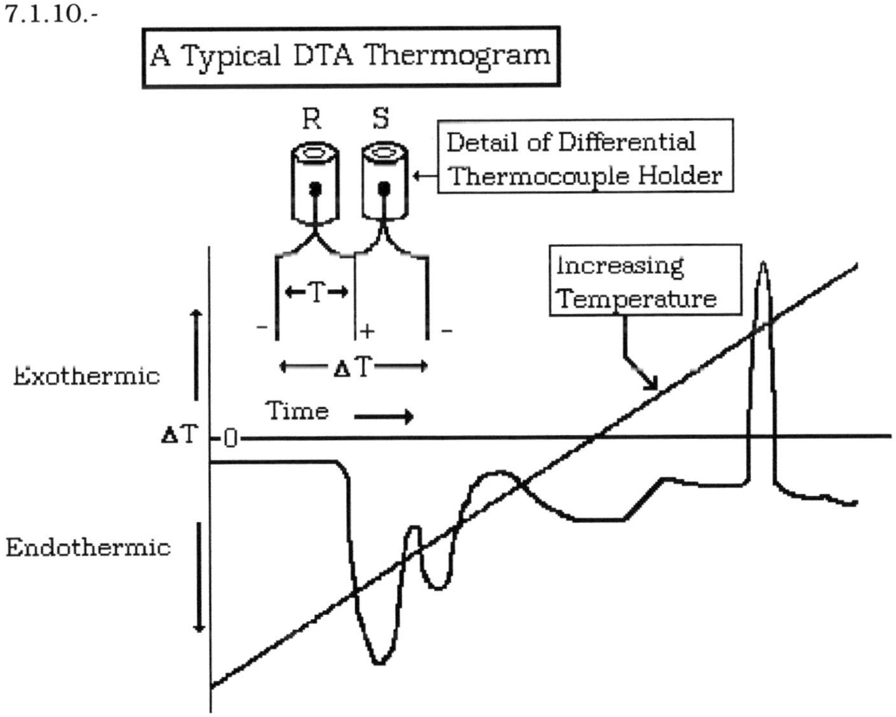

In this diagram, the arrangement of the sample, S, and the reference, R, across the differential TC is shown, and a typical DTA analysis is also given. Note that at low temperatures, the DTA peaks are endothermic. That is, heat is absorbed. Such peaks are similar to those obtained when water-of-hydration is lost, or when the solid state reaction undergoes a loss of water.

The above DTA curve is similar to that of the following reaction:

7.1.11.- $CaHPO_4 \cdot 2\,H_2O \Rightarrow CaHPO_4 + 2\,H_2O \Uparrow$

The first two peaks involve the loss of 2 waters of hydration. The **broad** endothermic peak following in the above diagram is similar to that we might see for a change in composition such as:

7.1.12.- $2\,CaHPO_4 \Rightarrow Ca_2P_2O_7 + H_2O \Uparrow$

The **exothermic** peak may be a change in structure such as:

7.1.13.- $\beta - Ca_2P_2O_7 \Rightarrow \alpha - Ca_2P_2O_7$

In general, solid state decomposition reactions occur as endothermic peaks ($\Delta H$ is negative and heat is absorbed) while phase changes, i.e.- changes in structure, occur as exothermic peaks ($\Delta H$ is positive and heat is evolved). Note that temperature change is programmed to be linear with time, in the above diagram. The components of a simple DTA Apparatus are shown in the following diagram, given on the next page as 7.1.14.

The first thing to note is that the furnace surrounds the sample-holder containing the differential thermocouples. A separate control thermocouple controls the furnace temperature and should be placed as close as possible to the position of the sample holder. Some commercial manufacturers use the Reference leg of the differential thermocouple to control the temperature. However, if you were to build a DTA using the components as shown in 7.1.14,

7.1.14.-

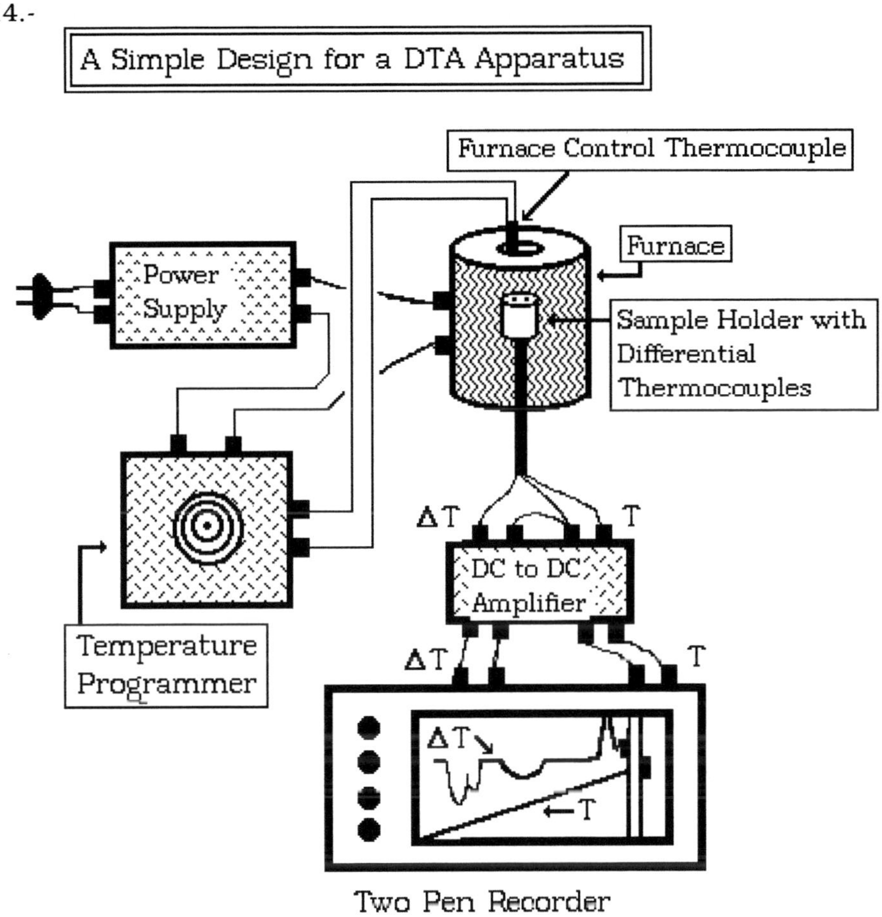


the separate furnace control thermocouple works just as well (even though it is not in exact juxtaposition with the sample and reference. A power supply is controlled by the temperature programmer. The temperature-programmer can be either mechanical or electrical in nature. A DC to DC amplifier is, in general, necessary to amplify the thermocouple signal to the recorder. It is possible to obtain a recorder with sufficient sensitivity to directly record the TC signal. However, it is generally better to amplify the TC signal in order to avoid spurious electrical "noise" that sometimes occurs. A two-pen recorder is superior to a one-channel recorder. In the latter case, it is necessary to switch from the differential TC signal to the temperature TC signal. However, in cases where the 2-channel recorder is not available, a suitable DTA chart be still be obtained. You will notice, on the above DTA graph shown, that the low temperature DTA peaks are endothermic. That is, heat is absorbed. Such peaks are similar to those obtained when water-of-hydration is lost, or when the solid state reaction undergoes a loss of water.

To use this apparatus, one follows a simple procedure. The method that one follows to obtain a DTA run is given as follows:

### 7.1.15.- GENERAL PROCEDURE FOR OPERATING A DTA APPARATUS

a. Load both sample and standard material into the DTA holder.

b. Put the DTA holder in place in the furnace.

c. Set rate for temperature programming.

d. Set DC Amplifier gain.

e, Set recorder gain and time drive

f. Begin temperature program.

g. Record Temperature changes

The sample size (a function of its crystallinity) is usually determined by trial and error. About 500 milligrams is usually sufficient.

It has been determined that the programmed temperature rate has a major effect upon the shape of the peaks observed, as shown in the following diagram, given as 7.1.16. on the next page.

The reason for this is practical. One **must** maintain a constant heat flow across the DTA head. For inorganics, a rate of 2 °C. spreads out the peak whereas a rate of 20 ° C. per minute seems to be about correct for most systems.

7.1.16.-

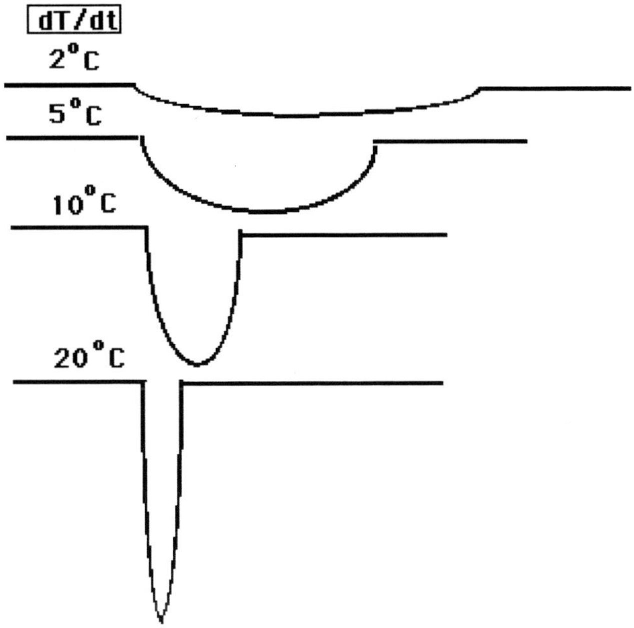


One can program at higher rates, even up to 100 °C./ min. However, the inorganic system cannot adjust fast enough to maintain a constant heat flow. This is, of course, a matter of system-design. However, for organics, a rate of 100 °C./min. may be required. Most commercial instruments have the correct heating rate programmed into the control circuits and one usually does not have to adjust heating-rates. Only when the DTA peaks are not sharp does one need to adjust heating rates to separate any near-lying peaks from one another.

One **can** quantify the heat flow involved within the system in terms of the DTA peak produced. Consider the following, given as 7.1.17. on the next page. In this diagram, the sample, S, is within the furnace which is at a temperature, $T_S$. The heat flow is dQ/dt and r is the **thermal resistance**. The sample undergoes an enthalpy change , ΔH, at its solid state reaction temperature.

7.1.17.-

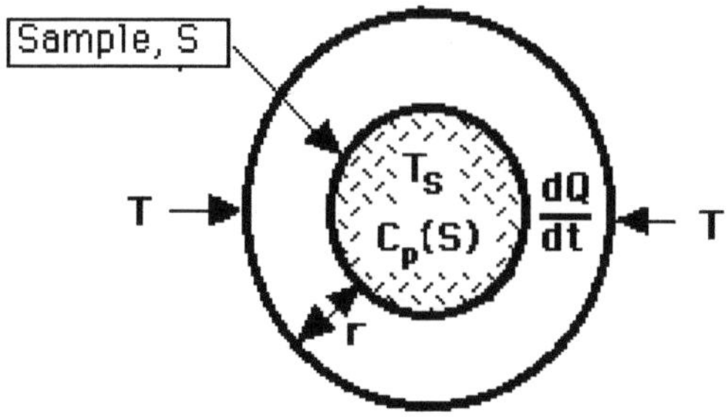


Thus, the heat flow is a function of the differences in temperature of the sample and that of the furnace, i.e.- $T > T_S$ . Then:

7.1.18.- $$dQ/dt = T - T_S / r$$

where t is the time. And the enthalpy change is the difference between that of the sample and the heat flow is:

7.1.19..- $$dH/dt = C_{p(S)}\ dT/dt - dQ/dt = C_{p(S)}\ dT/dt - [T - T_S / r]$$

If we set up the same equations for the reference material, combine these equations and rearrange, we obtain:

7.1.20.-

$$r[dH/dt] = \underbrace{[T_S - T_R]}_{\textbf{I}} + \underbrace{[r(C_{p(S)} - C_{p(R)})]\ dT_R/dt}_{\textbf{II}} + \underbrace{[rCp(S)\{d(T_S - T_R)/dt\}]}_{\textbf{III}}$$

Note that we have divided the equation into three (3) parts, each surrounded by a bracket, i.e. [ ].

This allows us to interpret a DTA peak as shown in 7.1.21., given on the next page. The deviation from the base line (at 0) is a function of both I and II, that is- the difference between sample and reference temperatures (**I**) and differences in heat capacities of sample and reference.

7.1.21.-

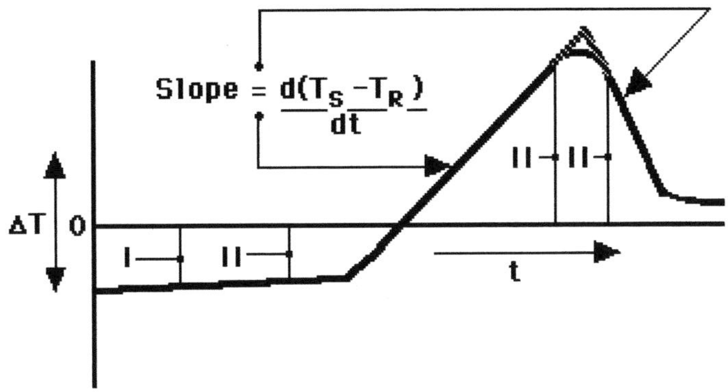


Thus, if the apparatus is properly designed, **I** is not a problem, but **II** cannot be controlled. It is **II** that causes the deviation from linearity which results in a peak. The slopes are a function of the heat capacity differences between **sample plus reference** and **product plus reference.** The slopes (III) obtained are a function of differences between $T_S$ and $T_R$. Note that at the top of the peak, we still have approximately 1/2 sample (as reactant) and 1/2 product.

Referring to 7.1.20. and 7.1.21., it would seem that the **area** of the DTA peak should be proportional to ΔH. This is indeed the case and a comparison of peak areas does yield an experimental value for $\Delta H_{(S)}$, vis:

7.1.22.- $$\Delta H_{(S)} = \Delta H_{Std.} \cdot A_{(S)} / A_{(Std)}$$

This equation assumes a relation between A and ΔH. Actually, this is not too hard to prove, as we can show in the following.

Consider a heat sink as a block containing both S and R (sample and reference). This is illustrated in the following diagram, given as 7.1.23. on the next page. In this diagram, we define three temperatures, $T_B$, $T_S$ & $T_R$, where each subscript refers to block, sample and reference, respectively. We also have two heat capacities, $C_{p(S)}$ and $C_{p(R)}$.

7.1.23.-

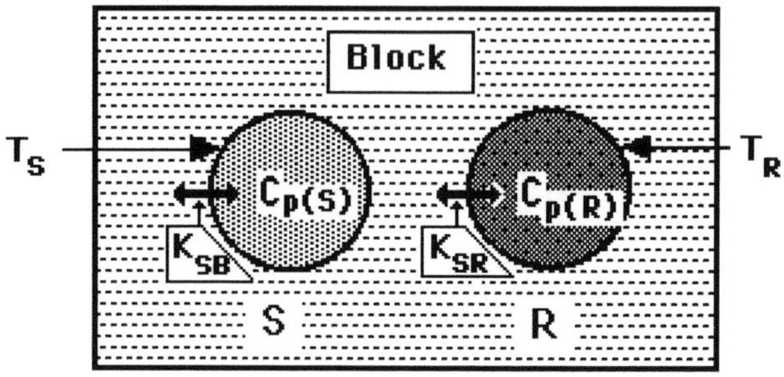


What we wish to prove is:

7.1.24.- $\Delta H = AREA/a$

where a is the **thermal diffusivity**. For convenience, we define the total heat flow in terms of heat flow between S & B, R & B, and R & S. We also use a STANDARD SAMPLE and run it against the reference, so that we can determine how many calories per gram are required for a given transition so as to calibrate the system.

The following table, given on next page, shows some materials suitable for such calibration, along with the temperature at which the solid state change occurs. Note that the reference material, $\alpha$-$Al_2O_3$, is thermally **inert** whereas the standard reference materials **are not.** Appropriate values of $\Delta H$ are available for the standard reference materials given in Table 5-2 on the next page.

Returning to our description of an analysis of the heat flows present in DTA, as shown in 7.1. 17., this gives us the following equations (Note that $\equiv$ is "defined as") :

7.1.25.-

$$dT/dt \equiv K$$
$$K_S \equiv (dQ/dt)_{SB}$$
$$K_R \equiv (dQ/dt)_{RB}$$
$$k \equiv (dQ/dt)_{RS}$$

TABLE 5-2

THERMOMETRIC FIXED POINTS

| FIXED POINTS* | TEMPERATURE °C. | ° F. |
|---|---|---|
| B. P. of $O_2$ | - 183.0 | - 297.3 |
| Sublimation Point of $CO_2$ | - 87.4 | - 109.2 |
| F.P. - Hg | - 38.9 | - 38.0 |
| Triple Point of Water | 0.01 | 32.0 |
| Ice Point | 0.00 | 32.0 |
| B.P. - Water | 100.0 | 212.0 |
| Triple Point of Benzene | 122.4 | 252.4 |
| B. P. of Naphthalene | 218 | 424.3 |
| F.P. of Sn | 231.9 | 449.4 |
| B.P. of Benzophenone | 305.9 | 582.6 |
| F.P. of Cd | 321.1 | 610 |
| F.P. of Pb | 327.5 | 621.5 |
| F.P. of Zn | 419.6 | 787.2 |
| B.P. of S | 444.7 | 832.4 |
| F.P. of Sb | 630.7 | 1167.3 |
| F.P. of Al | 660.4 | 1220.7 |
| F.P. of Ag | 961.9 | 1763.5 |
| F.P. of Au | 1064.4 | 1948 |
| F.P. of Cu | 1084.5 | 1984.1 |
| F.P. of Pd | 1554 | 2829 |
| F.P. of Pt | 1772 | 3222 |

* F.P. = Freezing Point, M.P. = Melting Point, and B.P. = Boiling Point

We can immediately write( see 7.1.2., 7.1.18. & 7.1.19.) :

7.1.26.- $C_{p(S)}\ dT_S / dt = K_S (T_B - T_S) + k ((T_R - T_S) + d(\Delta H)/dt$

$C_{p(R)}\ dT_R / dt = K_R (T_{SB} - T_R) + k (T_S - T_R)$

The equation for R is simplified because R is thermally inert and there is no change in enthalpy involved. To simplify matters further, we define:

7.1.27.- $\Phi_S \equiv C_{p(S)} / K_S$

$\Phi_R \equiv C_{p(R)} / K_R$

$H_S \equiv k / K_S$

$H_R \equiv k / K_R$

Making these substitutions, we get:

7.1.28. - $\Phi_S \; dT_S / dt + (1 + H_S)T_S - H_S T_R = T_B + (1 / K_S) \, d(\Delta H)/dt$

and:

$\Phi_R \; dT_R/dt + (1 + H_R) T_R - H_R T_S = T_B$

Now **if** $k = 0$, i.e.- there is no heat exchange between R & S (as in a properly designed apparatus), and **if** :

7.1.29.- $T_S = T_o + t \, (dT/dt)$

**as it will be** if we are programming the temperature. Then we can define $T_o$ as being equal to zero, so as to obtain the following equations:

7.1.30.- $\Phi_S \; dT_S / dt + T_S = t \, (dT/dt) + (1 / K_S) \, d(\Delta H)/dt$

and:

$\Phi_R \; dT_R/dt + T_R = t \, (dT/dt)$

If we are not in a region where a solid state reaction is taking place, then $d(\Delta H)/dt = 0$. The **change** in baseline temperature (see above) is now:

7.1.31.- $\Delta T_B = dT/dt \, (\Phi_S - \Phi_R) = T_S - T_R$

In other words, the change in baseline temperature is caused by a difference in the relative temperatures of sample and reference, which is related to their relative heat capacities. Thus, one needs to choose the reference material very carefully. If we subtract the equations in 7.1.30., we can get:

7.1.32.- $\Phi_S \; d(T_S - T_R)/dt + (T_S - T_R) - (dT_R/dt) = (1 / K_S)\; d(\Delta H)/dt$

This can be rearranged to:

7.1.33.- $\Phi_S \; d\,\Delta T_S/dt + \Delta T - (dT/dt) = (1 / K_S)\; d(\Delta H)/dt$

$\Delta T$ is the difference in temperature between the reference and the sample. In other words, it is $\Delta T$ that creates the DTA peak. Since we are not measuring **absolute** values of the temperatures, we can define a **relative** temperature:

7.1.34.- $\Delta T_{Rel} = \Delta T - \Delta T_B = \Delta T - (T_S - T_R)$

Using this relation, we get:

7.1.35.- $\Phi_S d\Delta T_{Rel}/dt + \Delta T_R + (dT/dt - dT_S/dt)(\Phi_S - \Phi_R) = (1/K_S)\; d(\Delta H)/dt$

If we choose a suitable reference material (such as $Al_2O_3$ for inorganics), then $dT_R/dt$ will be **equal** to $dT/dt$. Our equation is thus simplified to:

7.1.36.- $\Phi_S \int (d\,\Delta T_{Rel}/dt) + \Delta T_{Rel} = (1/K_S) \int d\,\Delta H/dt$

However, the first term is equal to zero, so:

7.1.37.- $\int d\,\Delta H = K_S \int \Delta T_{Rel}\; dt$

This is what we started out to prove, i.e.- $\Delta H$ equals the area of the peak times the total heat flow to the sample.

## B. DIFFERENTIAL SCANNING CALORIMETRY

Since $\Delta H$ is proportional to the area of the DTA peak, one ought to be able to measure heats of reaction directly, using the equation: 7.1.22. Indeed we can and such is the basis of a related method called Differential Scanning Calorimetry (DSC), but only if the apparatus is modified suitably. We find that it is difficult to measure the area of the peak obtained by DTA accurately. Although one could use an integrating recorder to convert the peak to an electrical signal, there is no way to use this signal in a control-loop feed-back to produce the desired result.

A more practical way to do this is to control the rate of heating, i.e.- dT/dt, and provide a **separate** signal to obtain a heating differential. One such way that became the basis of DSC is shown in the following diagram:

7.1.38.-

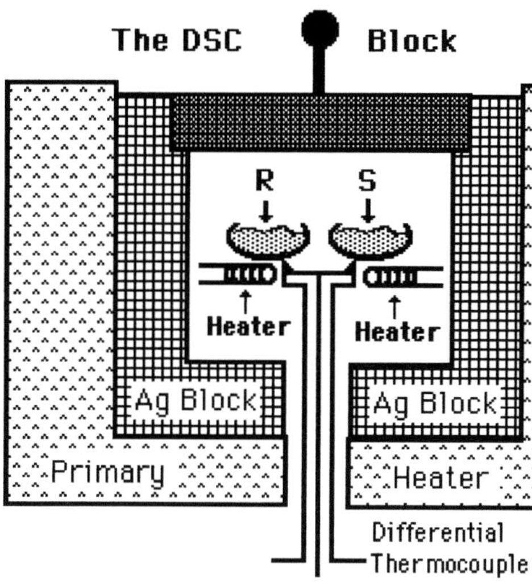


The apparatus consists of a DSC-head within a furnace, like the DTA apparatus. However, there is also a silver block which encloses the DSC head as well. This ensures complete and even heat dispersion. There are individual heaters for both the reference (R) and sample (S) pan-holders. What is measured is the **current required** to keep the differential thermocouple balanced, i.e. $\Delta T = 0$. This signal can be amplified and recorded. We use the same approach for DSC as we did for DTA. We start with the thermal heat flow equation which is similar to Ohm's Law :

7.1.39. $$dQ/dt = T_B - T_S / r$$

We can define the heat change involved with the sample as dh/dt so as to get the equation:

7.1.40.- $$dh/dt = C_{p\,(S)}\,[dT_S/dt] - dQ/dt = C_{p\,(S)}\,[dT_S/dt] - [T_S - T_B] \,/\, r$$

We are using the same terminology for DSC as we did for DTA. Following the methods given above, we arrive at:

7.1.41.- $$dq/dt = (C_{p(S)} - C_{p(R)})\, dT_B/dt + 1/r\{dT_B/dt \cdot t\}$$

We can thus "interpret" a DSC peak in terms of this equation, as we did for the DTA peak, as shown in the following diagram:

7.1.42.

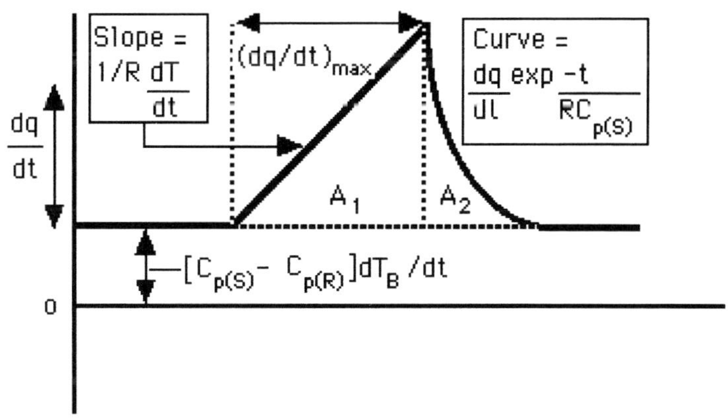

As can be seen, the initial part of the reaction up to $(dq/dt)_{max}$ is linear whereas the curve becomes exponential past the peak. This is due to the difference in $C_p$ between the reactant and product. Thus, $A_1 \neq A_2$.

Actually, the reaction peak in 7.2.5. is an idealized curve since the baseline is a function of the **difference** in heat capacities between reference to sample and reference and product. Usually, we have a different baseline, vis:

7.1.43.-

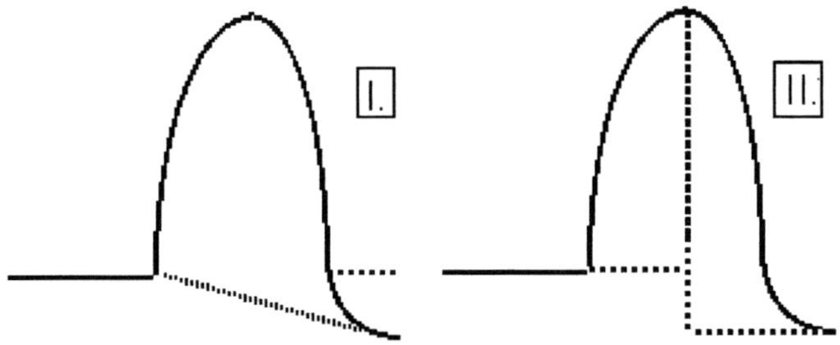


This presents a problem since it is difficult to estimate the area of the peak. One **cannot** simply extend the baseline as in Case I. A much better solution is that shown in Case II, wherein the two very asymmetrical peaks, i.e.- the initial and final parts of the overall thermal reaction taking place, are delineated. This problem has not been satisfactorily answered as yet and represents a challenge to anyone using DCS methods to characterize a solid state reaction.

## 7.2. - UTILIZATION OF DTA AND DSC

### A. Applications of DTA

One of the major uses of DTA has been to follow solid-state reactions as they occur. All decomposition reactions (loss of hydrates, water of constitution, decomposition of inorganic anions, e.g.- carbonate to carbon dioxide gas, etc.) are *endothermic and irreversible.* Likewise are the *synthesis* reactions such as

CaO reacting with $Al_2O_3$ to form calcium aluminate, $CaAl_2O_4$. Phase changes, on the other hand, are **reversible,** but may be endothermic or exothermic.

Thus, if we follow a solid state reaction by DTA and obtain a series of reaction peaks, it is easy to determine which are phase changes by recording the peaks obtained during the **cooling** cycle. Whereas DTA data are qualitative, those from DSC are quantitative and give information concerning the heat change (change in enthalpy) accompanying the exothermic or endothermic reaction. For example, one can obtain a value for melting of a solid state reaction product in terms of calories/gram or Kcal./mole.

DTA is especially suited in the construction of unknown phase diagrams of binary compounds. A hypothetical phase diagram and the DTA curves which would be used to construct it are shown in the following diagram, given as 7.2.1. on the next page.

In this diagram, the endothermic (Endo) peaks point to the right while the exothermic (Exo) peaks point to the left. Consider a system with two components, A and B (see 7.2.1.). They form an incongruently melting compound, AB. The compound, AB, forms only a limited solid solution with A. Most of the composition range is a two-phase region, with a eutectic. The DTA runs are superimposed on the specific composition points of the diagram. Thus at (1) on the diagram (about 10% A and 90% AB), we see one Exo and two Endo peaks. The Exo peak is the point where the two-phase mixture, A + AB, changes to a single phase, i.e. - a solid state solution of A in AB. Further on, an Endo peak indicates the melting point of AB, and finally that of A. At (2), we see only the two melting points, first that of AB and then A. But at (3), only the melting point of the eutectic is seen, that is, both A and AB melt at the same temperature.

In our Phase Diagram, the compound, AB, melts **incongruently,** that is - it decomposes at its melting point. Therefore. at (6), a double peak is seen representing the decomposition of AB and the melting of A. However, B melts at

7.2.1.-

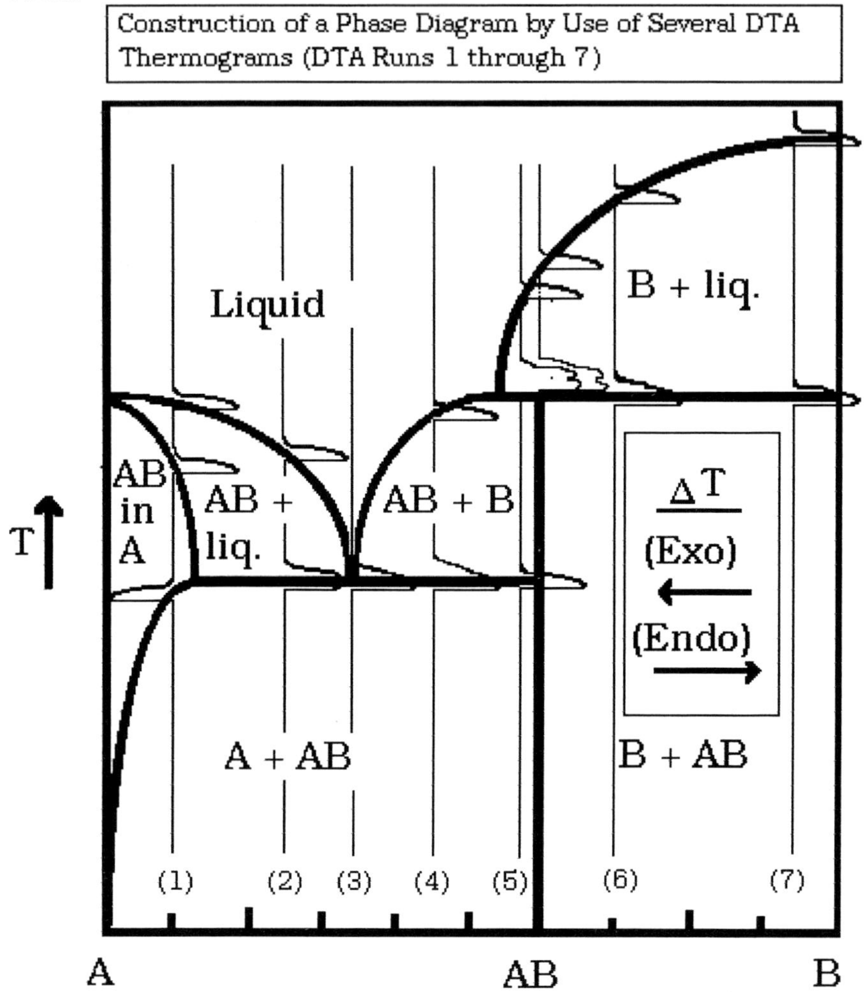


a later time. Note that one can pinpoint changes in the phase diagram quite accurately by running a DTA thermogram at specific composition points. For the most part, the thermal changes observed are specific, but it is wise to cool reversibly, while observing the DTA peaks in cooling so as to be sure exactly what the original peak represents.

Still another use to which DTA has been employed is the characterization of amorphous materials. The following shows a typical DTA thermogram obtained when a powdered sample of glass is run as shown in the following diagram:

7.2.2.-

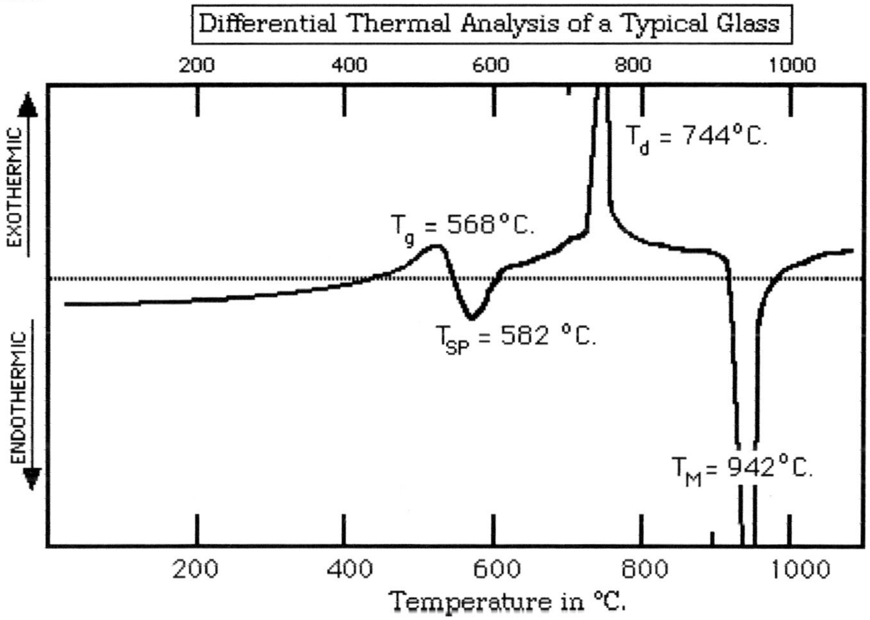


Note that nearly all of the characteristic "glass points" can bc dctcrmincd, i.c.-

$T_g$ = Glass transition temperature
$T_{SP}$ = Glass softening point temperature
$T_D$ = Glass devitrification temperature
$T_M$ = Melting temperature of crystallized product

The only one that is not readily accessible by DTA is the expansion coefficient. It is determined by use of a thermal expansion apparatus, i.e.- a dilatometer. Methods and uses of thermal expansion will be described in a succeeding section.

B. Uses of DSC

The greatest use for DSC has turned out to be for characterization of organic polymers. It has been found that most polymers are amorphous and have a characteristic $T_g$ , i.e.- a "glass" transition temperature which leads to a "crystalline" phase. Other common uses include determination of melting points, boiling points, $T_g$ , % crystallinity and oxidative stability.

In obtaining boiling points by DSC, it is necessary (14) to use a closed pan having an extremely small hole to allow the vapor to escape. The hole is made by use of a laser and should not be more than 50-80 μ in diameter. When a semi-volatile material is heated, an equilibrium will be established between material in the gas phase and in the condensed phase. As the material is being heated, the pressure exerted by the volatile phase, i.e.- the vapor pressure, increases. The rate of heating is important and should be kept between about 6-10 °C. per minute. It is important to have the two phases in equilibrium as the sample increases in temperature, hence the use of an escarpment in the sample pan to retard escape of the volatile material. This arrangement allows the vapor produced to sweep out the air and replace it. At the temperature where the vapor pressure of the sample exceeds the total pressure of its surroundings, the material boils. If the outside pressure is kept constant, there will be an endothermic heat flow associated with condensed phase material entering the vapor phase. As the temperature increases, the rate of boiling also increases. When all of the material is in the vapor stage, it remains in that state until all of the material has boiled off. If the hole is not small enough, then all of the material will be evaporate and be lost before the equilibrium condition is attained. If an equilibrium between vapor and material is not achieved, then the boiling point measured will not be the true boiling point.

As an example of how the data are obtained, the following diagram is presented as 7.2.3. on the next page. This diagram shows the behavior of water at 1.0 atmosphere when it is subjected to the above conditions. Note the design of the pan used to heat the water. It consists of an oval shape, something like an "egg"

7.2.3.-

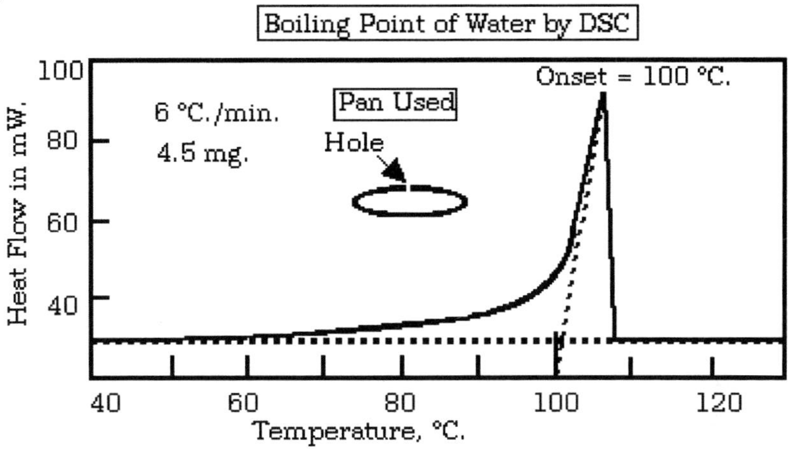


with a small hole on the top to limit the amount of water escaping at the boiling point. Keep in mind that the heat flow (which is related to the degree of vapor change achieved) is low in the beginning, but rises rather fast as the boiling point is reached. The flat **leading edge** of the endotherm represents the point where the sample temperature is constant at the boiling point.

If one measures the boiling points at several pressures, including that of atmospheric pressure, one can then extrapolate to obtain the vapor pressure of a material at ambient temperature. This is done using the Clausius-Clapeyron equation, i.e.-

7.2.4.-

$$E_o \equiv RT^2 \,(d \ln k_1 / dt)$$
$$E_o = \Delta H^* + RT$$

## 7.3.- Thermogravimetry

Thermogravimetric analysis (TGA) measures changes in weight of a sample being heated. A typical Thermogravimetric analysis (TGA) apparatus is shown in the following diagram:

7.3.1.-

A SIMPLE DESIGN FOR A THERMOGRAVIMETRIC APPARATUS

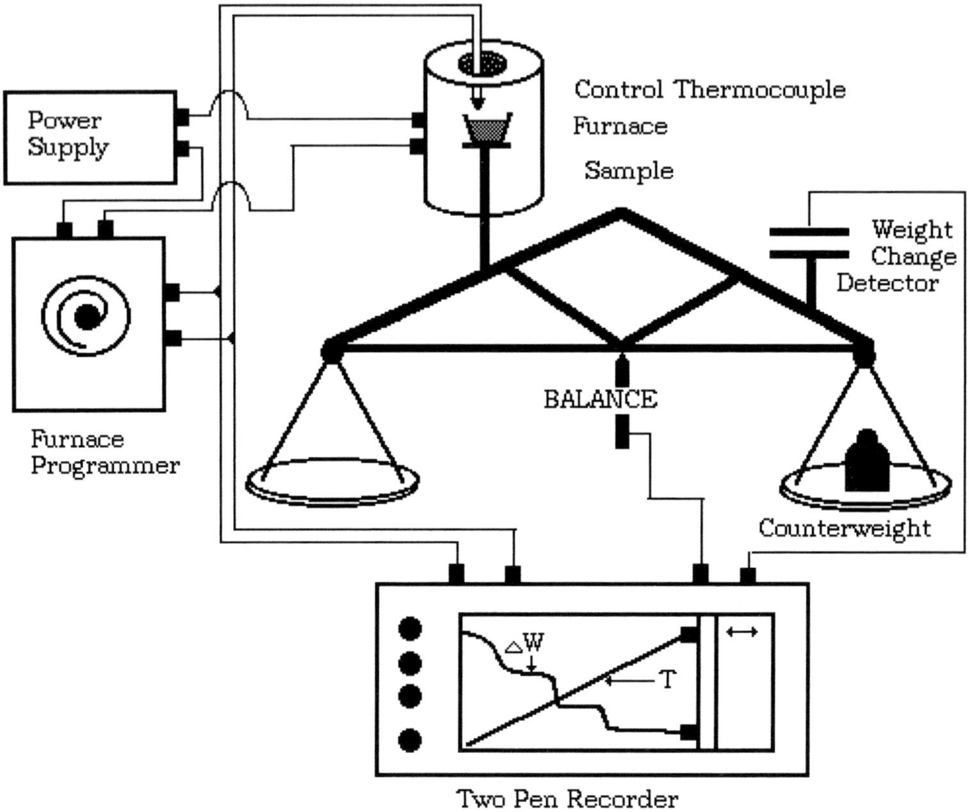


This instrument consists of an analytical balance having a weight-change detector on one side of the balance. Although current commercial TGA designs no longer use such a balance, the original ones did so. Current TGA designs employ piezoelectric crystals or similar types of crystals sensitive to gravitational force to measure changes in weight as the temperature is raised. As in the DTA design, a temperature programmer is needed along with a furnace temperature TC. A power supply is controlled by the temperature programmer. The temperature-programmer can be either mechanical or electrical in nature. It is possible to obtain a recorder with sufficient sensitivity to directly record the TC signal. However, it is generally better to amplify the TC signal in order to avoid spurious electrical "noise" that sometimes occurs. A two-pen recorder is superior to a one-channel recorder where it would be necessary to switch from the differential TC signal to the temperature TC signal. However, in cases where the 2-channel recorder is not available, a suitable TGA chart be still be obtained.

The actual weight is monitored in real time. Changes in weight, either gains or losses, are evident immediately. The apparatus itself consists of the sample situated within a crucible, which is enclosed within a temperature-controlled furnace. The sample, plus crucible, is counterbalanced on a sensitive balance. Weight changes are directly plotted on a two-pen recorder. Weight readout is usually accomplished by one of two methods, a linear transducer or a capacitance change between two flat plates, one of which is free to move with the balance swing. A resistance-capacitance tank circuit completes the electronics, producing a readable voltage. We usually employ a crucible to hold the powder sample, although flat pans are also suitable. Furnace temperature is controlled in a linear manner and recorded. In some cases, the TC is mounted directly at the sample position. It is important that sample and furnace temperatures be nearly equal, so as to record accurate weight losses and gains. Operational parameters for TGA are:

7.3.2.-

1. Sample Size (buoyancy)
2. Sample Closure
3. Heating Rate
4. Heating Mode
5. External Atmosphere

The SAMPLE SIZE is important because most balances have a limited range of weighing, as well as a limited sensitivity, i.e.- milligrams per gram of weight detectable. Many balances feature automatic counter-weight loading. If a sample is **fluffy** and a large crucible is used, then the <u>buoyancy factor</u> must be accounted for. At high temperatures, i.e.- > 800 °C., the density of air within the furnace is sufficiently **lower** than that of the outside, so that the apparent weight of the sample plus crucible appears lower than it actually is. And, the larger the crucible, the more air is displaced within the furnace. For 10.000 grams of total weight, the buoyancy factor will be about 0.002, enough that a correction needs to be made for precision work.

SAMPLE CLOSURE is important since it affects the *rate* of solid state reaction. Consider the following solid state reaction:

7.3.3.- $CaCO_3 \leftrightharpoons CaO + CO_2 \Uparrow$

**If** the gas is restricted from escaping, then the equilibrium is shifted **to the left** and the $CaCO_3$ does not decompose at its usual temperature. A **higher** temperature is required to effect decomposition. Likewise, *an external atmosphere*, such as $CO_2$ in the above reaction, restricts the apparent temperature of decomposition, the rate of reaction, and sometimes, the mode of decomposition (depending upon the nature of the compound under investigation). As an example, consider the following: $CaCO_3$ normally decomposes at about 860 °C. In a 1.0 atmosphere pressure of $CO_2$, the decomposition temperature is raised to about 1060 °C.

HEATING RATE is important from a practical aspect. Usually, we are measuring furnace temperature and cannot program the temperature too fast, for fear that the sample temperature will lag the furnace temperature by too great a degree. It is also for this reason that we use as small a sample as is practical to obtain a weight loss or gain which the balance can discriminate. This again depends upon the sensitivity of the balance and the nature of the reaction we are examining. In general, weight gains (due to oxidation) require more sensitivity and larger sample sizes than those of decomposition. Usually, we restrict heating rates to ≥ 15 °C and use a heating rate of about 6-8 °C/ min. at most. The HEATING MODE to be used depends upon the results we wish to achieve. A typical TGA run is given in the following diagram:

7.3.4.-

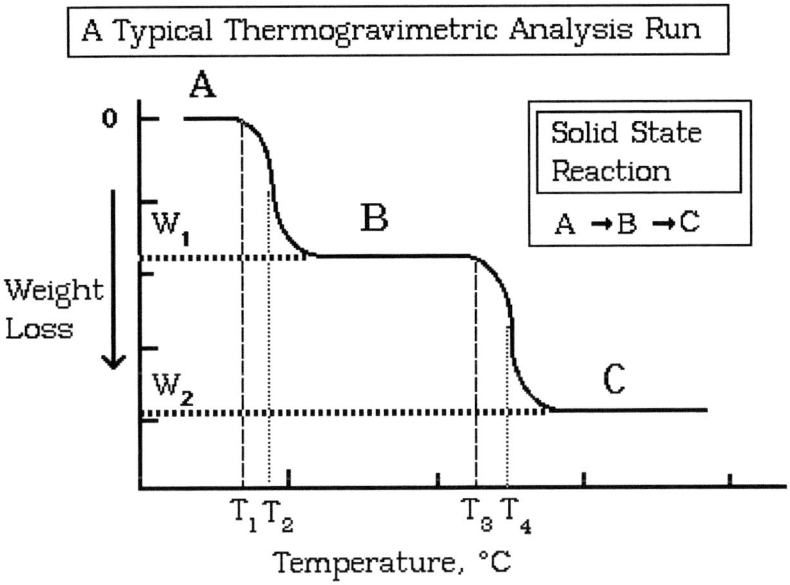


In the solid state reaction depicted, A begins to decompose to B at $T_1$ and the reaction temperature for decomposition is $T_2$ , with a weight loss of $W_1$. Likewise, the reaction of B to form C begins at $T_3$ and the reaction temperature (where the rate of reaction is maximum) is $T_4$ . Note that the weight loss becomes constant as each reaction product is formed and the individual reactions are completed. If we program the temperature at 6 °C/min., we would obtain the results in 7.3.4. This is called **dynamic thermogravimetry.**

However, if we set the furnace temperature just slightly greater than $T_2$ , we would obtain a reaction limited to that of A - B, and thus could identify the intermediate reaction product, B. This technique is called **isothermal thermogravimetry.** Thus, we can follow a solid state reaction by first surveying via dynamic TGA. If there are any intermediate products, we can isolate each in turn, and after cooling (assuming each is stable at room temperature) can identify it by x-ray analysis. Note that we can obtain an **assay** easily:

7.3.5.- Assay ≡ final weight / original weight

If there is a gaseous product, we can also identify it by converting weight loss to mols/mol of original reactant. In 7.3.3. above, 1.00 mol of $CO_2$ is expected to be lost per mol of reacting $CaCO_3$. The actual number of mols lost depends upon the original sample size. The most recent TGA apparatus includes what is called EGA, i.e.- effluent gas analysis. Most often, this consists of a small mass spectrograph capable of identifying the various gasses most often encountered in TGA. Gaseous weight losses can be classified according to the nature of the effluent gases detected. These include the following:

7.3.6.- Gaseous Products = $CO_2$ , $N_2O_4$ , $H_2O$ , $SO_2$ , $SO_3$ , CO.

Water - of hydration

- of constitution
- adsorbed

Regardless of how well a sample has been dried, it will always have an adsorbed monolayer of water on the surface of the particles. If we run the DTA carefully, we will see a small endothermic peak around 100 °C. Additionally, if we run the TGA properly, we will see a small loss plateau before the major losses begin. As a matter of fact, **if we do not see the loss of adsorbed water**, either the apparatus is not operating properly, or we do not have sufficient sensitivity to observe the reactions taking place. The different types of water which can be present during any inorganic solid state reaction is easily illustrated by the following example. Dibasic calcium orthophosphate forms a dihydrate:

7.3.7.- $CaHPO_4 \cdot 2\,H_2O$ - brushite

$CaHPO_4$ - monetite

Brushite reacts to form monetite which then reacts to form pyrophosphate:

7.3.8.- $2\,CaHPO_4 \cdot 2\,H_2O = 2\,CaHPO_4 + 2\,H_2O \Uparrow$

$2\,CaHPO_4 = Ca_2P_2O_7 + H_2O \Uparrow$

We can illustrate the type of calculations needed in order to determine the parameters involved in TGA runs.

In the following discussion, we present how one confronts this problem and the calculations needed to produce the desired results. The following Table, presents a typical problem that one encounters in TGA and the calculations needed to produce the desired results.

Table 7- 3

A TYPICAL PROBLEM IN TGA

| METHOD | ORIGINAL | FIRED PRODUCT |
|---|---|---|
| X-ray Analysis: | brushite | $Ca_2P_2O_7$ |
| Weight: | 17.311 gram | 12.705 gram |
| Molecular Weight: | 172.09 | 254.11 |

In this case, we start with a known material for which we have already used x-ray analysis to determine the nature of the fired product. We start with the reactions given in 7.3.8. for the reactions of calcium phosphate, since this also illustrates how assays are calculated. The steps include:

1. Determine assay
2. By subtracting actual assay from theoretical assay, obtain amount of water actually adsorbed on particle surfaces
3. Determine losses incurred by stages, if one wishes to determine the actual reactions occurring during the solid state reaction

Our first step is to analyze the solid state reaction by means of the values determined in the TGA analysis run.The reaction products are given above, along with the requisite molecular weights.

The next part required in the TGA analysis is given as follows (this is a continuation of Table 7-3):

<u>OVERALL SOLID STATE REACTION</u>:

$$2\ CaHPO_4 \cdot 2\ H_2O \quad = \quad Ca_2P_2O_7 + 3\ H_2O \Uparrow$$

| | | |
|---|---|---|
| Molecular Weight: | 2(136.059) + 2(18.015) | 254.11 + 54.045 |
| WEIGHT: | 17.311 gram | 12.705 gram (Assay = 73.39%; Theor. = 73.83%) |
| ANALYSIS: | 12.705 gram $Ca_2P_2O_7$ | ≈ 17.208 gram of $2\ CaHPO_4 \cdot 2\ H_2O$ |

By calculation, we find that we have 99.41% of $2\ CaHPO_4 \cdot 2\ H_2O$ and 0.59% adsorbed water (by subtraction). We can also determine from the TGA run the amount of water loss by stages in the overall reaction:

| | | |
|---|---|---|
| Losses by Stages: | 1st loss = adsorbed $H_2O$ | = 0.102 gram |
| | 2nd loss = brushite to monetite | = 3.532 gram $H_2O$ = 3.92 mols |
| | 3rd loss = monetite to pyrophosphate | = 0.972 gram $H_2O$ = 1.08 mols |
| | Total wt. of water lost | = 4.606 gram |

Note that the second loss corresponds to 3.92 mol of water per mol of reactant, whereas the 3rd loss is 1.08 mol. This illustrates a serious problem that can be encountered in dynamic TGA, **If the rate of heating is too fast** and not enough time occurs during programming to achieve true equilibrium between successive solid state reactions, then the loss of water from one reaction **carries over** into the next succeeding reaction.

## 7.4.- Determination of Rate Processes in Solid State Reactions

We have presented two methods useful in following solid state reactions. In order to completely classify a reaction, we need to obtain an estimate of the reaction kinetics and order of the solid state reaction. Both DTA and TGA have been used to obtain reaction rate kinetics. But first, we must reëxamine kinetic theory in light of solid state reactions, which differ from those involving gaseous components usually quoted in books dealing with reaction rate kinetics. the following discussion reiterates what we have already presented, but is given again here to reëmphasize its importance in solid state chemistry.

## A. TYPES OF SOLID STATE REACTIONS

In general, we can classify solid state reactions as being either homogeneous or heterogeneous. The former involves reactions by a single compound whereas the latter involves reactions between two different compounds. There are at least four (4) types of solid state reactions (as we have already presented in a prior chapter):

7.4.1.- Types of Solid State Reactions

| | |
|---|---|
| 1. decomposition: | $A \Rightarrow B + C$ |
| 2. synthesis : | $A + B \Rightarrow C$ |
| 3. substitutional: | $A + B \Rightarrow C + D$ |
| 4. consecutive: | $A \Rightarrow B \Rightarrow C$ |

All four may be heterogeneous, but only #1 (and sometimes #4) will be homogeneous. Rate processes are defined in terms of a rate, $\boldsymbol{r}$ , and a volume, V, usually a molar volume. Thus, we have:

7.4.2.- Homogeneous: $\boldsymbol{r} = 1/V_t \cdot dn_i/dt$

Heterogeneous: $\boldsymbol{r} = 1/V_f \cdot dV_t/dt$

where $n_i$ is initial mols, $V_t$ is volume at time, t, and $V_f$ is final volume. The fraction decomposed at any time is:

7.4.3.- $x \equiv V_t / V_f$

and this gives us:

7.4.4.- $dx/dt = k_1 \ (T) \cdot f(x)$

where $k_1$ is the rate constant for the reaction in 7.4.4. This equation is the general form for kinetics equations.

From the Kinetic Theory, when a gaseous system goes from an initial state to a final state, the reacting species must come close enough to react. At the moment of "joining", we have the "activated complex". By using this concept, we can obtain some general equations useful to us. The concept of the "activated complex" is illustrated in the following diagram:

7.4.5.-

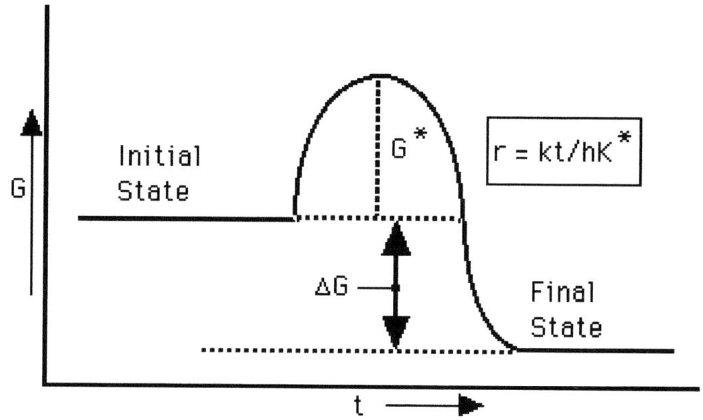


The free energy of the activated complex, G*, is higher than that of the initial state. K* is the equilibrium constant of the activated complex, and k is the Boltzmann constant. then, we can write thermodynamic equations as follows:

7.4.6.- $\Delta G \ = \ RT \ln K^*$

$$k_1 \ = kT/h \exp(-\Delta G^*/RT) \ = kT/h \exp(\Delta S^*/R) \exp(-\Delta H^*/RT)$$

where the star (*) refers to the activated species. These equations ought to be familiar to Physical Chemistry students.

From the Clausius-Clapyeron equation, we have:

7.4.7.- $E_o \equiv RT^2 \; d \ln k_1 / dt$

$E_o = \Delta H^* + RT$

where $E_o$ is an internal energy. By defining a frequency factor as:

7.4.8.- $Z = kT/h \exp(-E_o^*/RT)$

we can simply by means of 7.4.6. to:

7.4.9.- $k_1 = Z \exp(-E_o^*/RT)$

which is the ARRHENIUS EQUATION for the rate constant, $k_1$. We can transform the above equations into *general* equations as well, and use them to define the various types of reaction given above. This is given in the following table:

7.4.10.- RATE EQUATIONS FOR SOLID STATE REACTIONS

1. Simple nth Order — Rate Equation

$nA \overset{k_1}{\Rightarrow} B + C \qquad -dx_A/dt = k_1 x^n_A$

2. Parallel Reactions — Rate Equation

$A \overset{k_1}{\Rightarrow} B$

$-dx_A/dt = k_1 x_A + k_2 x_A$

$\{A \overset{k_2}{\Rightarrow} C\}$

3. Consecutive

$A \overset{k_1}{\Rightarrow} B \overset{k_2}{\Rightarrow} C \qquad -dx_A/dt = k_1 x_A - k_2 x_B$

These rate equations can be used for quite complicated reactions, but a specific method or approach is needed. Many authors have tried to devise methods for obtaining rate constants and orders of reaction for given solid state reactions. None have been wholly successful, except for Freeman and Carroll (1948).

The Freeman and Carroll method has been shown by Fong and Chen to be the only one which gives satisfactory answers to **known** reactions, whether zero order, 1st order, 2nd order, or even higher. Even fractional orders of reaction may be determined. This method can be used with either DTA or TGA data.

### B. THE FREEMAN-CARROLL METHOD APPLIED TO DTA DATA

The following is a description of the Freeman-Carroll method applied to DTA data. Consider the following DTA peak in which ΔT is plotted vs: time, t, a representation of which given as follows:

7.4.11.-

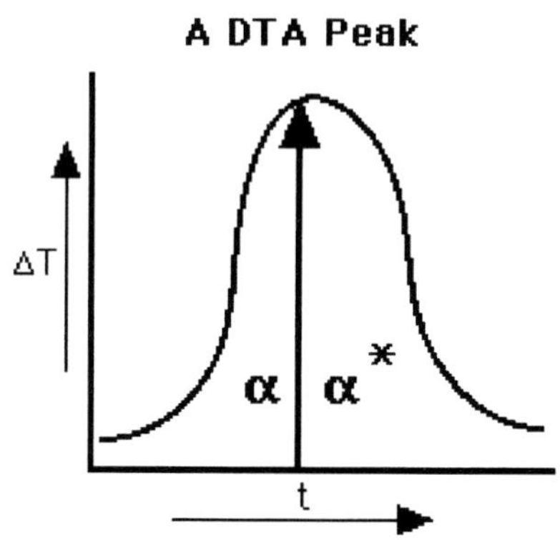


At any time, $\alpha$ is the fraction decomposed while $\alpha *$ is the fraction which has yet to react. We can set up equations as before:

7.4.12.- $$d\alpha / dt = k_1 (1 - \alpha)^n = Z/\phi \exp -E/RT (1 - \alpha)^n$$

The first part of the equation is the *general* kinetic equation, from which it is easy to obtain the last part (as we have shown).If we perform the mathematical operations of : 1) take the naperian log; 2) differentiation, and then 3) integration, we obtain the following equation:

7.4.13.- $\Delta \ln d\alpha / dt = n (\Delta \ln (1-\alpha) - E/R \Delta (1/T)$

If we now set: $\Delta T/A = d\alpha / dt$ and $\alpha * = A - \alpha$ (where A is the area under the peak) , we can then obtain:

7.4.14.- $\Delta (\ln \Delta T) = n (\Delta \ln \alpha *) - E/R \Delta (1/T)$

**This allows us to plot:** $\Delta (\ln \Delta T) / \Delta (\ln \alpha *)$ vs: $\Delta (1/T \ln \alpha *)$ so as to obtain a straight line. The **slope** is E/R and the intercept is **n** , the order of reaction. Having thus obtained the activation energy and order directly, we can then calculate the reaction **rate.**

C. THE FREEMAN- CARROLL METHOD APPLIED TO TGA DATA

In this case, we define our equation in terms of **weight,** w :

7.4.15.- $-dw/dt = k_1 f(w)$

As in the DTA method, we define weight reacted in time, t, as $w_t$ and final weight, $w_f$ . This allows us to define **weight to be reacted** as: $w = w_t - w_f$ .
Our weight function is then to be defined:

7.4.16.- $f(w) = w^n$

where n is a simple integer, representing the order of reaction. Thus, we obtain in the same manner as for the DTA data:

7.4.17.- $- dw/dt = Z \exp(-E/RT) w^n$

We now do the mathematical manipulations in the order given above to obtain:

7.4.18.- $\Delta$ (ln (-dw/dt)) - E/R ($\Delta$ (1/T)) + n $\Delta$ ln w

This allows us to plot, as before: $\Delta$ (ln - dw/dt) / $\Delta$ ln w vs: $\Delta$(1/T) / $\Delta$ ln w .

However, in many cases, it is easier to use **a**, the original weight, and calculate: (a - x) and dx/dt , where x is the fraction decomposed at time, t. Then we plot:

7.4.19.- $\Delta$ (ln dx/$\Delta$t) / $\Delta$ (ln (a-x)) vs: $\Delta$ (1/T) / $\Delta$ ln (a-x)

Thus if we take a well defined curve from either DTA or TGA and find points on the curve, we can calculate **all** of the kinetic parameters. Note that the main concept is to use the **amount left to react** at any given instant. In DTA, this was $\alpha^*$ while in TGA, it was (a- x). Both thermal methods give equally satisfactory results.

## 7.5.- DILATOMETRY

Measurement of the thermal expansion of solids is called "dilatometry". When energy is applied to a solid, one of the results is that the solid expands in space. Such expansion results because the energy generally ends up as "heat". That is- the lattice of the solid acquires increased vibrational modes which expands the lattice. On a practical scale, dilatometry is usually obtained in linear fashion wherein the change in unit length of a material caused by one-degree change in temperature is measured. Both volume change and linear change is expressed by:

7.5.1.- $$\alpha_V = \frac{1}{V_0}\left(\partial V/\partial T\right)_p \quad \text{and} \quad \alpha_L = \frac{1}{L_0}\left(\partial L/\partial T\right)_p$$

where V is the volume and L is the length and $V_0$ is the original volume and $L_0$ is the original length. $\alpha_V$ is the volume thermal expansion coefficient and $\alpha_L$ is the linear thermal expansion coefficient. Obviously, 3 $\alpha_L = \alpha_V$ (but only if the material is cubic. That is, the structure of the material is uniform in three dimensions). Both coefficients are ordinarily linear functions of temperature, that is- a plot of expansion vs: temperature yields a straight line. By using the *average* value of the linear coefficient of expansion, we get:

7.5.2.- $$\alpha_L = \frac{1}{L_0} \cdot \frac{\Delta L}{\Delta T}$$

where $\Delta L$ and $\Delta T$ are changes in length, L, and temperature, T, and $L_0$ is the original length of the sample being measured.

If a volume expansion is required, then measurements in three simultaneous dimensions are needed, a result experimentally difficult to achieve, to say the least. Even a slab of a single crystal does not completely solve the problem since thermal expansion in three dimensions is needed for the volume thermal expansion coefficient. The crystal has three (3) crystallographic axes and may have three (3) linear coefficients of expansion. Only if the crystal is cubic does one have the case where all three values of $\alpha_L$ are equal.

Thus, most thermal expansion measurements use the linear expansion coefficient method. This requires a rod of solid material and powders cannot be measured by this method. Either the material is melted and cast into rod form or it is cut from a slab of material. Glass and metals are ideally suited to such measurement, whereas inorganic and organic compounds are difficult to measure at best. Single crystals of inorganic or organic compounds are required for the most part, but the work of obtaining such is sometimes a daunting task.

A suitable apparatus is shown on the following following page as 7.5.3. Note that only a means of holding one end of the sample bar is needed while the sample is being heated. Most materials expand only about a small fraction of their actual dimensions. Because they expand less than a micrometer, i.e.- one millionth of a meter, with a one-degree increase in temperature, measurements

7.5.3.-

Simple Apparatus for Thermal Expansion Measurements

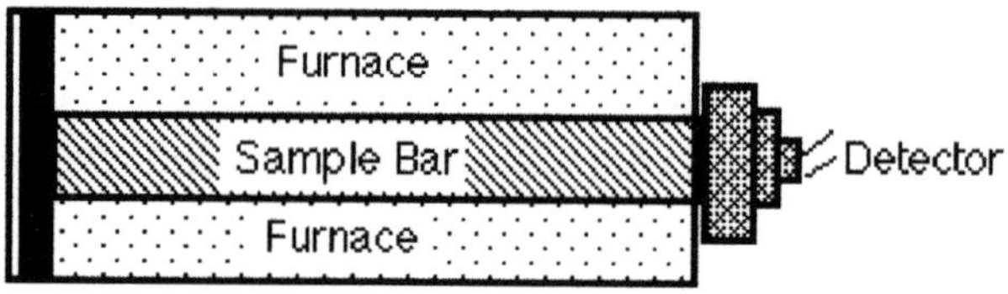


must be made by use of a highly sensitive device as the detector. Suitable measurements are usually made by the following methods:

7.5.4.- Suitable Detecting Systems for Dilametric Measurement of Solids

a. Microscope
b. Dial gauge
c. Telescope with Micrometer
d. Strain gauge
e. Interferometer
f. Mirror reflection
g. Lattice constants via x-ray diffraction
h. Sample density

The MICROSCOPE method (a) entails the use of a hot-stage which fits on the microscope. A calibrated eyepiece is needed as well. By setting up and controlling the temperature of the bar under investigation, a series of lengths can be obtained using the equation of 7.9.2. When these points are plotted against temperature, a straight line will be obtained. The slope of the line will be the linear coefficient of expansion, $\alpha_L$.

Both the DIAL GAUGE (b) and TELESCOPE WITH MICROMETER (c) methods are variants of the microscope-method (a). A temperature-controlled furnace is needed and a physical set-up where the DIAL-GAUGE or TELESCOPE WITH

MICROMETER can observe the change of length as a function of several settings of temperature. $\alpha_L$ is obtained as described before.

Both the INTERFEROMETER method (d) and the MIRROR REFLECTION (e) methods use optical means to detect changes in linear expansion of the sample under test. The instrumentation is more complex and will not be described here. The other two methods, X-RAY LATTICE CONSTANTS (f) and SAMPLE DENSITY (g) have not been employed to any great extent for determination of $\alpha_L$ and are only included for the sake of completeness.

Modern commercial dilatometric instruments use the electrical output from a "strain-gauge". The heart of a strain gauge consists of a crystal (or sometimes a disc pressed from a powder) which is piezoelectric. The term, piezoelectric, is defined as the appearance of positive electric charge on one side of certain nonconducting crystals and negative charge on the opposite side when the crystals are subjected to mechanical pressure. Piezoelectricity was discovered in 1880 by Pierre and Paul-Jacques Curie, who found that when they compressed certain types of crystals (including those of quartz, tourmaline, and Rochelle salt) along certain axes, a voltage was produced on the surface of the crystal. The next year, they observed the converse effect, the elongation of such crystals upon the application of an electric current.

Pressure on certain electrically neutral crystals--those not having a center of structural symmetry- polarizes them by slightly separating the center of positive charge from that of the negative charge. Equal and unlike charges on opposite faces of the crystal result. This charge separation may be described as a resultant electric field and may be detected by an appropriate voltmeter as a potential difference, or voltage, between the opposite crystal faces. This phenomenon, also called the piezoelectric effect, has a converse, i.e.- the production of a mechanical deformation in a crystal across which an electric field or a potential difference is applied. A reversal of the field reverses the direction of the mechanical deformation. This effect has been used in microphones.

The advantage of using a strain gauge when making thermal expansion measurements lies in the fact that a continuous electrical signal can be obtained as the sample bar is being heated. A typical dilatometric run for a glass sample is shown in the following:

7.5.5.-

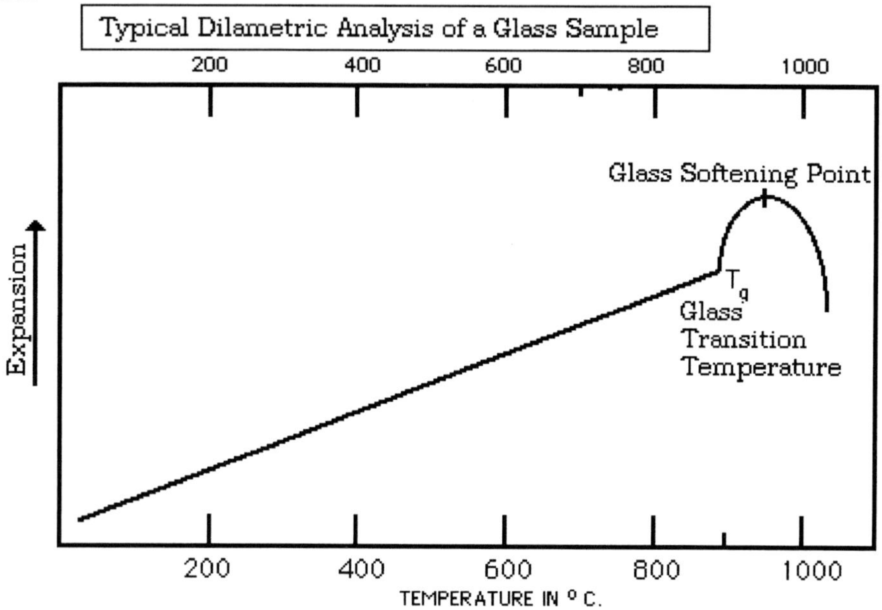


Note that the linear coefficient of expansion, $\alpha_L$ is obtained from the slope of the straight line. The glass softening point is also easily observed as is the glass transitional temperature (which is the point where the amorphous glassy phase begins its transition to a crystalline phase. These glass-points can also be used to cross-check values obtained by the DTA method.

It is possible for a glass to have more than one linear coefficient of expansion. One such case is shown in the following diagram, given as 7.5.6. on the next page.

7.5.6.-

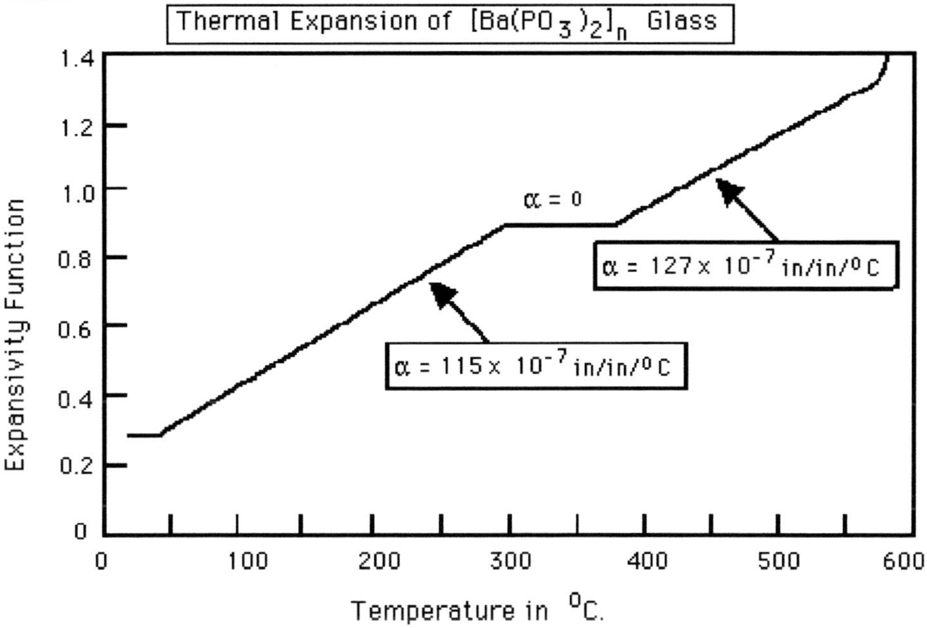


This glass is polymeric and consists of barium meta-phosphate units, i.e.- $[Ba(PO_3)_2]_n$ , polymerized to form long chains in the amorphous state. Note that three values are shown. The first is the expected linear expansion, typical of most glasses. In this case, there is a temperature range where the glass ceases to expand (between 300 and 380 °C) and is flat. Above about 380 °C., a third value of $\alpha_L$ can be seen. Although this behavior is not typical for most glasses, it does illustrate the fact that an amorphous phase can have more than one coefficient of expansion. Note also that the softening point is not indicated in the diagram. Its value lies above 550 °C. The same is true for other solids as well.

To illustrate this point, consider a typical metal. When a bar of such metals are heated, they expand in a linear manner, but may undergo a phase transition as they reach a critical temperature. This is shown in the following diagram:

7.5.7.

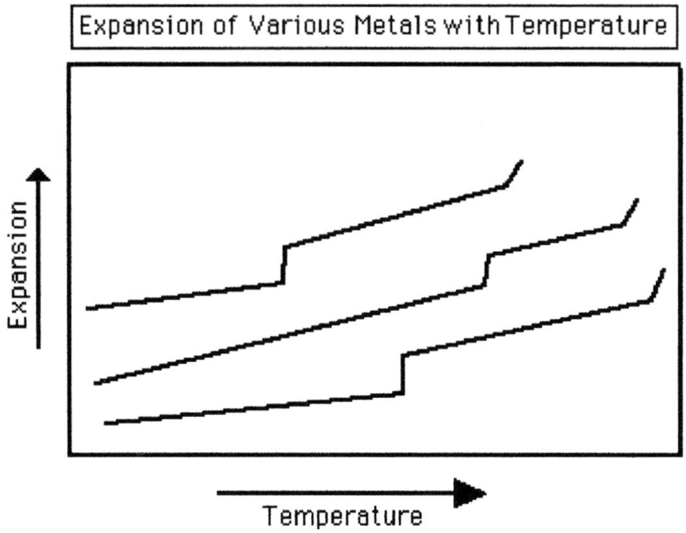


In this case, a change in structure occurs. Many metals are elemental in nature and when refined to a pure state have a cubic structure. At some critical temperature (defined by the number of metal electrons per atom present and the type of metallic bonding), a change to a hexagonal form occurs.

This is shown in the following diagram. Note that a simple shift in unit cell dimension is all that is required for the crystal structure change to take place.

7.5.8.-

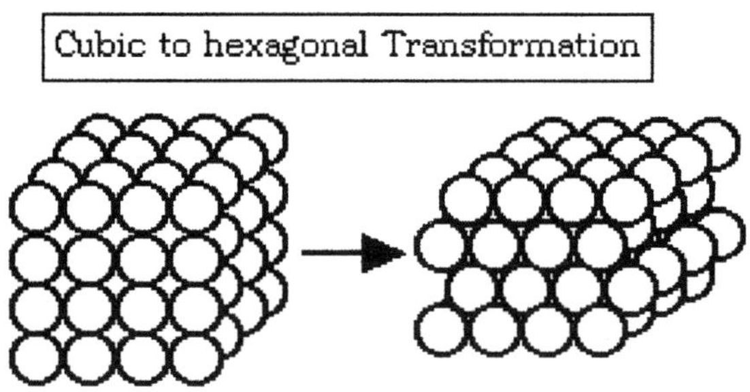

Observe also that increased thermal energy is all that is required for the shift to take place. Such change occurs well below the melting point of the metal.

Dilatometric measurement is the only way that such change was originally discovered and the use of x-ray analysis was needed to confirm the exact nature of the change measured.

**7.6.-** Thermometry

Thermometry is the science of measuring the temperature of a system or the ability of a system to transfer heat to another system. Temperature measurement is important to a wide range of activities, including manufacturing, scientific research, and medical practice. Thermometry relates to the dilatrometric measurement in that the expansion of a gas, liquid or solid is used to determine temperature.

The accurate measurement of temperature has developed relatively recently in human history. The invention of the thermometer is generally credited to Galileo. In his instrument, built about 1592, the changing temperature of an inverted glass vessel produced the expansion or contraction of the air within it, which in turn changed the level of the liquid with which the vessel's long, open-mouthed neck was partially filled. This general principle was perfected in succeeding years by experimenting with liquids such as mercury and by providing a scale to measure the expansion and contraction brought about in such liquids by rising and falling temperatures.

By the early 18th century as many as 35 different temperature scales had been devised. The German physicist, Daniel Gabriel Fahrenheit, in 1700-30 produced accurate mercury thermometers calibrated to a standard scale that ranged from 32, the melting point of ice, to 96 for body temperature. The unit of temperature (degree) on the Fahrenheit scale is 1/180 of the difference between the boiling (212) and freezing points of water. The first centigrade scale (made up of 100 degrees) is attributed to the Swedish astronomer Anders Celsius, who developed it in 1742. Celsius used 0 for the boiling point of water and 100 for the melting point of snow. This was later inverted to put 0 on the cold end and 100 on the hot end, and in that form it gained widespread use. It was known simply as the centigrade scale until in 1948 the name was changed to honor Celsius. In 1848 the British physicist William Thompson (later Lord Kelvin) proposed a system that used the degrees that Celsius used, but was keyed to absolute zero (-273.15 °C); the unit of this scale is now known as the kelvin, i.e.- °K. The Rankine scale employs the Fahrenheit degree keyed to absolute zero (-459.67 °F) , i.e.- °R.

Any substance that somehow changes with alterations in its temperature can be used as the basic component in a thermometer. Gas thermometers work best at very low temperatures. Liquid thermometers are the most common type in use. They are simple, inexpensive, long-lasting, and able to measure a wide temperature span. The liquid is almost always mercury, sealed in a glass tube with nitrogen gas making up the rest of the volume of the tube.

Electrical-resistance thermometers characteristically use platinum and operate on the principle that electrical resistance varies with changes in temperature. Thermocouples are among the most widely used industrial thermometers. They are composed of two wires made of different materials joined together at one end and connected to a voltage-measuring device at the other. A temperature difference between the two ends creates a voltage that can be measured and translated into a measure of the temperature of the junction end. The bimetallic strip constitutes one of the most trouble-free and durable thermometers. It is simply two strips of different metals bonded together and held at one end. When heated, the two strips expand at different rates, resulting in a bending effect that is used to measure the temperature change.

Other thermometers operate by sensing sound waves or magnetic conditions associated with temperature changes. Magnetic thermometers increase in efficiency as temperature decreases, which makes them extremely useful in measuring very low temperatures with precision. Temperatures can also be mapped, using a technique called thermography that provides a graphic or visual representation of the temperature conditions on the surface of an object or land area. Phosphors are now being used in this application.

## 7.7.- APPLICATION OF DILATOMETRY TO PLASTIC MATERIALS

The term "plastic" refers to that class of materials which is organic in nature. That is, plastics are composed of hydrogen, carbon, nitrogen and sometimes fluorine, i.e.- they are hydrocarbons. Such materials are actually "polymers" composed of monomeric units which are caused to join to form rather large uniform molecules having molecular weights as high as 100,000 or more. Plastics or polymers have found a variety of end-uses in our economy. In addition to containers such as bottles for beverages, plastics are used in all sorts of products and polymers have been developed that are stronger and more flexible than many metals.

For the most part, plastics are man-made since very few plastics are natural, i.e.- nature-made. Natural plastics include large molecular-weight proteins and similar molecules. Man-made plastics can be classified as either thermoplastic or thermosetting. Each class derives its physical properties from the effects of application of heat, the former becoming "plastic" (that is- it becomes soft and tends to flow) while the latter becomes less "plastic" and tends to remain in a softened state. This difference in change of state derives from the actual nature of the chemical bonds in the polymer. Thermoplastic polymers generally consist of molecules composed of many monomeric units. A good example is that of polyethylene where the monomeric unit is: -($CH_2$-$CH_2$)- . The molecule is linear and the polymer consists of many units tied together in one long string. In contrast, thermosetting polymers consist of **cross-linked** units where the cross-linking is three-dimensional. That is, the molecules are linked together in three dimensional-space:

```
-(CH-CH)-
  |   |
-(CH-CH)-
```

This difference in spatial characteristics has a profound effect upon the polymer's physical and chemical properties. In thermoplastic polymers, application of heat causes a change from a solid or glassy (amorphous) state to a flowable liquid. In thermosetting polymers, the change of state occurs from a rigid solid to a soft, rubbery composition. The glass transition temperature, $T_g$, and the coefficient of expansion have a profound effect on the performance and reliability of many polymer applications. Although $T_g$ is usually quoted and accepted as a single value, the transition usually occurs within a range of temperatures. Factors such as intra-chain stiffness, polar electromagnetic forces and co-polymer compatibility (when two or more polymers are blended together to improve physical properties) can affect the size of the glass-transition region. As shown in the following diagram, property changes occur throughout a temperature region which depends upon the type of polymer(s):

7.7.1.-

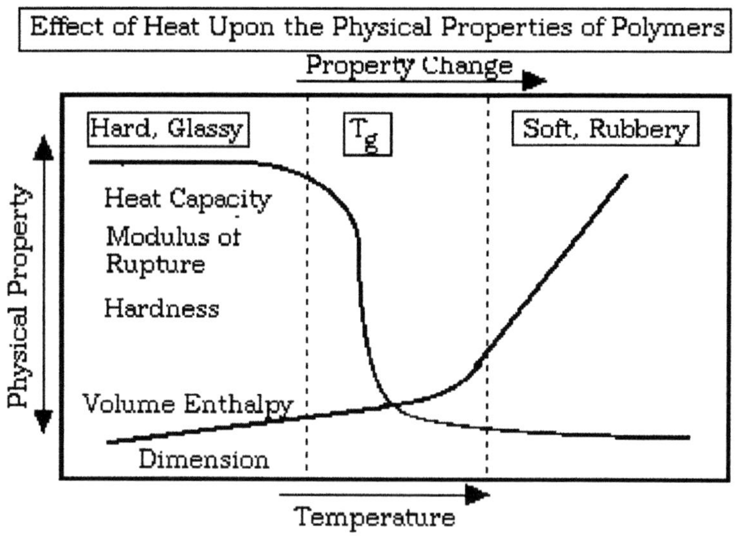


It should be clear that the point where a polymer shifts from a glassy, hard state to a soft, rubbery one is not well defined but occurs within a band of temperatures. In contrast, it is easy to define $T_g$ as a single temperature point for glasses and most ceramic materials. We have already described how $T_g$ is measured. Similar methods are used when plastics or organic materials are involved. DSC is the method most often employed since it gives both enthalpy and heat capacity data from the same analysis.

## **7.8-** OPTICAL MEASUREMENTS OF SOLIDS

In terms of their optical properties, all solids fall into one of two classes. Either they are transparent to light (here we are restricting the term "light" to visible radiation) or they are opaque. In the latter case, all of the radiation may be reflected. However, most solids reflect some wavelengths and absorb others. This is the condition that we call "color". If all visible wavelengths are absorbed, the solid is said to be "black" while reflectance of all visible wavelengths results in a "white" solid. We intend to show how "color" is measured but first must define the nature of "light".

### A. DEFINING LIGHT

Light is composed of photons (which are individual energy bundles) that propagate through space. The correct term for light is electromagnetic radiation. We use the term "light" to refer to those photons which we can see. "Dark" refers to the absence of visible photons (You may be interested to know that a photon is now regarded as a energy carrier between fundamental particles, i.e.- leptons such as protons and neutrons). Light travels at a constant speed, i.e.- $3.0 \times 10^{10}$ meters/second, through a vacuum (a space where no matter exists). When matter is present, its speed is diminished, but is still constant. Since the speed of light is constant, individual photons can vary only in energy, a state which results in differences in their wavelength. That is, a photon's energy is manifested as a specific wavelength. Our main concern will be that of "color", which is the science of measuring what type of photons are reflected and those that are absorbed by a solid. What this means is that color is determined by which photons, in a stream of photons having various energies, are either reflected or absorbed by the solid. This mechanism implies some sort of interaction between each individual photon and the electrons composing the solid. Photons (i.e.- electromagnetic energy) can vary in wavelength from:

$\gamma$ –rays = $10^{-12}$ meters
x-rays = $10^{-10}$ meters
ultraviolet = $10^{-8}$ meters
visible = $10^{-6}$ meters
infrared radiation = $10^{-4}$ meters
microwaves = $10^{-2}$ meters
radio = $10^{2}$ to $10^{4}$ meters

where the values given are averages of the spread of wavelengths. Note that radio waves can be several miles long. But, since light travels at 186,000 miles/sec., transmission to any point on Earth is nearly instantaneous. The frequency varies from about $10^{5}$ cycles/second (radio waves) to $10^{20}$ cycles/second ($\gamma$ –rays).

The relationship between electromagnetic radiation and matter (solids) is intertwined in the so-called "space-time" phenomenon. All solids *emit* photons, even yourself. The concept of "absolute zero" lies in the fact that no photons are emitted at 0° K. As the temperature rises, a spectrum of photon energies is emitted, as shown in the following diagram, given as 7.8.1. on the next page.

This diagram shows the radiation emitted by "black-bodies" at specific temperature. A black-body is one that has a uniform temperature over all of its surface. One way to make a black-body is to form an hollow enclosure and to heat it to a given temperature. If a small hole is made in the side of the enclosure, radiation characteristic of the temperature will be emitted.

It should be clear that all bodies (even your own) radiate photons in the infra-red range of energies. Yours is similar to that of the earth and probably peaks near to 10.0 microns or 10,000 Å. If you place your hand on your face, you feel warmth because the emitted photons are reabsorbed by your hand.

7.8.1.-

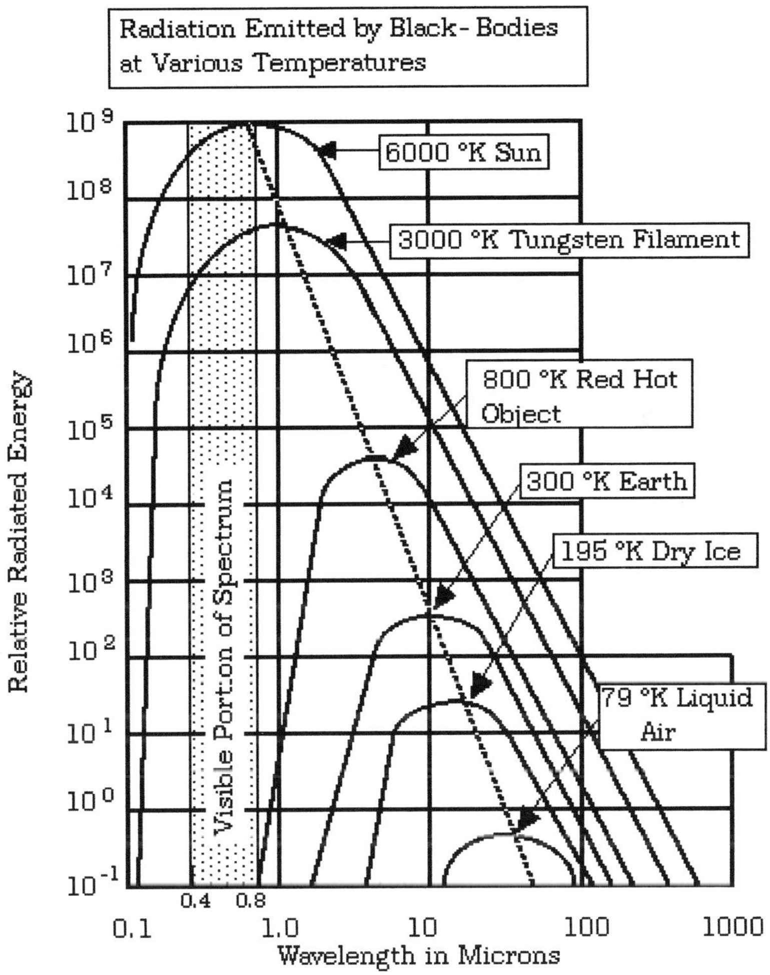


Nonetheless, our primary interest lies in the 0.4 to 0.7 micron range, which we call the visible part of the electromagnetic spectrum. Note that even bodies at liquid-air temperatures emit photons between 10 and 100 microns in wavelength, i.e.- 100,000 and $10^6$ Å in wavelength. The earth itself at a temperature of 300 °K. has an emission between about 20,000 and 300,000 Å in wavelength, i.e.- 2 μ. and 300 μ. Light itself has a specific wavelength measured in Ångstroms (1Å = 0.0000000001 meters = 0.000000000003 feet).

Sir Isaac Newton first demonstrated the separation of wavelengths (colors) of sunlight in 1675 by use of a glass prism. The wavelengths to which the human eye responds is called the "visible spectrum". We see each of these wavelengths of light as a perceived color. Additionally, it was found that "white" is a compilation of all of the colors while "black" is an absence of said colors. Photons having the highest energy are violet in color while the lowest energy ones are those we call deep-red. The intensity, or "brightness" of light is simply the numbers of photons reaching the eye at any given moment. On either side of the visible spectrum are "ultraviolet" and "infrared" wavelengths. The Sun itself radiates throughout the visible spectrum and far into the ultraviolet and infrared ranges as well.

The following diagram, shows the Sun's radiation, including the visible spectrum which encompasses the 4000 Å to 7000 Å range of wavelengths.

7.8.2.-

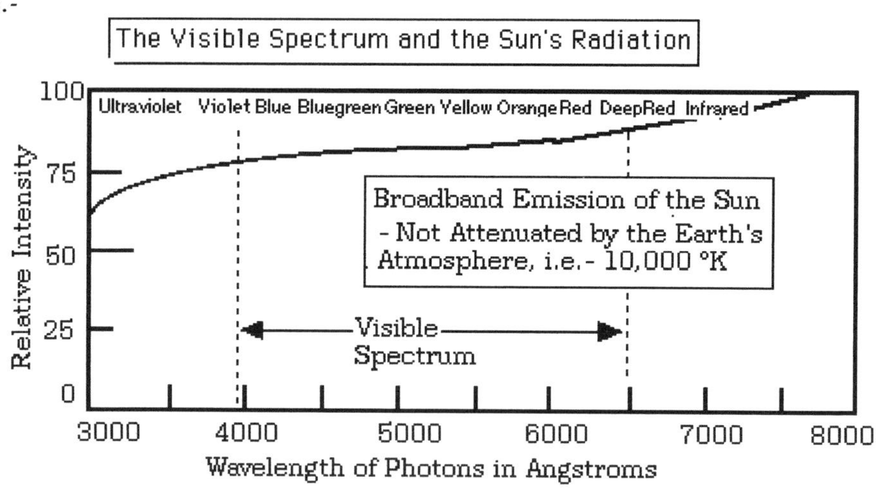


It is this range that we usually call "daylight". Note the progression of hues as we move from violet to deep red colors. Although the Sun radiates wavelengths as small as 20 Å and as large as 25,000 Å, most of the energy is concentrated between 3000 and 10,000 Å. Attenuation by the Earth's atmosphere (ozone in particular) causes the energy to peak near to 6500 Å. (What this means is that if we have a hot "body" at 6500 °K (°K = °C + 273) it would look like the Sun at noon).The actual color temperature of the Sun is about 10,000 °K but it appears to us to be about 6500 °K due to scattering of light within the atmosphere.

It is **not** happenstance that the human eye responds exactly to the same wavelength intensities as those of the noonday Sun. In other words, mankind was born, and has evolved, under sunlight and his eyes have adapted to sunlight intensity, also known as "Daylight". The relative response of the human eye is shown in the following diagram:

7.8.3.-

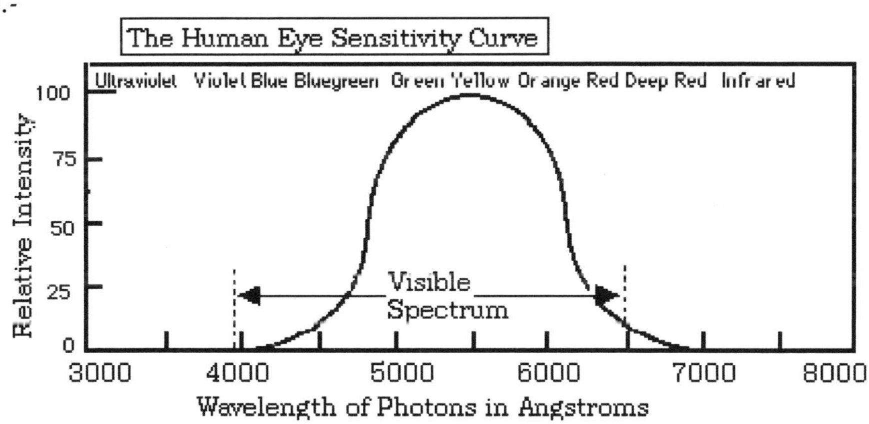


Note that the eye response is greatest in the green and yellow regions of the spectrum. and that response to blue and red wavelengths is much lower.

## B. MEASUREMENT OF COLOR

In order to measure color, we first need to define how color is perceived, what is perceived, and the factors involved in perception of color. First of all, "color" is a result of photons striking the human eyeball, specifically the retina, and causing a response which we interpret as color. Color arises because each photon has its own energy, i.e.- its own wavelength. Color is perceived when photons strike the retina of the eye and cause electrical activity in the optic nerve. In order to define the nature of a photon, we need to consider exactly what light is and how it interacts with matter. This problem has been studied for many centuries by various investigators and continues to be studied by contemporaneous scientists.

## C.-THE NATURE OF LIGHT

Nowadays, we know that light is comprised of "photons", which are quantized waves having some of the properties of particles. The concept of photons with wave properties has its roots in the study of optics and optical phenomena. Until the middle of the 17th century, light was generally thought to consist of a stream of some sort of particles or *corpuscles* emanating from light sources. Newton and many other scientists of his day supported the idea of the corpuscular theory of light. It was Newton in 1703 who showed that "ordinary" sunlight could be dispersed into its constituent colors by a prism, but the phenomenon was not clearly grasped at that time. Significant experiments on the nature of light were carried out by:

1) Fresnel and Thomas Young (1815) on interference and diffraction respectively
2) Maxwell in 1873 who postulated that an oscillating electrical circuit should radiate electromagnetic waves
3) Heinrich Hertz in 1887 who used an oscillating circuit of small dimensions to produce electromagnetic waves which had all of the properties of light waves
4) Einstein in 1905 who explained the photoelectric effect (He did so by extending an idea proposed by Planck five years earlier to postulate that the energy in a light beam was concentrated in "packets" or *photons*..

The wave picture was retained in that a photon was considered to have a frequency and that the energy of a photon was proportional to its frequency).

We can summarize all of the above research carried out over the last two centuries in that a photon is a quantum of radiation and a carrier of force between particles, whereas an electron is a quantum of matter. Now, let us examine the more mundane aspects of light measurement including color measurement.

D.- ABSORBANCE, REFLECTIVITY AND TRANSMITTANCE.

When a beam of photons strikes a solid, specific interactions take place which can be related to the Quantum Theory. These interactions have been measured in the past and certain formulas have been found to apply. According to HUYGHEN'S principle of electromagnetic radiation **scattering,** when photons come into close contact with a solid, the electric and magnetic field vectors of the incident photons couple with those of the electrons associated with the atoms comprising the solid. The following presents some of these formulas applicable to concepts given here:

### 7.8.4.- TERMINOLOGY APPLICABLE TO OPTICAL PROPERTIES

a. Absorbance : $A \equiv \log 1/T = \log I_0 / I$

b. Transmittance: $T \equiv I / I_0$ ( I is measured intensity; $I_0$ is original intensity)

c. Absorptivity: $\mathfrak{A} \equiv A / bc$ (A is measured absorption; b is the optical path-length; c is the molar concentration)

d. Reflection: Reflectivity $\equiv$ spectral reflection, i.e.- at specific angles

Reflectance $\equiv$ diffuse reflection, i.e.- scattered radiation

e. Intensity: I is defined as the energy / unit area of a beam of electromagnetic radiation.

This interaction leads to at least four (4) components, namely: R- the radiation **reflected**, A- the radiation **absorbed**, T- the radiation **transmitted**, and S - the radiation **scattered.** A depiction of these interactions is given in the following diagram:

7.8.5.-

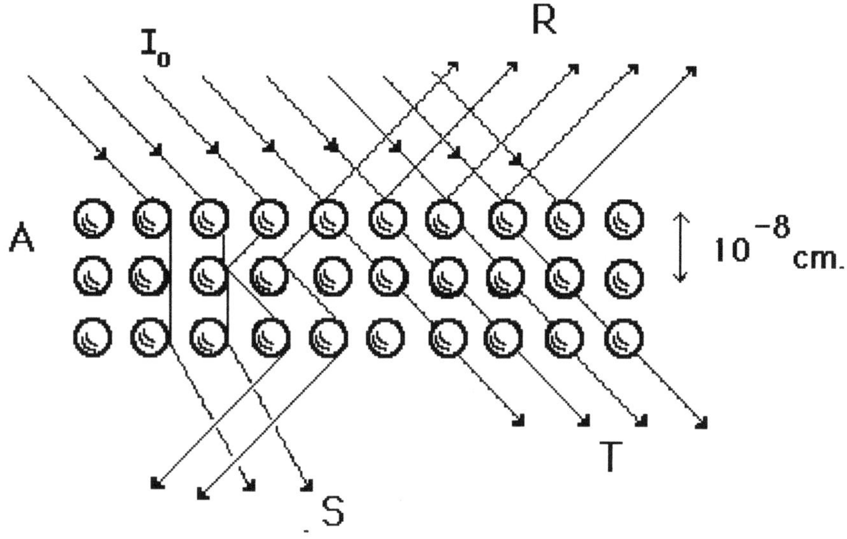


The original intensity of the radiation is defined as $I_0$ . A part of the intensity is absorbed, another part is transmitted, still another part is scattered, and a part of the total intensity is reflected. The components, S and T, are processes which are **independent** of the wavelength (frequency) of the incident photons, whereas R and A are primarily wavelength dependent. It is here that the factor of "color" arises. The **exact amount** of energy extracted from $I_0$ by each process is a complex set of variables depending upon the type and **arrangement** of atoms composing the solid.

To simplify these concepts, consider an optically homogeneous thin film. By optically homogeneous, we mean one that is thin enough so that no scattering can occur. If a beam of photons is incident to the surface at a given angle (but less than that where all of the beam is transmitted- the so-called Brewster angle), part of the beam will be **reflected** and part will be **absorbed**, as shown in the following:

7.8.6.- SINGLE REFLECTION IN A THIN FILM

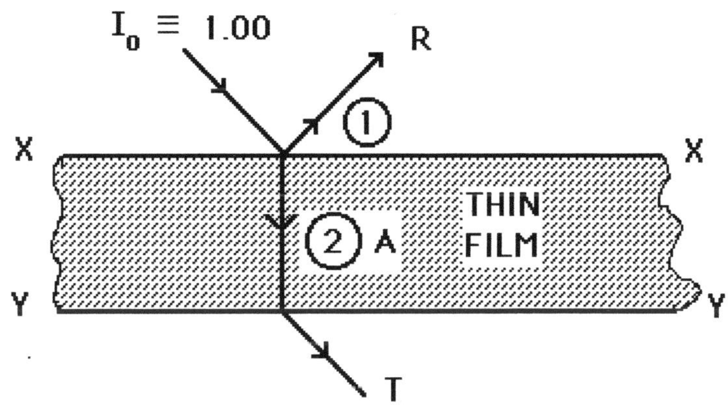


The reflectance, R, is a consequence of the difference in refractive indices of the two media, (1) - air, and (2)- the semi-transparent thin film. The amount of absorption is a function of the nature of the solid. Obviously, the amount transmitted, T, is determined by both R and A. In this case, A, the **absorbance,** is defined as:

7.8.7.- $A \equiv (1 - R) - T$

and the original beam intensity, $I_o \equiv 1.00$, is diminished according to the **Beer-Lambert law**, vis-

7.8.8.- $A = \ln I_o / I = \varepsilon c l$

where l is the pathlength (depth of film traversed), $\varepsilon$ is a **molar extinction coefficient** of the absorbing species, and c is the concentration of the absorbing species. Although we have shown but one reflection at X, another is equally likely to occur at Y as well. This problem of multiple reflections was worked out by Bode (1954). We will show only the results of his work and use it to prove the validity of the equation of 7.8.7. for the case of the homogeneous thin film.

Consider the energy interactions which occur when a **photon** strikes a solid. The timeframe of **interaction** is about $10^{-18}$ seconds:

7.8.9.- Speed of photon = v = $10^{10}$ cm./sec.

Distance between lattice planes = d = $10^{-8}$ cm.= ~ 1 Å

Time for photon to traverse lattice = d/v = ~ $10^{-18}$ seconds

The fact that the photon does traverse the lattice planes does not mean that the photon will be absorbed or even scattered by the solid. The reflectance of the photon is a function of the nature of the compositional surface, whereas absorption depends upon the interior composition of the solid. A "resonance" condition must exist before the photon can transfer energy to the solid (absorption of the photon). In the following, we show this resonance condition in general terms of both R & A.

7.8.10.- ENERGY TRANSFER TO A SOLID BY A PHOTON

| R | A | ENERGY TRANSFER | EXAMPLES |
|---|---|---|---|
| High | High | Moderate | "Colored" solid |
| Low | High | Very High | "Black" solid |
| High | Low | Nil | "White" solid |
| Low | Low | Nil | Heat transfer |

This result defines absorption of light in terms of both reflectance and absorption. It is well to note that either one or the other (or both) are required phenomena in order for a photon to interact with a solid.

Thus, when both R & A values are high, we say the solid is colored. If either R or A is high, and the other value is low, then we have either a white solid or a black solid. As an example of controlled absorption, consider the case of a pigment. It is quite common to add controlled amounts of a transition metal to a transparent solid to form an inorganic pigment. In one such case, we add ~1% of chromium oxide to aluminum oxide to obtain a pink solid, i.e.- "ruby". The $Cr^{3+}$ ion in the $Al_2O_3$ lattice absorbs blue and green light and reflects or transmits mostly the red wavelengths. It should be clear, then, that both processes, R & A, are wavelength dependent. That is, they depend upon the energy of the photon(s) striking the solid.

E.- <u>MEASUREMENT OF COLOR</u>

Having defined the nature of a photon and how it interacts with matter, we can now proceed to describe a method to measure color. We find that even though our vision is sufficient to categorize color as a gross feature, we still need to be able to measure small differences of color. One method, used in the past, was to assemble a series of samples having small but observable differences in color, and then compare an unknown to these. Because the human eye is a superb color instrument, this approach was perfectly feasible. This method used the so-called the Munsell Color Tree as the standard samples. But, it was based on being able to provide reproducible color swatches from reproducible pigments. You can imagine the problems associated in doing so.

However, the retina does not respond equally to all wavelengths. For equal energies, a yellow-green light produces a much stronger response in the human eye than a red or blue light. Thus, we say that the yellow-green light is "brighter" than the red or blue lights. This is called the luminosity response of the eye. However, the dark-adapted human eye does not respond to light in the

are much different. Thus, we need a method that can discriminate between such cases.

It was Newton, using a glass prism plus slits, who first demonstrated that sunlight consisted of colors or chroma. Subsequent work then showed that colors could be duplicated by mixing the three primaries, red, green and blue to obtain the various chroma, including shades of "white". Actually, these shades involved the luminous intensity, that is- the amount of light falling upon a surface. Let us first define some of the terms of measurement of luminosity and then proceed to determine how to measure the chroma, or chromaticity. The actual theory is complicated and long so that we shall present only a simplified version of the whole theory.

### III. Intensity and Scattering

Consider a thin flat plate which has no absorption, only reflectance and transmittance. We will find that light falling upon its surface with a certain intensity has a luminance, L, while the **light transmitted is defined as H**, the exittance, or emittance. H will equal L if no absorption takes place or if there is no scattering at the surface of the thin plate. However, even at atomic distances, a certain amount of **scattering** does take place. We find that two types of scattering are possible. If the surface is perfectly smooth on an atomic level, then a light wave would be back- scattered along the same exact path, and we would have a **perfect** diffuser, which we call $S_o$ . However, there is always an angle associated with the scattering, which we call $S_\alpha$ (where $\alpha$ is defined as the scattering angle), and we have an imperfect diffuser. This is shown in the following diagram, given as 7.8.13. on the next page.

If we view the thin plate from the left where it is illuminated with intensity, L, what we see is the back-scattered light, or light diffusion from the surface. If the plate is a perfect diffuser, then we will see an exact amount of L scattered back along the same plane as L as a diffuse component. Note that we are not speaking of reflection (which is an entirely different mechanism where the

7.8.13.-

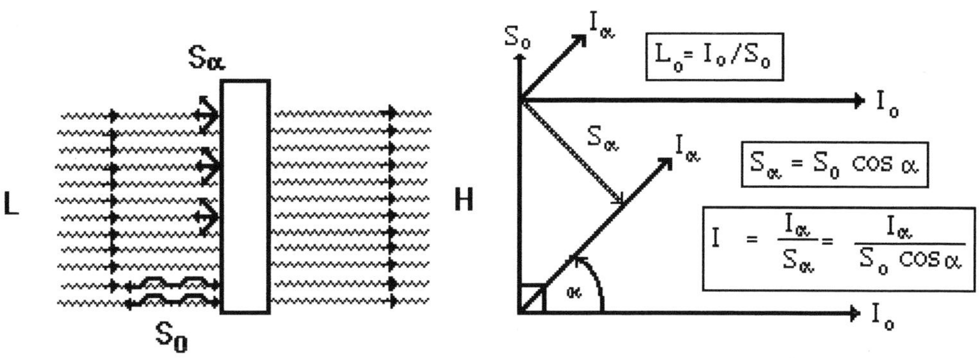


wavelength of the light is affected) but of scattering (where the light is absorbed, then reëmitted at the same wavelength). For scattering by a perfect diffuser, $L_o$ should equal $I_o$ / $S_o$. However, this is never the case. What we find is that there is an angular dependence of scattering, and that:

7.8.14.- $I = I_\alpha / S_\alpha = I_\alpha / (S_o \cos \alpha)$

where $\alpha$ is the angle of scattering. Therefore, H, the light transmitted, does not equal L. If we define $\phi$ as the flux of light, i.e.- the number of photons incident per second, we find:

7.8.15.- $H = \phi / S_\alpha = \phi / S_o \cos \alpha$ where: $\phi = 4\pi I$ (point source)

$\phi = \pi I$ (flat surface)

The intensity units for $\phi$ are related to a primary radiation source, the **candela**, Cd. The definition of a candela is:

7.8.16.- CANDELA: *a unit of luminous intensity, defined as 1/60 of the luminous intensity per square centimeter of a black-body radiator operating at the temperature of freezing platinum (1772 °C), Formerly known as a candle. The unit is abbreviated as: Cd*

This gives us the following intensity units:

7.8.17.- INTENSITY UNITS

$$1.0 \text{ Cd (at a one-foot distance)} \approx L/\pi \text{ (foot-lamberts)} \approx H/\pi \text{ (lumens)}$$

Thus, we have two units of measurement of intensity. One is related to scattering from a surface, L, i.e.- in foot-lamberts and the other is related to emittance, H, i.e.- in lumens per square foot. Although we have assumed "white" light up to now, either of these two can be wavelength dependent. If either is wavelength dependent, then we have a pigment (reflective- but more properly called scattering) with intensity in foot-lamberts, or an emitter such as a lamp or phosphor (emittance) with intensity in lumens.

## IV. Color Processes and Color Matching Systems

In terms of color, we can have one of two processes:

1. Additive processes (emittance)
2. Subtractive processes (reflectance).

If we wish to match colors, the primary colors are **quite different,** as shown in the following:

7.8.18.-

| ADDITIVE PRIMARIES | SUBTRACTIVE PRIMARIES |
|---|---|
| red | magenta |
| green | yellow |
| blue | cyan |

Let us now consider how to set up a color matching system. One way to do so is to average the eye responses of a large number of individuals so as to eliminate the individual "quirks" of the human eye (some people can "see" into the violet or ultraviolet region whereas others cannot). The resultant average is that we call "The Standard Observer".

### F. <u>The Standard Observer and The First Color-Comparator</u>

Since color matching is meant for humans, it is natural to define color in terms of an average, or "Standard Observer". Our first step is to build an instrument which contains three colored lamp sources, a place for the individual observer, intensity detectors, and a monochromator, as shown in 7.8.19. on the next page.

Essentially, such an instrument is called a color-comparator. The design is simple and was used many years ago to obtain the experimental results needed to fully define the true nature of color measurements. In this instrument, there are two (2) sources of light to be compared. One is from a set of three lamps whose emission is modified by means of suitable filters to give a red beam , a green beam and a blue beam. These are mixed at the screen to form a single spot (although we have not illustrated it in that way, so as to be more discernible). The other source comes from a monochromator so as to obtain a monochromatic beam of light. There are controls to adjust the individual beams of red, green and blue light, as well as that of the monochromatic beam. In this way, the mixed beams of light can be directly compared to the monochromatic spot. In the back of the screen are photoelectric detectors to measure the energy intensity of the beams of light being compared. We need about 5000 observers to obtain a satisfactory average, both for the dark-adapted and the light-adapted human eye. There are three (3) things that need to accomplished:

1. Define eye response in terms of color at **equal energy**
2. Define shades of "white" in terms of % red, % green and % blue, at equal energies of those "whites".
3. Define "color" in terms of % red, % green and % blue, as compared to monochromatic radiation

7.8.19.-

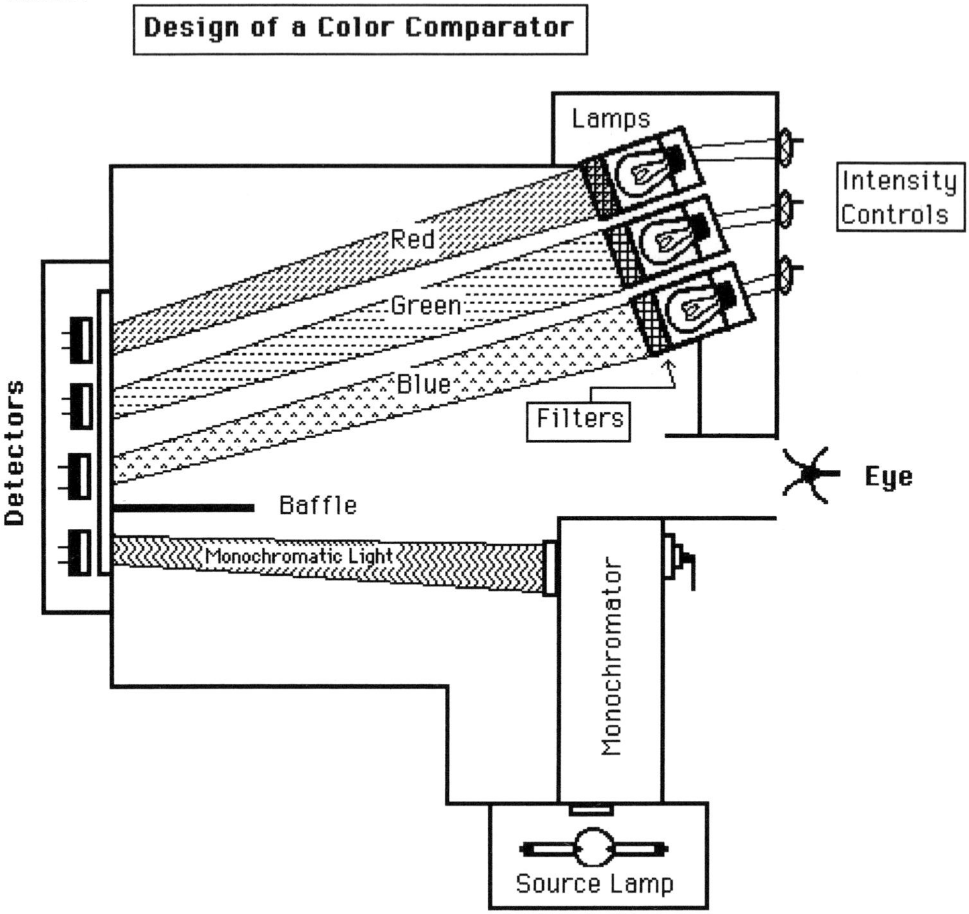


The difficulty in setting up the initial system for color comparisons cannot be underestimated. The problem **was** enormous. Questions as to the suitability of various lamp sources, the nature of the filters to be used, and the exact nature of the primary colors to be defined occupied many years before the first attempts to specify color in terms of the standard observer were started. As we said previously, the Sun is a black-body radiator having a spectral temperature of about 10,000 °K (as viewed directly from space). Scattering and reflection within the Earth's atmosphere is sufficient to lower the effective black-body radiation perceived to 6500 °K. Thus, the Sun is a 6500 °K. source which we call "daylight". The direct viewed brightness of the Sun at the Earth's surface is about 165,000 candela/$cm^2$ , that of the Moon - 0.25 candela/$cm^2$ , and a clear sky is about 0.8 candela/$cm^2$.

For these reasons, DAYLIGHT has been defined as: "The northern skylight at 11:30 am. at Greenwich, England on October 31, 1931". This is also the definition of ILLUMINANT- C. The other standard illuminant that we use is ILLUMINANT- A, which is the radiation emitted from an incandescent tungsten filament operating at 3250 °K. These are shown in the following diagram:

7.8.20.-

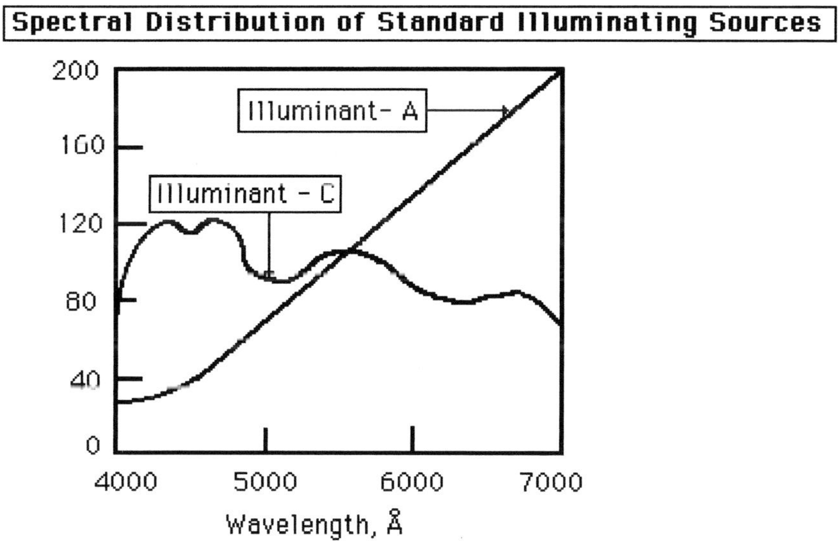


Referring back to our Color Comparator of 7.8.19., we use these concepts to calibrate our lamps in terms of spectra and relative energy in terms of these standard sources. Illuminant - B , by the way, was originally defined as "average sunlight" but it was soon determined that "average" is not the same at all parts of the Earth's globe.

Our next step in using the Color Comparator is to set up proper filters so as to obtain and use "primary" color lamp sources. We find that by using the monochromator of the Color Comparator, we can approximate the wavelength response of our so-called "Standard Observer", in the red region, the green region and the blue region of the visible spectrum. But we also find that we need a band of wavelengths for each color, since a monochromatic beam is not at all suitable. This is where the choice becomes subjective, since we are relying upon the perceived response of individuals. We find that by choosing a blue filter peaking at 4400 Å, a green filter peaking at 5200 Å, and a red filter peaking at 6200 Å (but having a lesser peak in the blue), we have a "Blue" -blue, a "Green"-green and a "Red"-red which will satisfy most observers. Note that the original single color was chosen by approximating the human eye response in the three-color regions, using monochromatic light to obtain a brightness response, and then adjusting the broad-band transmission properties of the three filters of the lamps to obtain the proper colors. A final check of these lamp filters would be to mix all three colors additively, and then to evaluate the "white" thereby produced. We can then substitute a Standard Lamp for the monochromator and see if we can reproduce its exact color temperature. If not, then we need to modify the transmission characteristics of our filters used on the source lamps.

Once we have done this, we now have our three primary colors in the form of standard lamps, and can proceed to determine Items 1,2 & 3, given above on page 421. To do this, we vary the wavelength of the monochromatic light, and determine relative amounts of red, green and blue light required to match the monochromatic color. This is done, as stated before, for about 5000 observers.

The result is finalized response curves for the Standard Observer, also called "Tristimulus Response curves", and is shown as follows:

7.8.21.-

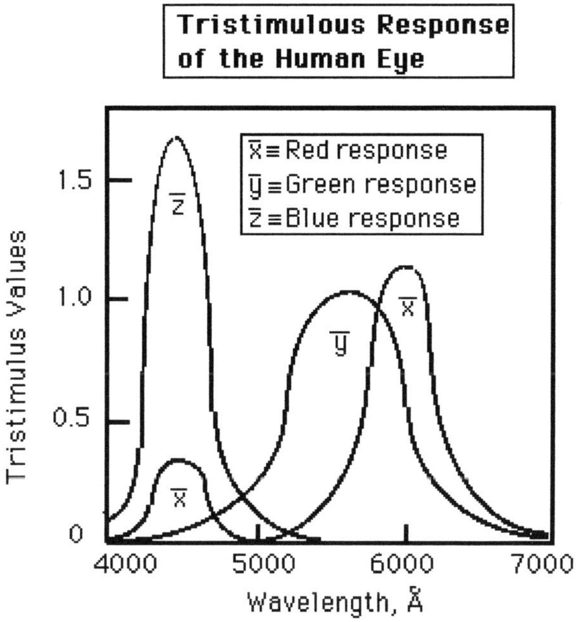


We finally arrive at the result we want, since we can now set up "Tristimulus Filters" to use in defining colors. We can now define $\bar{y}$ as our standard luminosity curvc for thc human cyc (photopic vision). Notc that $\bar{x}$ , the red tristimulus value, has a certain amount of blue in it in order to duplicate the response of the red preceptor in the retina.

I. Tristimulus Coefficients

Our next step is to define colors in terms of tristimulus responses. We know that we can define the energy of any spectral curve as a summation of intensities times wavelengths, i.e.-

7.8.22.-

$$E_R = \sum (Id\lambda)_R$$
$$E_G = \sum (Id\lambda)_G$$
$$E_B = \sum (Id\lambda)_B$$

Therefore if we take the spectral curve, and multiply it by the **overlap** of each tristimulus response curve, we get TRISTIMULUS VALUES, i.e.-

7.8.23.-

| Emittance (additive) | Reflective (scattering) (subtractive) |
|---|---|
| $X = \bar{x} \sum I_R \, d\lambda$ | $X = \bar{x} \sum I_R R_R \, d\lambda$ |
| $Y = \bar{y} \sum I_G \, d\lambda$ | $Y = \bar{y} \sum I_G R_G \, d\lambda$ |
| $Z = \bar{z} \sum I_B \, d\lambda$ | $Z = \bar{z} \sum I_B R_B \, d\lambda$ |

However, we find that these values are difficult to use since each color give a set of tristimulus values, but each set does not have a specific relation to any other. The reason for this is that the intensity of $I_R \neq I_G \neq I_B$. Therefore, we define a set of chromaticity coordinates, as shown in the following, where the tristimulus values are used to define what we now call "Chromaticity Coordinates".

7.8.24.- CHROMATICITY COORDINATES:

$$x + y + z \equiv 1.00$$

$$x = X / X + Y + Z \qquad y = Y / X + Y + Z \qquad z = Z / X + Y + Z$$

Note that we now have three (3) different sets of values related to color specification, and three different types of symbols. In order not to get confused, we reiterate them again. They are:

7.8.25.- VALUES RELATED TO COLOR SPECIFICATION

| Color | Tristimulus Response | Tristimulus Values | Chromaticity Coordinates |
|---|---|---|---|
| red: | $\bar{x}$ | X | x |
| green: | $\bar{y}$ | Y | y |
| blue: | $\bar{z}$ | Z | z |

The advantage of chromaticity coordinates is that we now have a set of **normalized** values which we can use to compare colors having different intensity values (and thus different energy values as well). Furthermore, we need only specify x and y since x + y + z = 1.00. This allows us to specify **monochromatic** radiation in terms of our chromaticity coordinates. Actually, we can plot a three-dimensional value on a flat (2-dimensional) surface.

II. Chromaticity Coordinate Diagrams

Since monochromatic radiation is a **boundary** of color-mixing, then we can construct a CHROMATICITY COORDINATE DIAGRAM in terms of x and y :
7.8.26.-

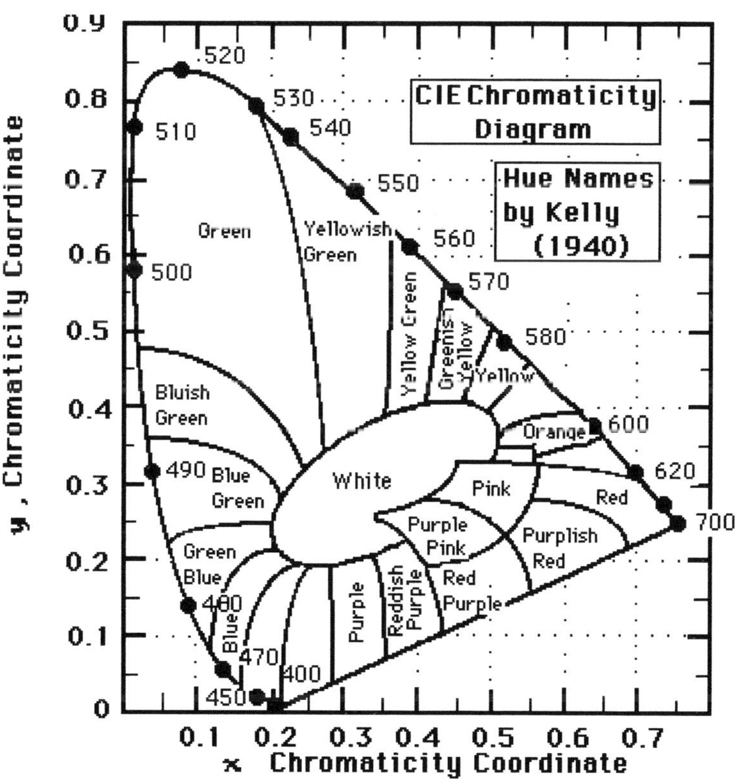

Note that the diagram is bounded, as we have already stated, by the values of monochromatic light. Thus, we can find any color, be it monochromatic or polychromatic, in terms of its x and y coordinates.

We have delved into the methods used by previous investigators in an effort to quanticize color measurement and used the same methods that they did. Once this was done, color specifications became standardized and were not subject to vagaries of the color-method used or the deviations caused by the human eye.

It would be more dramatic to print the various hue areas in color but it is difficult to accurately print reflectance hues. It is better to name the colors directly. This was done by Kelly (1940). Note also that we do not use the term "color" anymore but use the term "hue".

Any hue can be specified by x and y. For example, we can specify the locus of black-body hues and even Illuminants A, B & C. This is summarized as follows:

7.8.27.- <u>x and y Chromaticity Coordinates</u>

| | x | y |
|---|---|---|
| 6000 Å | 0.640 | 0.372 |
| 5200 | 0.080 | 0.850 |
| 4800 | 0.140 | 0.150 |
| | | |
| Illuminant A (3250 Å) | 0.420 | 0.395 |
| Illuminant B (4500 Å) | 0.360 | 0.360 |
| Illuminant C (6500 Å) | 0.315 | 0.320 |

These points on the chromaticity diagram are shown in the following diagram, given as 7.8.28. on the next page.

Nonetheless, while determining color by the color- comparator mode was a distinct improvement over prior methods of color comparison, the method was slow and not easy to accomplish. What was needed was an instrumental method of color measurement and matching.

7.8.28.-

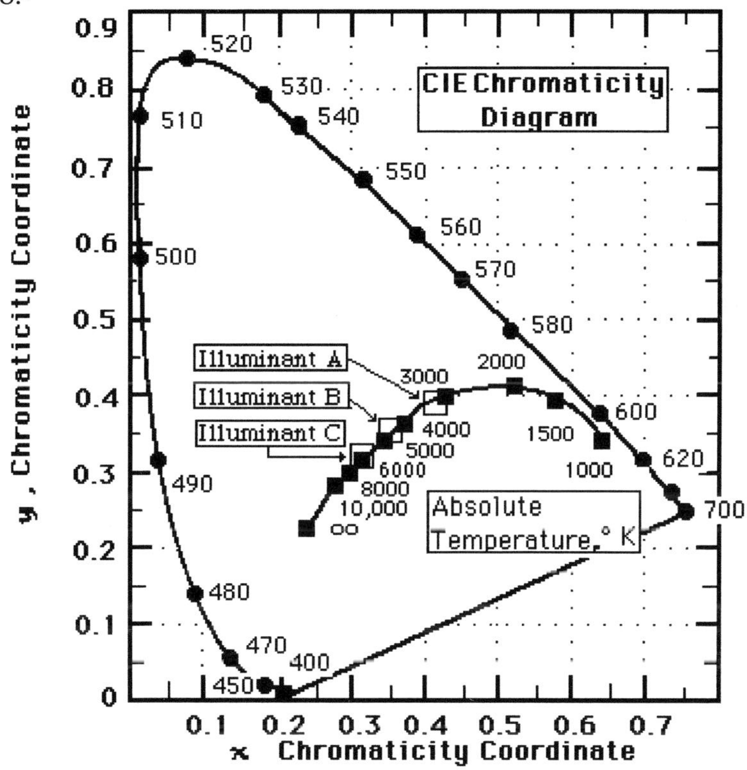


The requirements for an instrumental method of specifying reflected color include a light source, the colored object and a detector. What this means is that all we need is a source, an object and a detector. However, since the response characteristics of these optical components are not linear, nor flat, we need an analogue system in order to be able to measure color.

The analogue system simply corrects for the non-linearity of the source and detector. The reflectance analogue system, given as 7.8.29. on the next page, shows schematically how the spectral power distribution of a CIE source, the spectral reflectance, R, of an object, and the spectral color-matching functions, x, y and z, combine by multiplication (each wavelength by each wavelength), followed by summation across the spectrum, to give the CIE tristimulus values.

7.8.29.- Reflectance Analogue System for Specifying Color

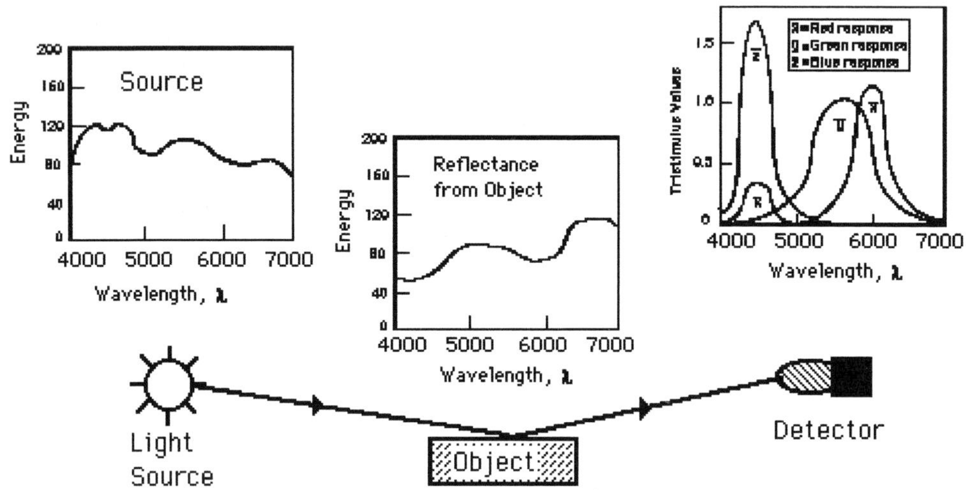


However, these analogues are actually hypothetical. The reason for this is that it is nearly impossible to obtain optical measurement components, such as the source and the detector, whose response to light across the visible spectrum is flat (or nearly so). However, this is not an impossible task and we find that an excellent match can be obtained to the transmission functions of 7.8.21., i.e.- those of the Standard Observer. This is typical for commercially available instruments. Now, we have an instrument, called a Colorimeter, capable of measuring reflective color.

The actual analogue values we need to measure reflectance are given on the next page as 7.8.30. as follows. Note that the optical response curves of the measuring parts, i.e.- the non-linearity of the source and detector, are now corrected in the response of the overall instrument.

7.8.30.-

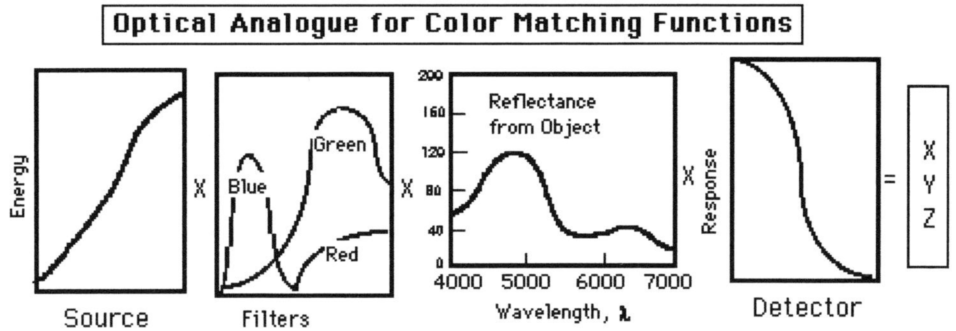


We find that an excellent match can be obtained to the transmission functions of 7.8.20. This is typical for commercially available instruments. Now, we have an instrument, called a Colorimeter, capable of measuring reflective color.

A system for emittance color matching is given in the following :

7.8.30.-

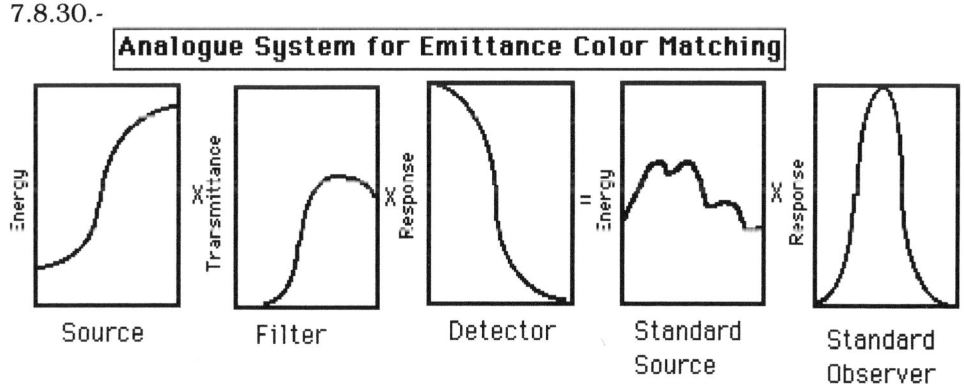


This diagram shows the energy spectrum of a given source, coupled with a filter of defined transmittance, which is established by a detector of known spectral response, as modified by a standard source and modified to that of a Standard Observer. Once an instrument has been set up properly with the proper optical response modifications, it will produce chromaticity coordinates that accurately reflect those needed to define the "color" of the emitter or reflector being measured.

The process for actually measuring emissive color is somewhat different and more challenging. First, we must obtain an emission spectrum by means of a spectrofluorimeter. We can now integrate I $d\lambda$ to obtain the energy and then specify this in terms of $x$ and $y$. There are special methods which have been developed to do so wherein selected wavelengths are used, depending upon the nature of the emission spectrum. We will not delve further into this method other than to state that it does exist.

## **7-9** COLOR SPACES

If we have a certain color, a change in intensity has a major effect on what we see (in both reflectance and emittance). For example, if we have a blue, at low intensity we see a bluish-black, while at high intensity we see a bluish-white. Yet, the hue has not changed, only the intensity. This effect is particularly significant in reflectance since we can have a "light-blue" and a "dark-blue", without a change in chromaticity coordinates.

The concept of "lightness" involves the reflective power of materials. If the reflectance approaches 100%, we say the material is "white", whereas complete absorption (0.00% reflectance) produces a "black" material. Let us examine exactly what is meant by these terms, particularly those of **subtractive** color mixing. When colors are prepared by mixing dyes or pigments, the resultant reflective hue is controlled by a subtractive process of the three (3) primaries, whose reflectance spectrum is given above. When these are mixed, the resulting hue is that where the curves **overlap,** as is easily seen in 7.9.1., given on the next page.

When we mix these primary colors, their reflectances remove more of the incident light, and we see the part where the reflectances are reinforced. Thus,

7.9.1.-

## THE PRINCIPLE OF COLOR MIXING BY SUBTRACTION

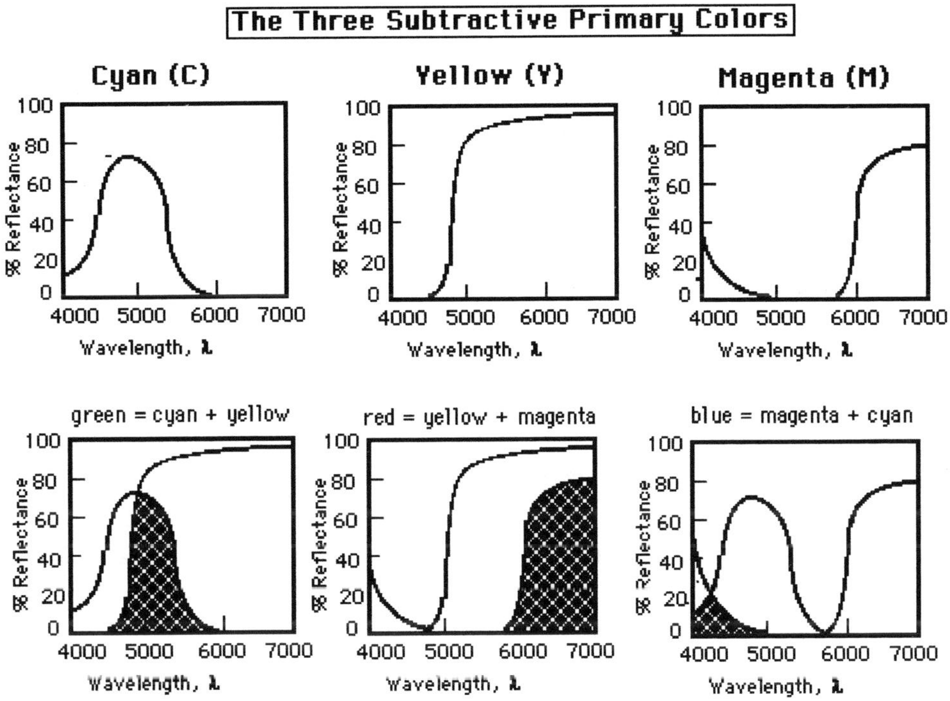


we can get red, green and blue, but they are not primary colors in the subtractive system. Intermediate hues can be obtained also in this process when the subtractive primaries are used in less than full concentration. That is, they are "lightened". Although we can explain this effect on a spectrophotometric basis, we do not have a way of specifying hues in terms of **saturation**, using the CIE system.

### A. The Munsell Color Tree

One of the first attempts to specify reflective colors and color mixing was accomplished by Munscll (1903). Hc dcviscd a color system based on factors he called **hue, chroma and value.** Munsell set up a three-dimensional arrangement based upon *minimum perceptual color difference* steps. He based these upon direct observation since he did not have the instrumental means to do so. Therefore, his results are not the same as the system that we use today.

The advantage of the Munsell system was that in addition to being able to specify hue, we could also specify Chroma, which is the degree of saturation of a specific hue. Value is the relative amount of "blackness" or "whiteness" (equal parts of black and white produce gray) in the particular reflective color being specified. It is this system that gives us access to the degree of "grayness" and/or saturation of any hue.

On a practical basis, if we wish to set up this system, we would assemble a set of "color-chips". Each color-chip would be specified by two factors, H = hue, and V/C, which is value (grayness) modified by chroma (saturation). The actual number of layers in the Munsell Color Tree was determined by "minimum perceptual difference". That is, the minimum change that produces a visual perceptible difference. This arrangement specifies all light colors as well as the dark ones. To use such a system, one would choose the color-chip closest to the hue and saturation of the test color and thus obtain values for H and V/C. However, it was soon discovered that the system was not perfect. Reasons for this include the facts that the hues defined by Munsell are not those of the primaries of the human eye. Furthermore, Munsell was somewhat subjective in his definitions of hues.

In 1920, Priest showed that if the Munsell-Chips were viewed on a white-background, the "brightness", i.e.- lightness as viewed by the human eye, could be related to the Munsell system by:

7.9.3.- $V = 10\,Y^{1/2}$

where Y was defined as "brightness" or eye response. Over the years, the Munsell System has been modified to agree with actual human perception, using a "middle-gray" background .

Nevertheless, it is well to note that it was the subjective observation of the lack of correction for luminosity in the Munsell System that gave impetus to the development of the CIE Color System. The major problem with the Munsell system was that each person attempting to match colors did not produce the exact same result. So color matching became dependent upon the person.

B. Color Matching and MacAdam Space

When Munsell devised his color space, he did so on the basis of minimum observable color perception steps. But the problem with the Munsell System was one of reproducibility, which the CIE Standard Observer cured. In formulating a color match, one wants to be able to predict the correct concentration of colorants required, whose scattering and absorption properties are known, i.e.- the lightness, so as to match the sample submitted, starting with their spectrophotometric curves. In practice, this is not so simple, since two colors must have identical spectrophotometric curves to be **exactly** equal. It turns out that the human eye will identify the two colors to be equal if their spectrophotometric reflectances are reasonably close. Two colors may appear to be equal under Daylight illumination, but quite different under incandescent lamp illumination. These colors are known as "metamers" and the phenomenon "metamerism".

It was soon determined that the 1931 CIE chromaticity diagram, and luminance function, Y, are not representative of **equal** visual spacing. that is, equal changes in Y do not represent equal changes in visual perception for all values of Y. Nor do equal increments of $x$ and $y$ represent the same visual effect for all locations on the chromaticity diagram. In other words, there is a minimum perceptual difference on both $x$ and $y$ (i.e.- $\Delta x$ and $\Delta y$). But, the size of $\Delta x$ and $\Delta y$ is not the same at all parts of the chromaticity diagram. This is the same problem that Munsell encountered and is due to the fact that the human eye is an integrating instrument, not a dispersive one. MacAdam (1942) studied this problem and established limits of Minimum Perceptual Difference steps (MPD) in the form of ellipses on the CIE diagram. This is shown as follows:

7.9.4.-

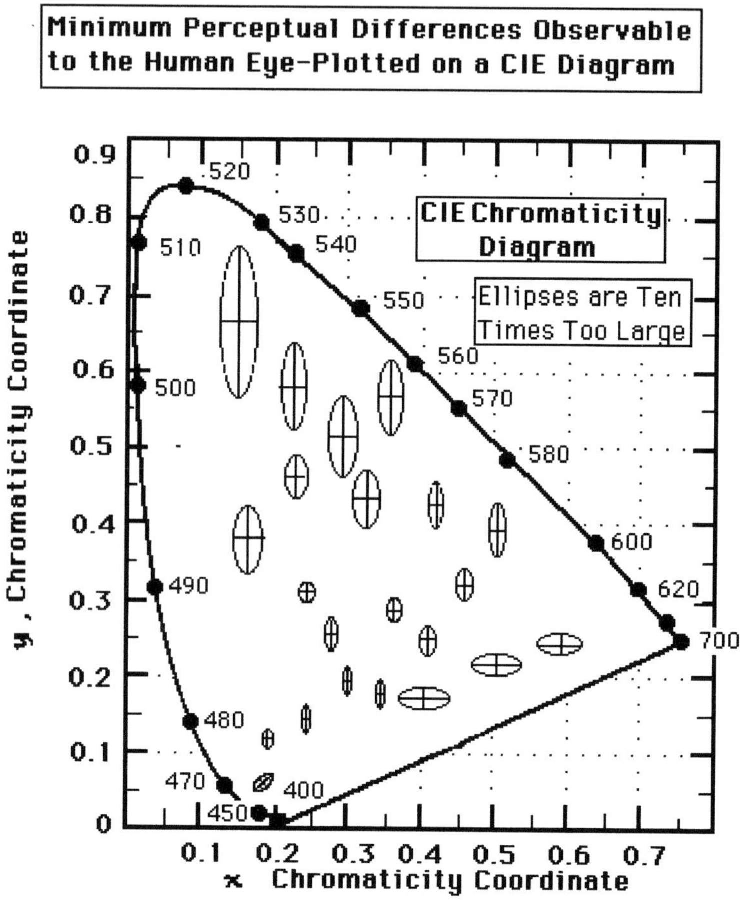


It turns out that a linear transformation of x and y coordinates to a new set of coordinates, u and v, is sufficient to do the job.

MacAdam's transformation equations were:

7.9.5.- $u = 4x / (-2x + 12y + 3)$

$v = 6y / (-2x + 12y + 3)$

Plotting these values gave the following chromaticity diagram:

7.9.6.-

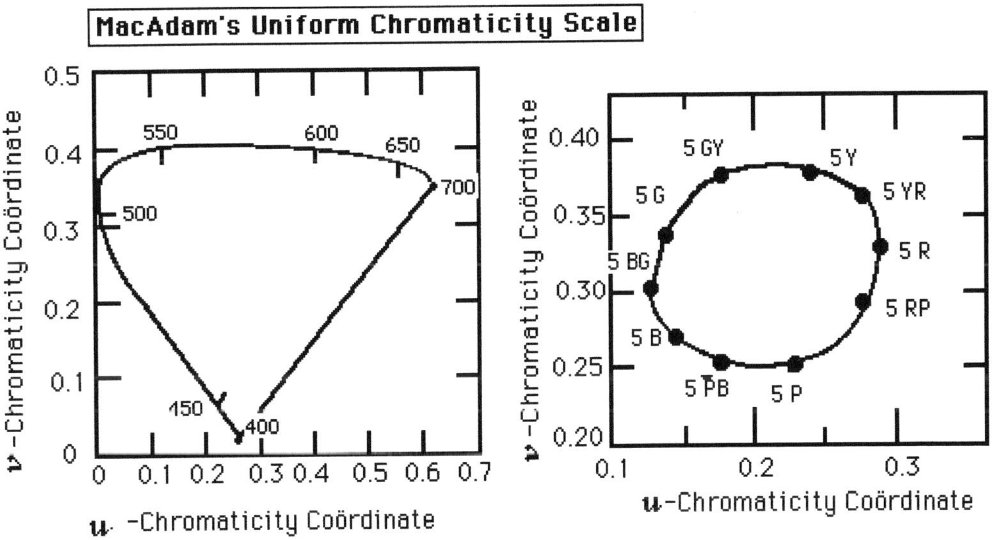


The true test is how well the Munsell hues plot out on the CIE diagram. As can be seen on the right hand side of the diagram, the Value 5 - Chroma 8 hues do construct a nearly perfect circle. Thus, the MacAdam transformation is a definite improvement over the 1931 CIE system. In 1960, the CIE adopted the MacAdam System, having defined the equations (with MacAdam's help):

7.9.7.- $u = 4X / (X + 15Y + 3Z)$

$v = 6Y / (X + 15Y + 3Z)$

The following diagram, shows MacAdam's Uniform Chromaticity Scale, and has Minimum Perceptible Color Difference steps plotted as ellipses:

7.9.8.-

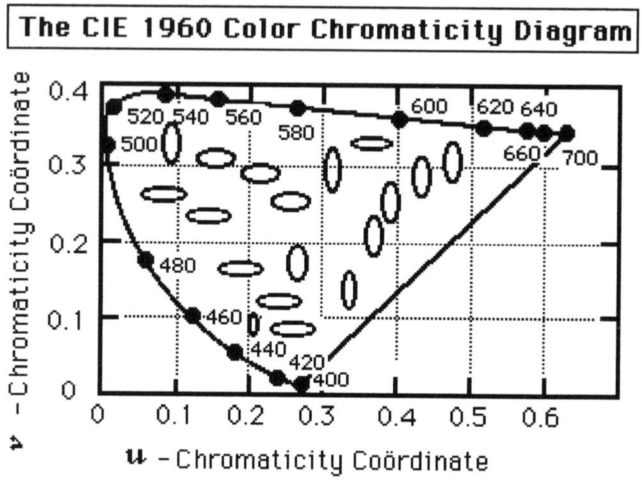


It should be clear that the CIE Color Chromaticity diagram, adopted in 1960 is a superior method for delineating color differences. Indeed, it is the one specified for just such usage.

<u>Recommended Reading</u>

1. Robert E. Reed-Hill, "Physical Metallurgy Principles"- Van Nostrand, Princeton, New Jersey (1964).

2. A.J. Dekker, "Solid State Physics" - Prentice-Hall, Englewood Cliffs, New Jersey (1958).

3. Perelomova & Tagieva, "Problems in Crystal Physics"- MIR Publ., Moscow, Eng. Transl. - (1983).

4. W.W. Wendlandt, *Thermal Methods of Analysis*, Interscience-Wiley, New York (1964).

5. P.D. Garn, *Thermoanalytical Methods of Investigation*, Academic Press, New York (1965).

6. W.J. Smothers and Y Chiang, *Handbook of Differential Thermal Analysis*, Chemical Publishing Co., New York (1966).

7. E.M. Barral and J.F. Johnson, in *Techniques and Methods of Polymer Evaluation*, P. Slade & L. Jenjins- Ed., Dekker, New York (1966).

8. W.W. Wendlandt, *Thermochim. Acta* **1** , (1970)

9. D.T.Y. Chem, *J. Thermal Anal.*, **6** 109 (1974) - Part I

10. Chen & Fong, loc. cit. **7** 295 (1975) - Part II.

11. Fong & Chen, loc. cit., **8** 305 (1975) - Part III

12. C. Duval, *Inorganic Thermogravimetric Analysis*, 2nd Ed., Elsevier, Amsterdam (1963).

13. C. Keatch, *An Introduction to Thermogravimetry*, Heydon, London (1969).

14. "Vapor Pressure Determination using DSC", A. Brozens, R.B. Cassel, C.W. Schaumann & R. Seyler, *Proceeding. N. Am. Therm. Analysis Soc.* **22nd Conf.** Denver, Colo. (1993).

## Problems for Chapter 7

1. You have added a solution of calcium chloride to a solution of sodium metaborate. A crystalline precipitate resulted. In attempting to analyze your product,

an unknown x-ray pattern was obtained. You then used DTA and TGA to attempt to characterize this product. Using the following DTA and TGA data,

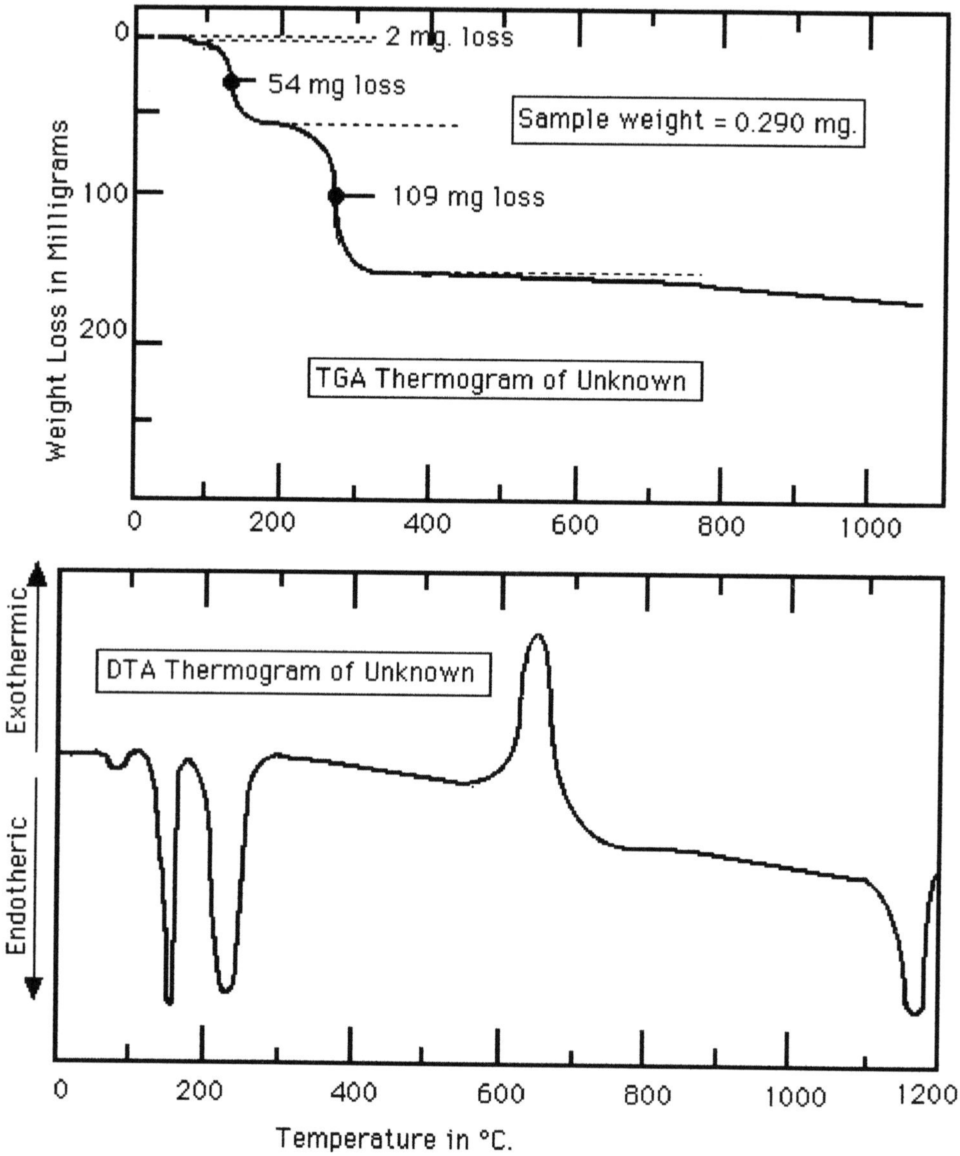

identify all of the products formed during the reactions by writing the reactions which occur during heating. Be sure to characterize both the initial and final product. Write also the precipitation reaction. As an aid, x-ray diffraction analysis was able to identify a product obtained by calcination at 1000 °C as $\alpha$- calcium metaborate. Be sure to identify all of the peaks in the thermograms.

2. Construct your own DTA apparatus, using a "building-block" approach. The desired parameters for your DTA are: max temp = 1200 °C; Atm = A, N & R. Be sure to specify the particulars of your apparatus: e.g.- temp programmer = 1 to 30 °C per minute, etc.

3. Using your DTA, draw the expected curve for decomposition of:

a. $SrHPO_4$
b. $CaCO_3$
c. $Y(NO_3)_3 \cdot 9\ H_2O$

You may wish to look up decomposition temperatures in a Handbook of Chemistry.

4. Draw an expected TGA curve for all 3 compounds, using the temperatures that you used in your DTA curves.

5. Given the color coordinates of: x= 0.34 & y = 0.51. Identify the color having this property. If x = 0.34 & y = 0.61, identify the difference between these two colors. Identify the amounts of red, green and blue in each of these.

6. Do the same for the following chromaticity coordinates (use all 6 combinations possible):

x = 0.25 , 0.45, 0.65

y = 0.20 , 0.40, 0.60

There are nine combinations to be made. Identify each of the three at a time in terms of increasing green, or red.